Manufacturing and Processing of Advanced Materials

Edited By

Amar Patnaik

Mechanical Engineering Department
Malaviya National Institute of Technology Jaipur,
Rajasthan,
India

Albano Cavaleiro

Mechanical Engineering Department
University of Coimbra,
Coimbra, Portugal

Malay Kumar Banerjee

Research Chair, SGVU Suresh Gyan Vihar University,
Jaipur, Rajasthan, India

Ernst Kozeschnik

Head of the Institute
Materials Science and Technology, TU Wien,
Austria

&

Vikas Kukshal

Mechanical Engineering Department
National Institute of Technology,
Uttarakhand,
India

Manufacturing and Processing of Advanced Materials

Editors: Amar Patnaik, Albano Cavaleiro, Malay Kumar Banerjee, Ernst Kozeschnik and Vikas Kukshal

ISBN (Online): 978-981-5136-71-5

ISBN (Print): 978-981-5136-72-2

ISBN (Paperback): 978-981-5136-73-9

need for a court order if at any point you breach any terms of this License Agreement. In no event will any delay or failure by Bentham Science Publishers in enforcing your compliance with this License Agreement constitute a waiver of any of its rights.

3. You acknowledge that you have read this License Agreement, and agree to be bound by its terms and conditions. To the extent that any other terms and conditions presented on any website of Bentham Science Publishers conflict with, or are inconsistent with, the terms and conditions set out in this License Agreement, you acknowledge that the terms and conditions set out in this License Agreement shall prevail.

Bentham Science Publishers Pte. Ltd.
80 Robinson Road #02-00
Singapore 068898
Singapore
Email: subscriptions@benthamscience.net

BENTHAM SCIENCE

CONTENTS

FOREWORD

I am greatly honoured to write this foreword for the edited book titled "**Manufacturing and Processing of Advanced Materials**" in the field of Mechanical and Materials Engineering published by Bentham Science. The advanced materials and manufacturing processes are revolutionizing all the specialized areas in different fields such as electrical, electronics, mechanical and civil. A high degree of novelty is needed in both structural and functional materials to achieve the goal of sustainability.

This book will provide scientists, researchers, and academicians working in the field of advanced materials and manufacturing to gain a better understanding of information and the necessary components of material science and mechanical engineering. The book chapters published cover recent innovations and emerging methods, materials, and optimization techniques, such as those related to advanced materials, composite materials, manufacturing technologies, industrial tribology, material characterization, *etc*.

I appreciate the authors for their technical input to complete the book. I would like to acknowledge the editors of the book and wish the authors good luck in their future endeavors.

Alok Sathpathy
Department of Mechanical Engineering
National Institute of Technology, Rourkela,
India

PREFACE

The application of advanced engineering materials and manufacturing techniques is currently capturing the attention of prominent researchers, scientists, and academicians in a variety of high-tech fields. There are large numbers of challenges that can be solved by improving the compatibility of the materials and by adopting proper manufacturing processes. Hence there is a continuous need for research and development in the field of materials and manufacturing. This book is dedicated to current research activities in the field on advance materials and manufacturing and engineering applications.

The book entitled "**Manufacturing and Processing of Advanced Materials**" embrace innovations in the field of materials, manufacturing processes, and distinguished applications proposed by numerous researchers, highly qualified professionals, and academicians. This book focuses on the properties and end applications of a wide range of engineering materials, including metals, polymers, composites, fiber composites, ceramics, and other similar materials, as well as their characterization and prospective applications. This book is an updated reference of research activities that bring together various theories, methods, and technologies for a wide range of manufacturing processes adopted for industrial application. All the chapters were subjected to a peer-review process by the researchers working in the relevant fields. The chapters were selected based on their quality and their relevance to the title of the book.

Successful completion of this book includes the efforts of many people. It is very imperative to acknowledge their contribution to shaping the structure of the book. Hence, all the editors would like to express special gratitude to all the reviewers for their valuable time spent in reviewing process and completing the review process on time. Their valuable advice and guidance helped in improving the quality of the chapters selected for publication in the book. We would like to thank all the authors of the chapters for the timely submission of the chapter during the rigorous review process. Finally, we would like to thank the publisher for the invaluable input in the organizing and editing of the book.

Amar Patnaik
Mechanical Engineering Department
Malaviya National Institute of Technology, Jaipur
Rajasthan,
India

Albano Cavaleiro
Mechanical Engineering Department
University of Coimbra,
Coimbra, Portugal

Malay Kumar Banerjee
Research Chair, SGVU
Suresh Gyan Vihar University, Jaipur
Rajasthan,
India

Ernst Kozeschnik
Head of the Institute
Materials Science and Technology, TU Wien
Austria

&

Vikas Kukshal
Mechanical Engineering Department
National Institute of Technology,
Uttarakhand,
India

List of Contributors

Aditya Purohit	Department of Mechanical Engineering, Malaviya National Institute of Technology Jaipur, Rajasthan, India
Aparna Duggirala	School of Laser Science and Engineering, Jadavpur University, Kolkata, 700032, India
Akshay Dvivedi	Department of Mechanical and Industrial Engineering, Indian Institute of Technology, Roorkee, India
Apurbba Kumar Sharma	Department of Mechanical & Industrial Engineering, IIT Roorkee, Roorkee-246174, India
Ankit Kumar Maurya	Mechanical Engineering Departmen, Madan Mohan Malaviya University of Technology Gorakhpu, Uttar Pradesh, India
Anjani Kumar Singh	Mechanical Engineering Departmen, Madan Mohan Malaviya University of Technology Gorakhpu, Uttar Pradesh, India
Abhishek Pothina	Department of Mechanical Engineering, National Institute of Technology Patna, India
Amit Kumar	CSIR-Central Mechanical Engineering Research Institute, Durgapur-713209, India
Bappa Acherjee	Department of Production and Industrial Engineering, Birla Institute of Technology, Mesra, Ranchi, 835215, India
Botcha Appalanaidu	Department of Mechanical and Industrial Engineering, Indian Institute of Technology, Roorkee, Uttarakhand, India
Bipul Kumar Singh	Mechanical Engineering Departmen, Madan Mohan Malaviya University of Technology Gorakhpu, Uttar Pradesh, India
Chandra Kant	Department of Mechanical Engineering, National Institute of Technology Srinagar, Jammu, and Kashmir, India
Dasari Sai Naresh	R&D Mechatronics, Design and Engineering, VEM Technologies PVT Ltd, Hyderabad, Telangana, 502321, India
Dileep Chekkaramkodi	Department of Mechanical Engineering, National Institute of Technology Calicut, Calicut- 673601, India
Gaurav Kumar	Department of Mechanical Engineering, NIT Uttarakhand, Srinagar-246174, India
Ganesh S. Kadam	Babasaheb Ambedkar Technological University, Lonere, Maharashtra, India SIES Graduate School of Technology, Navi Mumbai, Maharashtra, India
Ghulam Ashraf Harmain	Department of Mechanical Engineering, National Institute of Technology Srinagar, Jammu, and Kashmir, India
Jinu Paul	Department of Mechanical Engineerin, National Institute of Technology Calicut, Calicut- 673601, India
Jyoti Behera	Department of Physics GIET University, Gunupur, Rayagada, Odisha, India

Kriti Srivastava	Department of Mechanical Engineering, National Institute of Technology, Patna-800005, Bihar, India
K. Ravi Prakash Babu	Department of Mechanical Engineering, Prasad V Potluri Siddhartha Institute of Technology, Kanuru, Andhra Pradesh, India
Mahaveer Prasad Sharma	Malaviya National Institute of Technology Jaipur, Rajasthan, India
Mukund Kumar	Department of Mechanical Engineering, BIT Mesra, Ranchi - 835215, India
Mudit K. Bhatnagar	Department of Mechanical Engineering, JSS Academy of Technical Education Noida, Uttar Pradesh, India
Mamatha Theetha Gangadhar	Department of Mechanical Engineering, JSS Academy of Technical Education Noida, Uttar Pradesh, India
Mohit Vishnoi	Department of Mechanical Engineering, JSS Academy of Technical Education Noida, Uttar Pradesh, India
Murahari Kolli	Lakireddy Bali Reddy College of Engineering, Mylavaram, Krishna District, Andhra Pradesh, India
Mohammed Yusuf A. Yadwad	Department of Mechanical Engineering, PES University, Bangalore-560085, India
Muhammed Hunize Chuttam Veettil	Department of Mechanical Engineering, National Institute of Technology Calicut, Calicut- 673601, India
Murali Kodakkattu Purushothaman	Department of Mechanical Engineering, National Institute of Technology Calicut, Calicut- 673601, India
N. Rajesh Mathivanan	Department of Mechanical Engineering, PES University, Bangalore-560085, India
Pankaj Kumar Gupta	Department of Mechanical Engineering, Malaviya National Institute of Technology Jaipur, Rajasthan, India
Rajendra Kumar Arya	Department of Mechanical Engineering, Indian Institute of Technology, Mumbai, Maharastra, India
Raju Shrihari Pawade	SIES Graduate School of Technology, Navi Mumbai, Maharashtra, India
Rohit Kumar Babberwal	Department of Mechanical Engineering, Army Institute of Technology, Pune-411015, Maharashtra, India
Raosaheb Bhausaheb Patil	Department of Mechanical Engineering, Army Institute of Technology, Pune-411015, Maharashtra, India
Spruha Aniket Dhavale	School of Mechanical Engineering, Vishwanath Karad MIT World Peace University, Pune, Maharashtra, India
Shivprakash Bhagwatrao Barve	School of Mechanical Engineering, Vishwanath Karad MIT World Peace University, Pune, Maharashtra, India
Souradip Paul	School of Laser Science and Engineering, Jadavpur University, Kolkata, 700032, India
Souren Mitra	Department of Production Engineering, Jadavpur University, Kolkata, 700032, India
Shivnandan Bind	Mechanical Engineering Department, National Institute of Technology Patna, India

Sanjay Mishra Mechanical Engineering Departmen, Madan Mohan Malaviya University of Technology Gorakhpu, Uttar Pradesh, India

Siddharth Srivastava Department of Mechanical Engineering, JSS Academy of Technical Education Noida, Uttar Pradesh, India

Satendra Singh Department of Mechanical Engineering, Malaviya National Institute of Technology Jaipur, Rajasthan, India

Saroj Kumar Sarangi Department of Mechanical Engineering, National Institute of Technology Patna, India

Saurabh Mishra CSIR-Central Mechanical Engineering Research Institute, Durgapur-713209, India

Surendra Kumar CSIR-Central Mechanical Engineering Research Institute, Durgapur-713209, India

Simadri Priyanka Achary Department of Physics GIET University, Gunupur, Rayagada, Odisha, India

Sanjukta Mishra Department of Physics GIET University, Gunupur, Rayagada, Odisha, India

Tapas Bajpai Microfluidics & MEMS Centre, CSIR-Advanced Materials and Processes Research Institute (AMPRI), Hoshangabad Road, Bhopal 462026, India

Tapan Kumar Patnaik Department of Physics GIET University, Gunupur, Rayagada, Odisha, India

Upama Dey School of Laser Science and Engineering, Jadavpur University, Kolkata, 700032, India

Velavali Sudharshan Department of Mechanical Engineerin, National Institute of Technology Calicut, Calicut- 673601, India

Vansh Malik Department of Mechanical Engineering, JSS Academy of Technical Education Noida, Uttar Pradesh, India

Vishwas G. Department of Mechanical Engineering, PES University, Bangalore-560085, India

Yogesh Kumar Department of Mechanical Engineering, National Institute of Technology, Patna-800005, Bihar, India

CHAPTER 1

A Review on the Joining of Dissimilar Materials with Special Context to Laser Welding

Aditya Purohit[1], Tapas Bajpai[1,*], Pankaj Kumar Gupta[1] and **Arpana Parihar[2]**

[1] *Department of Mechanical Engineering, Malaviya National Institute of Technology Jaipur, Rajasthan, India*

[2] *Microfluidics & MEMS Centre, CSIR-Advanced Materials and Processes Research Institute (AMPRI), Hoshangabad Road, Bhopal 462026, India*

Abstract: In recent times, there has been an increasing demand for dissimilar metal fabrication, as this weldment utilizes the specific benefits of different metals for a particular application. In this paper, the recent trends evolving in the field of dissimilar material joining, which introduces residual stresses, distortions and formation of brittle intermetallics within the structure is discussed. As these are highly undesirable, therefore various techniques were studied by the researchers, which reduce the distortions and formation of brittle intermetallics. The use of numerical techniques in this field was also studied as they provided the researchers with an insight into the process. Mostly, the joining of dissimilar material is done using friction stir welding and laser welding, but the use of friction stir welding has constraints in terms of material temperature thus, the joining of dissimilar weldment is discussed by giving a special context to laser welding technology.

Keywords: Dissimilar material joining, Laser welding, Optimization techniques.

INTRODUCTION

Laser welding is a state of art technique which is normally performed in keyhole mode. Unlike conventional welding techniques, this welding technique enables deeper penetration in the material combined with lower heat-affected zone (HAZ) since the focal diameter of the laser beam can be adjusted according to need. Application of laser welding varies from structural applications, automotive sector to sophisticated applications such as dentures, implants and complex electrical circuits. Dissimilar material welding is one of the fields to be explored. It's a phenomenon concerning the current industrial trend because dissimilar material welding uses the specific properties of different materials for an application that

* **Corresponding author Tapas Bajpai:** Department of Mechanical Engineering, Malaviya National Institute of Technology Jaipur, Rajasthan, India; E-mail: tapas.mech@mnit.ac.in

Amar Patnaik, Albano Cavaleiro, Malay Kumar Banerjee, Ernst Kozeschnik & Vikas Kukshal (Eds.)

ranges from dissimilar metal electronic connection of Al-Cu in an electronic vehicle to large structures where dissimilar joints are required [1]. The problem in dissimilar welding is that because of dissimilar materials, the materials have different thermal expansion coefficients, which leads to stress application on the material due to fluctuation in temperatures, and if that stress is higher than the yield stress of the material, then the material might fail [2]. Apart from this, the other reason is the formation of intermetallic phases in the fusion zone because of less solubility between different metals. These phases have hardness variation in the weldment zone, and this variation of hardness might lead to the failure of the joint [3].

It's a well-known fact that in order to get a mechanically sound joint, it is required for the joint to be stress-free to avoid distortions and crack initiation. The major contributor to it is the different expansion coefficients of the material and phase transformation of the material. There are three components of stresses, mainly quenching, phase transformation and shrinkage. Shrinkage leads to tensile stresses, *i.e.,* the region which is last to cool, faces tensile stresses and vice versa. Another factor is phase transformation, so the region that faces phase transformation first is subjected to tensile stresses. Whichever factor dominates the most, its effect is shown on the weldment [4]. Due to high-temperature exposure on the weldment, the temperature rises, and subsequently, a thermal gradient is set up between the weldment zone and the base metals. Due to the gradient and the difference in various material properties, residual stresses are set up in the material, and subsequently, distortions are introduced. The tensile residual stresses in the welding zone lead to a compromise in the structural integrity of the joint. If stresses are above the material's permissible limit, it might lead to the failure of the material. The subsequent distortions introduced within the assemblies are also highly undesirable since they compromise on the tolerance front, which is not desirable in the manufacturing unit. Deformation introduced in the assembly can vary from longitudinal and transversal shrinkage to angular deformation. Therefore to address the above issues and mitigate these phenomena, the researchers are working in this direction. Different aspects of dissimilar material welding were examined by analysis of various fronts of this particular field. The effects of different phenomena which lead to the deterioration of the joint were analyzed, and different researchers tried to mitigate the problem in their own way. Classification on the basis of various paths adopted by the researchers [5].

OPTIMIZATION

Optimization techniques serve as an integral part of research because various optimization techniques give the researchers an idea about the variable parameters, which should be set in such a way that the optimum results are achieved. Various techniques for researchers are available to optimize the variables such as RSM (Response surface methodology), which plots a 3D curve of output variable with respect to input variables.

Researchers as presented in Table **1** used the following optimization techniques:

1. Design of experiments (DOE) or Taguchi optimization

2. ANOVA or Analysis of Variance

3. RSM or response surface methodology

4. Combination of the above techniques

Getting an idea of how various variables in a welding process interacted with each other and the final results, helps in giving a clear outlook of the process and also helps in predicting outcomes. Bhattacharya. *et al.* [6] tried to study the effect of various parameters such as power, frequency, and scanning speed on the weld width and HAZ of the weldments. Centrally composite design technique was used to design the set of experiments for the experiment, and RSM was used to develop a mathematical model, keeping weld width and HAZ as the output. Specimens of polycarbonate and acrylic were taken and welded using Nd-YAG. RSM results showed that with the increase of power, HAZ and weld width increased. After attaining good weld width and strong weld, it starts to decrease with the rise in power. Results also showed that weld width and HAZ do not depend on frequency. The results of ANOVA also validated the same. One way of predicting the quality of the weld is by analyzing the bead geometry, *i.e.,* the dimension of the bead should be kept minimum, with the weldment also serving its purpose. Juang and Tarng [7] analyzed and optimized the weld bead geometry of weldment of stainless steel prepared by gas tungsten arc welding (GTAW). Researchers narrowed down the large number of experiments down to a fixed number using Taguchi DOE technique. Since the bead geometry consists of different variables such as front height, back height and back width therefore instead of optimizing a single variable using a loss function, all the variables are optimized at the same time by assigning weighted residuals to each function according to the literature. The optimization criteria taken by the authors for that particular loss function was "Lower the Better."

Toughness in weldments is an essential property as far as bridge construction and shipbuilding are concerned. Anawa and Olabi [8], fabricated sheets of 316 SS and low-carbon steel. These dissimilar metals were joined using a continuous CO_2 laser. Firstly using design expert, Taguchi set of experiments was designed, keeping laser power, focus diameter, and speed as variables and impact strength or toughness as the response output. The analysis of the S/N ratio with "larger the better" characteristic gave us a particular set of variables for which the highest toughness of the specimen was achieved. Further analysis was done using ANOVA, which predicted the significance of the model and showed that LASER power is the most influential factor keeping in mind the impact strength of the weldment. Prabhakaran and Kanan [9] tried to weld two dissimilar metal plates of AISI 1018 and AISI316, properties were examined on the basis of central composite design (CCD), and they were analyzed using response surface methodology (RSM). The response surface plot showed the combined effect of parameters on weld strength, and it is quite evident that high power, low speed, and high focal length help increase the weldment strength.

Table 1. Optimization of weld parameters with respect to input parameters.

S. No.	Author	Technique	Material	Output	Conclusion
1.	Bhattacharya et al. [6]	RSM	Polycarbonate and acrylic	HAZ and weld width.	HAZ is not affected by frequency, but with an increase in power, initially, it increases, but after that, it decreases.
2.	Juang and Tarang [7]	Taguchi DOE	Stainless Steel (SS)	Bead geometry	Instead of optimizing a single parameter, different parameters were optimized at the same time.
3.	Anawa and Olabi [8]	ANOVA	316 SS - Low carbon steel.	Toughness	ANOVA results showed that power was the most influential variable on toughness.
4.	Vermanaboina et al. [12]	Taguchi DOE and ANOVA	SS316 and Inconel	Residual stresses	A combination of minimum residual stresses was found, and the root gap was found as the most influential factor.

Kalins et al. [10], in their work, tried to weld titanium sheets using Nd-YAG lasers. Using Taguchi L9 orthogonal array, we attempted to study the effect of various parameters on the tensile strength of the joint, and finally, an optimum combination of parameters was found. ANOVA results concluded that laser power was the most influential parameter on the tensile strength. Predictive equations were also developed to predict the output.

Patil and Waghmar [11] studied the effect of current, voltage and speed on tensile strength has been analyzed using Taguchi designed experiments on AISI1030 mild steel. ANOVA results concluded that welding current and welding speed are the most influencing parameters which affects the tensile strength. Optimisation of parameters to keep residual stresses to the minimum, ultimately reducing distortions and giving the desired mechanical properties. Vemanaboina *et al.* [12], butt welded combination of SS316 and Inconel, Taguchi L9 orthogonal array was applied keeping residual stress as the output and a particular combination for which residual stresses are minimum were found out. ANOVA was also applied, and it concluded that the root gap was the most influencing factor in deciding the residual stress.

RESIDUAL STRESSES AND DISTORTIONS

The variation introduced within the base metals and weld path in terms of temperature both laterally and transversely introduces temperature gradient within the sample. This variation of gradient combined with the different expansion coefficients introduces residual stresses within the material. If not mitigated, they seriously compromise with the overall strength. There are various techniques to measure the residual stresses, such as X-ray diffraction (XRD) and blind hole drilling method. These techniques measure the lattice and bulk distortion by nondestructive and destructive techniques, respectively. As could be seen from the previous studies that the residual stresses over the weldment can be harmful for the weldment and its mechanical properties. Therefore efforts were made by the authors to mitigate the effect. Madhvan *et al.* [1] attempted welding of AA 6061- AZ31B Mg in lap joint configuration. They found that as the heat input is increased the solidification rate decreases consequently reduction in magnitude of tensile stress is observed. Roshith *et al.* [13] , joined thick plates of SMO 254SS by using Pulsed Continuous Gas Tungsten Arc Welding (PCGTAW) with (ErNiCrMo3) as filler and autogenous CO_2 laser welding with 3032 J/mm and 120 J/mm heat input, respectively. A variation in the pattern of residual stresses was observed in both the cases, as with PCGTAW due to external filler material and increased heat input tensile stresses are induced in the weldment and compressive stresses in the base material whereas in the weldment, welded by laser welding compressive stresses were induced. Tensile stresses in the PCGTAW lead to lower fatigue life, premature failure and crack initiation.

Apart from conventional preheating and post-heating, other techniques were also studied and introduced by the researchers to mitigate the residual stresses within the weldment. Mohanty *et al.* [14] welded a specimen of AISI 316 using continuous CO_2 laser welding. A Vibratory Stress Relieving (VSR) setup, which used vibrations of the order of 70 Hz over the workpiece, tried to relieve the induced residual stresses by vibration instead of by changing the microstructure. Results showed that after VSR, the value of residual stresses in the specimen decreased to a limit, which had a positive effect on the tensile strength of the joint, thus increasing UTS of the joint. It was observed that the hardness of the weldment decreased after the exposure of VSR, which, apart from saving energy compared to the conventional methods of increasing the temperature to refine the structure, the use of VSR also ensured no intermetallic compounds were formed in the weldment. Reddy *et al.* [15] studied the effect of filler wire on residual stresses distribution. Marging steel as a base material was joined using different filler materials namely maraging steel filler, austentic SS and medium alloy medium carbon steel. Peculiar residual stress behavior is observed between different filler materials as austenitic filler material showed tensile stress at the center compared to maraging steel and low alloy medium carbon steel which showed compressive stress at the center. It was also found that post weld aging results in minimizing the residual stresses because of over tempering.

Effect of welding speed on the residual stresses was analyzed by Sindhu *et al.* [16], in which lap joints were prepared using two speeds, namely 5.1 and 4.1 m/sec. Further they were subjected to fatigue and tensile testing. They found that higher static and fatigue strengths were observed in the samples welded with 4.1 m/sec, as compared to the samples welded at 5.1 m/sec weld speed. Also, higher residual stresses were found in specimens welded at 4.1 m/sec as compared to the samples welded at 5.1 m/sec weld speed. This is due to the higher heat input available at low welding speed. The hardness profile of the WZ, BMZ and HAZ was observed. It was found to have abrupt variation in the hardness profile. The effect of residual stresses on dissimilar material welding was analyzed by various researchers, in this chain work was done by Bajpei. *et al.* [17], they tried to study and mitigate the effect of residual stresses on thin sheets of Al alloys welded together by GMAW. In their work they experimentally welded the sheets using different cooling conditions. It was observed that apart from water cooled model, all the models had tensile residual stresses induced inside them which would ultimately compromise with the strength of the weldment joint.

Apart from inducing residual stresses, the difference in thermal expansion coefficient and thermal gradient would lead to distortions in the joining plates. Distortions can be of various kinds, such as transverse, longitudinal and angular. Distortions seriously compromise the tolerance in the weldment structure. Various researchers carried out their studies on this. To study or analyze an output parameter, it must be related to the input parameters of the process, therefore Shichun and Jinsong [18] related the material deformation with various criteria, namely laser parameters, geometric parameters and material properties, and concluded that bend angle increased with an increase in the number of passes, power, thermal expansion coefficient and the bend angle decreased with the increase in sheet thickness, scanning speed and beam diameter. Alternate ways of reducing the distortions were also studied by the researchers, Mochizuki *et al.*

[19] introduced the concept of riverside preheating to minimize distortions. In their work, before the MIG welding, a riverside TIG welding whose change in position is taken as a variable, and the power input (Q) is also taken as a variable, and the combined effect of all those on angular distortions was studied experimentally and numerically. Das and Biswas [5] studied the effect of parameter variation, namely thickness, power, number of scans, scanning speed and thickness. The results were analyzed, keeping output as deformation. A combination of parameters for minimum deformation was found, and ANOVA results showed us that the bending of sheets was significantly affected by the number of passes and sheet thickness. An overall idea about the distortions in welding, their kind and types were documented by O.P. Gupta [20], who studied the various kind of deformations occurring during the welding and concluded that longitudinal shrinkages showed a regular variation along the weld line and transverse shrinkages are shown to having a tendency of being less at the starting point which leads to angular deformation of the plates.

MISCELLANEOUS

Apart from the above points, researchers also investigated various other phenomena to mitigate the problems or thoroughly understand the problem. These included the effect of offsetting the laser beam on mechanical properties and microstructure. The simulation further emulates the welding process and studies the various outcomes graphically and numerically. Apart from that, there were some other variables that would be discussed in the next section.

Offsetting of Laser Beam

The offsetting of the laser beam while welding dissimilar material usually mitigates the effect of intermetallic formations. The offsetting is usually provided away from the material having a higher reflective index. Chen *et al.* [21] studied the effect of processing parameters on the characteristic of Cu-SS joint using laser welding and observed that the weld mode transformed from brazing to fusion as the weld offset was shifted from SS to the interface of the base metals, respectively. The effect of various parameters such as offset, oblique angle, welding speed, and power on mechanical behavior, microstructure, and appearance was studied. The optimal value of various input parameters is to get researchers to put forward the desired output parameters. According to Chen *et al.* [22], laser butt welded Ti-6Al-4V and Inconel 718 concluded that brittle intermetallic phases could be reduced by deviating the laser beam on the Inconel side of the weld and crack-free welds of the alloy combination could be obtained by the use of higher power and velocity. Reduction in porosity was found when high velocity was employed because less time was made available for solidification, which thus promoted uniformity in the density of the structure.

Numerical Modelling of Dissimilar Laser Welded Joints

Researchers employed various simulating packages such as SYSWELD, Abaqus and ANSYS to simulate and study, the welding process and welding outputs.The technique usually uses temperature loads as the input for the process of generation of mechanical outputs, such as residual stresses and distortion. Attar *et al.* [23] performed a simulation of the mathematical model developed for the welding of dissimilar materials such as 304SS and Copper. Various heat models were used to develop a heat source for the process. After observing the optical image of the weldment and comparing it with the various developed heat models, the volumetric double conical heat source model was found more suitable than the other models. It was also concluded that material with higher yield strength would store higher stress and thus have lower distortion. Bajpei *et al.* [24] studied the behavior of residual stresses in thin dissimilar Al sheets. Experimental results were verified using a fem-based simulation in which goaldak's volumetric heat source was used. The high temperature was observed in AA5052 compared to AA6061, because of the latter's high conductivity. Significant longitudinal stresses were observed in AA6061 compared to AA5052 since the ultimate strength of AA6061 was higher.

Monfared *et al.* [25] welded austenitic steel and compared the experimental analysis with the simulated results. SYSWELD was used to simulate the welding conditions, and various hypothetical conditions were proposed, which were necessary to propose the simulation and related residual stresses with the angular deformation. Firstly thermal analysis was done, which created input for further mechanical analysis, which predicted residual stresses and deformation, according to the boundary conditions employed. A thorough study on the nature of residual stresses on the top and bottom surface was studied both in the longitudinal as well as transversal direction.

Effect of Variation of Pulse

Modern-day lasers available to us are generally of two types, namely continuous and pulsed. The effect of change in pulse shape, frequency, and various parameters associated with the laser pulsation in welding of dissimilar material was analyzed by the researchers. Lerra *et al.* [26] studied the effects of pulse shape, pulse distance, and pulse energy on the weld seam's mechanical, thermal and electrical characteristics. Pulsed energy was varied, keeping shape constant for different pulse separation distances. Researchers took Al-Cu as their base metal. Weld seam dimensions increased until cutting was done for increased pulse energy. Decreasing pulse distance led to higher penetration. Optimal process parameters led to low weld depth. Mathivanan *et al.* [27], carried out their study on Cu and Al sheets, oscillation of laser source, and pulsed nature of power source on the weldment nature has been studied. When a tensile shear test was carried out on the sample without beam oscillation, the weldment broke abruptly, confirming its brittle nature. Whereas the weldment with beam oscillation showed behaviour that confirmed induced ductility. The microstructure analysis of the oscillated samples confirmed the results, which showed dimples, confirming induced ductility.

Variation of frequency of the power source and its effect on the weld geometry because of environmental effects such as air pressure was interestingly analysed by Ghosh and Sharma [28], who studied the variation of frequency and mean current on the nature of weldment. An alloy of Mg-Zn-Al was taken, increasing current and frequency increases the hardness of the weldment. The fluctuation in current may give rise to an air aspiration effect due to the change of pressure around the weld zone, which would lead to a loss in the material. Diametto *et al.* [29] welded sheets of Cu using a fiber laser and observed the effect of wobbling of the laser head on the weld bead parameters. It was observed that for lower power, voids and ripples were visible, which eventually got diminished at higher speeds. Increasing the rotational diameter decreases surface voids but also decreases surface penetration. When different trajectories were compared, it was concluded that weld parameters were mostly related to the frequency of rotation as higher frequency induced overlapping, which means less and less penetration. At lower speeds, due to high concentration of heat at certain points leads to maragonia effect leading to voids and spatter.

Other Techniques

Apart from the above-mentioned techniques, other techniques, such as the effect of water cooling, use of flux, variation in clamping forces, variation in temperature during testing, *etc.*, were employed by the researchers to analyze and study their effect on output parameters. Liu *et al.* [30] observed thermal expansions, and cooling contractions introduce variations in the clamping forces, which affects welding strength. Different levels of pre-set clamping forces were applied on the different thicknesses of sheets, and it concluded that samples welded with pre-set welding forces had higher UTS. The higher thickness of the sheet had a lower maximum temperature. Eventually, the force fades out as the expansion from the load cell moves away. Pankaj *et al.* [31] experimentally investigated butt joints of AISI304 and mild steel, prepared using continuous CO_2 welding. In contrast to similar joints, the dissimilar joint had lower elongation before fracture. The fracture occurred near the mild steel side due to the presence of equiaxed dimples, as observed through SEM. The grain structure was found to be coarse near the steel side because of lower heat conductivity. Sharma *et al.* [32], welded AHSS(Advanced High Strength Steel), namely TRIP780 and DP980. In their study, the effect of prestraining on the sample TRIP 780 and unstrained samples were also compared. It was observed that the energy absorption in the prestrained samples was comparatively lower than in the strained sample. Prestraining also converted the austenitic phase into martensite, thus inducing hardness in the sample.

Xu *et al.* [33] fabricated dissimilar Titanium –Al joints using pulsed Nd-YAG laser welding. Hot cracking susceptibility and shear fracture behaviour of the welded joint was examined by varying various parameters. The HCS of the weldment increases with the increase of power, but after a point, it starts to decrease as a higher amount of heat, increases the volume of the melted pool; this reduces the cooling from the pool and enables the melted metal to repair off the cracks. Wu *et al.* [34] studied the effect of water cooling on the weldments prepared using Ni-SS as the base metals, filler wire of ERNiCrMo-4 and concluded that adding cooling water refines the microstructure, increases load strength, increases hardness but also decreases weld depth and increases reinforcement. Zhang *et al.* [35] analyzed the effect of hybrid laser and arc weld welding on low carbon and austenitic steel and concluded that zones influenced by hybrid arc had wider HAZ and deeper penetration with large grain size. Formation of austenite and martensite in hybrid and laser zone, respectively, due to the addition of filler. Good tensile and ductile behaviour was observed in the hybrid zone compared to the laser zone. Antony and Rakeshnath [36], joined Cu with SS316L welded butt joints were prepared using a CO_2 laser. Two levels of samples were created with different power but the same speed. Results showed that samples made with more power showed higher tensile strength; also, a proper fusion was observed in that case. Mai and Spowage [37] fabricated Steel-Kovar, using Nd-YAG laser, using a 350 *W* power-driven laser without filler material used to join the material. EDX analysis showed a uniform mixture in the weld zone of steel and Kovar. Comparative analysis showed that weld depth decreased with an increase in welding speed, pores decreased, and their size increased.

CONCLUSION

The paper discussed various challenges encountered while joining dissimilar metals. Optimization methods associated with the different welding process to get the optimum output such as residual stress, tensile strength , bead geometry, toughness etc are covered in this survey. ANOVA was applied to the process to point out the factors significantly affecting the process. RSM methodology is helpful in graphically predicting the relation of various input parameters with respect to output parameters simultaneously. Application of simulation packages to study the laser welding process and various output parameters such as distortion, residual stresses and thermal distribution was studied. Various heat models were applied and the model which simulated the LASER welding process was also studied.

REFERENCES

[1]   S. Madhavan, M. Kamaraj, and L. Vijayaraghavan, "Cold metal transfer welding of aluminium to magnesium: Microstructure and mechanical properties", *Sci. Technol. Weld. Join.,* vol. 21, no. 4, pp. 310-316, 2016.

[2]   K. Fahlström, O. Andersson, U. Todal, and A. Melander, "Minimization of distortions during laser welding of ultra high strength steel", *J. Laser Appl.,* vol. 27, no. S2, p. S29011, 2015.

[3]   N. Kumar, S. Kataria, B. Shanmugarajan, S. Dash, A.K. Tyagi, G. Padmanabham, and B. Raj, "Contact mechanical studies on continuous wave CO_2 laser beam weld of mild steel with ambient and under water medium", *Mater. Des.,* vol. 31, no. 8, pp. 3610-3617, 2010.

[4]   K. Bandyopadhyay, S.K. Panda, and P. Saha, "Optimization of fiber laser welding of DP980 steels using RSM to improve weld properties for formability", *J. Mater. Eng. Perform.,* vol. 25, no. 6, pp. 2462-2477, 2016.

[5]   B. Das, and P. Biswas, "Effect of operating parameters on plate bending by laser line heating", *Proc. Inst. Mech. Eng., B J. Eng. Manuf.,* vol. 231, no. 10, pp. 1812-1819, 2017.

[6]   R. Bhattacharya, N. Kumar, N. Kumar, and A. Bandyopadhyay, "A study on the effect of process parameters on weld width and heat affected zone of pulsed laser welding of dissimilar transparent thermoplastics without filler materials in lap joint configuration", *Mater. Today Proc.,* vol. 5, no. 2, pp. 3674-3681, 2018.

[7]   S.C. Juang, and Y.S. Tarng, "Process parameter selection for optimizing the weld pool geometry in the tungsten inert gas welding of stainless steel", *J. Mater. Process. Technol.,* vol. 122, no. 1, pp. 33-37, 2002.

[8]   E.M. Anawa, and A.G. Olabi, "Effects of laser welding conditions on toughness of dissimilar welded components", *In Applied Mechanics and Materials Trans Tech Publications Ltd.,* vol. 5, pp. 375-380, 2006.

[9]   M.P. Prabakaran, and G.R. Kannan, "Optimization of CO_2 laser beam welding process parameters to attain maximum weld strength in dissimilar metals", *Mater. Today Proc.,* vol. 5, no. 2, pp. 6607-6616, 2018.

[10]   B. Kalins, V. Naveen, L. Girisha, U.N. Kumar, R. Subbiah, and S. Marichamy, "Numerical investigation on laser welding of titanium plates", *Mater. Today Proc.,* vol. 45, pp. 2224-2227, 2021.

[11]   SR Patil, and CA Waghmare, "Optimization of MIG welding parameters for improving strength of welded joints", *Int. J. Adv. Engg. Res.,* vol. 14:16, 2013.

[12]   H. Vemanaboina, G. Edison, and S. Akella, "Effect of residual stresses of GTA welding for dissimilar materials", *Mater. Res.,* p. 21, 2018.

[13]   P. Roshith, M. Arivarasu, N. Arivazhagan, and A. Srinivasan, "KV PP. Investigations on induced residual stresses, mechanical and metallurgical properties of CO2 laser beam and pulse current gas tungsten arc welded SMO 254", *J. Manuf. Process.,* vol. 44, pp. 81-90, 2019.

[14]   S. Mohanty, M. Arivarasu, N. Arivazhagan, and K.P. Prabhakar, "The residual stress distribution of CO_2 laser beam welded AISI 316 austenitic stainless steel and the effect of vibratory stress relief", *Mater. Sci. Eng. A,* vol. 703, pp. 227-235, 2017.

[15]   G. Madhusudhan Reddy, and P. Venkata Ramana, "Influence of filler material composition on residual

stress distribution of dissimilar gas tungsten arc weldments of ultrahigh strength steels", *Sci. Technol. Weld. Join.,* vol. 16, no. 3, pp. 273-278, 2011.

[16] R.A. Sindhu, M.K. Park, S.J. Lee, and K.D. Lee, "Effects of residual stresses on the static and fatigue strength of laser-welded lap joints with different welding speeds", *Int. J. Automot. Technol.,* vol. 11, no. 6, pp. 857-863, 2010.

[17] T. Bajpei, H. Chelladurai, and M.Z. Ansari, "Experimental investigation and numerical analyses of residual stresses and distortions in GMA welding of thin dissimilar AA5052-AA6061 plates", *J. Manuf. Process.,* vol. 25, pp. 340-350, 2017.

[18] W. Shichun, and Z. Jinsong, "An experimental study of laser bending for sheet metals", *J. Mater. Process. Technol.,* vol. 110, no. 2, pp. 160-163, 2001.

[19] M Mochizuki, and M Toyoda, "Weld distortion control during welding process with reverse-side heating", *Journal of Engineering Materials and Technology, Transactions of the ASME,* vol. 129, no. 2.

[20] O.P. Gupta, "A Study of Distortion in Welding", *AIMTDR Conference* 2007, pp. 265-270 .

[21] H.C. Chen, A.J. Pinkerton, and L. Li, "Fibre laser welding of dissimilar alloys of Ti-6Al-4V and Inconel 718 for aerospace applications", *Int. J. Adv. Manuf. Technol.,* vol. 52, no. 9, pp. 977-987, 2011.

[22] S. Chen, J. Huang, J. Xia, X. Zhao, and S. Lin, "Influence of processing parameters on the characteristics of stainless steel/copper laser welding", *J. Mater. Process. Technol.,* vol. 222, pp. 43-51, 2015.

[23] M.A. Attar, M. Ghoreishi, and Z.M. Beiranvand, "Prediction of weld geometry, temperature contour and strain distribution in disk laser welding of dissimilar joining between copper & 304 stainless steel", *Optik,* vol. 219, p. 165288, 2020.

[24] T. Bajpei, H. Chelladurai, and M.Z. Ansari, "Numerical investigation of transient temperature and residual stresses in thin dissimilar aluminium alloy plates", *Procedia Manuf.,* vol. 5, pp. 558-567, 2016.

[25] A.H. Monfared, "Numerical simulation of welding distortion in thin plates", *J. Engi. Phy. Thermo.,* vol. 85, no. 1, pp. 187-194, 2012.

[26] F. Lerra, A. Ascari, and A. Fortunato, "The influence of laser pulse shape and separation distance on dissimilar welding of Al and Cu films", *J. Manuf. Process.,* vol. 45, pp. 331-339, 2019.

[27] K. Mathivanan, and P. Plapper, "Laser welding of dissimilar copper and aluminum sheets by shaping the laser pulses", *Procedia Manuf.,* vol. 36, pp. 154-162, 2019.

[28] P.K. Ghosh, and Vijay Sharma, "Chemical composition and microstructure in pulse mig welded Al-Zn-Mg Alloy", *Mater. Trans.,* vol. 32, no. 2, pp. 145-150, 1991.

[29] V. Dimatteo, A. Ascari, and A. Fortunato, "Continuous laser welding with spatial beam oscillation of dissimilar thin sheet materials (Al-Cu and Cu-Al): Process optimization and characterization", *J. Manuf. Process.,* vol. 44, pp. 158-165, 2019.

[30] Q.S. Liu, S.M. Mahdavian, D. Aswin, and S. Ding, "Experimental study of temperature and clamping force during Nd: YAG laser butt welding", *Opt. Laser Technol.,* vol. 41, no. 6, pp. 794-799, 2009.

[31] P. Pankaj, A. Tiwari, R. Bhadra, and P. Biswas, "Experimental investigation on CO2 laser butt welding of AISI 304 stainless steel and mild steel thin sheets", *Opt. Laser Technol.,* vol. 119, p. 105633, 2019.

[32] R.S. Sharma, and P. Molian, "Yb: YAG laser welding of TRIP780 steel with dual phase and mild steels for use in tailor welded blanks", *Mater. Des.,* vol. 30, no. 10, pp. 4146-4155, 2009.

[33] W.F. Xu, M.A. Jun, Y.X. Luo, and Y.X. Fang, "Microstructure and high-temperature mechanical properties of laser beam welded TC4/TA15 dissimilar titanium alloy joints", *Trans. Nonferrous Met.*

Soc. China, vol. 30, no. 1, pp. 160-170, 2020.

[34]　D. Wu, B. Cheng, J. Liu, D. Liu, G. Ma, and Z. Yao, "Water cooling assisted laser dissimilar welding with filler wire of nickel-based alloy/austenitic stainless steel", *J. Manuf. Process.,* vol. 45, pp. 652-660, 2019.

[35]　X. Zhang, G. Mi, and C. Wang, "Microstructure and performance of hybrid laser-arc welded high-strength low alloy steel and austenitic stainless steel dissimilar joint", *Opt. Laser Technol.,* vol. 122, p. 105878, 2020.

[36]　K. Antony, and T.R. Rakeshnath, "Dissimilar laser welding of commercially pure copper and stainless steel 316L", *Mater. Today Proc.,* vol. 26, pp. 369-372, 2020.

[37]　T.A. Mai, and A.C. Spowage, "Characterisation of dissimilar joints in laser welding of steel–kovar, copper–steel and copper–aluminium", *Mater. Sci. Eng. A,* vol. 374, no. 1-2, pp. 224-233, 2004.

CHAPTER 2

A Study on Friction Stir Welding of Composite Materials for Aerospace Applications

Spruha Aniket Dhavale[1,*] and **Shivprakash Bhagwatrao Barve**[1]

[1] *School of Mechanical Engineering, Dr. Vishwanath Karad MIT World Peace University, Pune, Maharashtra, India*

Abstract: Composite materials define the new age for the development of technologies. In a recent study of aerospace structure, conventional materials are replaced by almost 75% by modern composites. The strength-to-weight ratio of aluminum alloys makes it an attractive choice among other materials for industries. Carbon fiber-reinforced composite materials are better alternatives to conventional materials to prepare lighter panels. In many applications, the joining of these composites becomes troublesome. The promising technique to join the advanced composites is still unrevealed. The proper joining method is an essential need for composites. With this aim, we carry out the study to check the feasibilities of Friction Stir Welding for light panels used in aerospace industries. Friction Stir Welding is a solid-state welding method. The external tool solidifies and mixes the metal at the edge line. We reviewed several research articles to understand the difficulties associated with the friction stir welding method.

Keywords: Aluminum metal composite, Friction stir welding, Light weight structures.

INTRODUCTION

The eco-system balance and sustainability concerns increase the demand for lightweight structures in aerospace. The reduction in fuel consumption is possible only through the overall weight reduction of an aircraft [1].

Aluminium alloys are the most suitable choice for structural parts as they are less costly and provide a good weight-to-strength ratio [2]. Scientists have achieved advances in aluminium alloys for these explicit requirements, such as high-temperature resistance, corrosion resistance, and high load-bearing capacities [3].

---
[*] **Corresponding author Spruha Aniket Dhavale:** School of Mechanical Engineering, Dr.Vishwanath Karad MIT World Peace University, Pune, Maharashtra, India; E-mail: spruha.dhavale@gmail.com

Fig. (**1**) shows the particular requirements and stresses encountered by a specific part of an aeroplane.

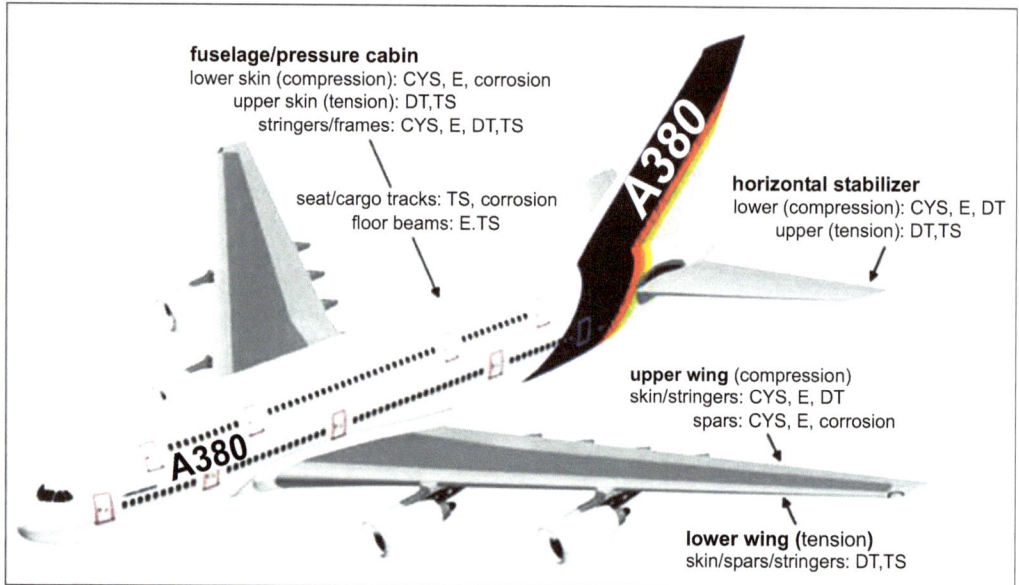

Fig. (1). Property requirements of the specific section concerning stresses encountered [2].

The composite materials are made out of these necessities [4]. Now, these materials define the new age for the development of spacecraft technologies. Several studies have been carried out on aluminum-based composites to get enhanced properties and energy-efficient materials [5, 6]. Wide varieties of aluminum-based composites are available which are used in the aerospace field. Conventional welding showed some limits to forming the solid joint in between the composites. Numerous problems are stated with the fusion welding of these materials, including thermal expansion, gas solubility and oxide formations [7] In many applications, the joining of these composites becomes troublesome. The promising technique to join the advanced composites is still unrevealed [8]. It's essential to get proper joining techniques to prepare firm joints between composites.

The FSW found an appropriate method to join aluminium-based composites. This opens new opportunities for preparing high-quality welds for composites [9, 10]. The FSW is a process where metals are solidified and mixed with the help of an external tool [11]. Fig. (**2**) shows the process and weld created by FSW. The tool in FSW is capable to generate sufficient frictional heat, which is below the melting point. The metals undergo plastic deformation due to stirring action and mix which result in strong weld formation. This welding process avoids formations for defects such as voids, porosities, shrinkages and segregation of reinforcements. The defect rate is reduced up to zero percent through extensive control of the process parameters.

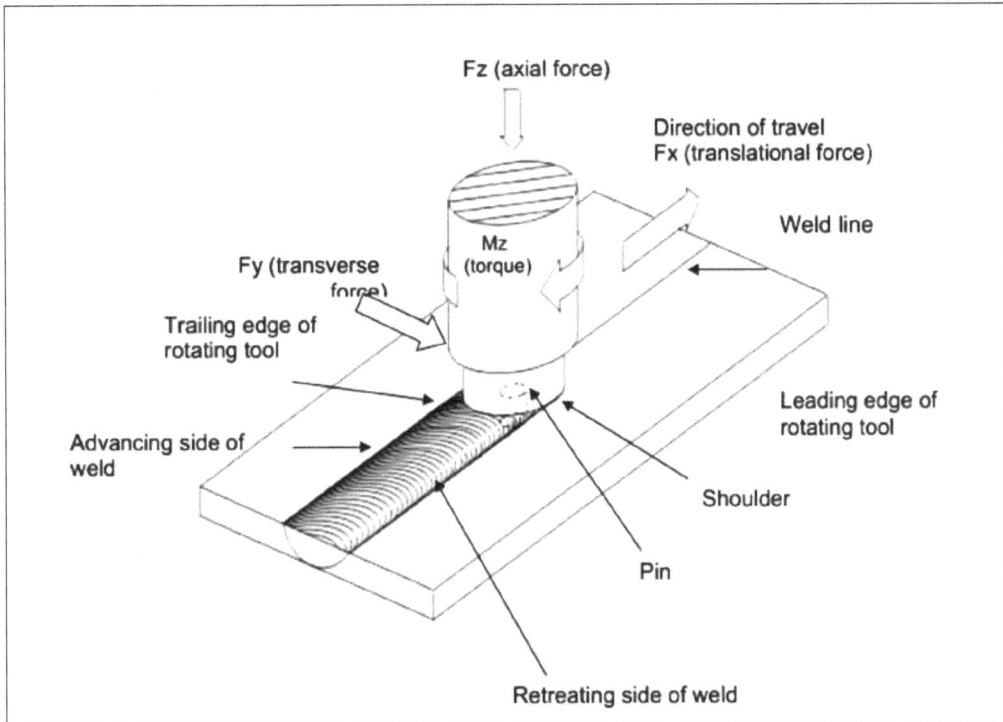

Fig. (2). Friction Stir Welding Process [11].

The aluminium alloys in aerospace industries have been explored by Rajan *et al.* The authors carried out the comparative analysis and discussed enhancement in the mechanical properties of aluminium's 2000 series, 7000 series and aluminium lithium alloys through FSW welding. In addition, a review of developments in aluminium alloy joining methods has been carried out. Their study indicated that FSW was the best choice for aerospace joining applications. FSW was proven to be a game-changer since the joints were seamless, had fewer defects, allowed low welding loads, and improved weld quality [12].

Aluminium Composite Materials

Aluminium Matrix Composites (AMCs) have properties that can improve a product's performance and efficiency [13]. It has an excellent strength-to-weight property, making it ideal for aerospace applications [14]. Table **1** shows the list of nanoparticles used for reinforcement in the Aluminium metal matrix. Liu *et al.* studied and summarised the research work performed on carbon nanotubes with metal matrix [15, 16]. Due to their high specific strength, these composites are expected to use in structural applications.

Table 1. List of nanoparticles for reinforcement [17].

Sr. No.	Nanoparticles Type	Subtype	Properties
1	Metal oxide nanoparticles	Iron oxide	Corrosion sensitive, unstable and reactive
		Silicon dioxide	Less toxic, stable
		Aluminium oxide	Highly reactive, highly temperature-sensitive
		Zinc oxide	Antibacterial, anti-corrosive
		Titanium oxide	Magnetic
2	Carbon nanoparticles	Carbon nanofiber	High electrical and thermal conductivity, good mechanical properties
		Carbon nano tubes	Flexibility, Elasticity, and considerable strength
		Graphene	Extremely high strength, Ultralight, immensely tough, Flexible, compact structure, Transparent, thinnest material
		Fullerenes	Transmit light. Can be a semiconductor and superconductor
		Carbon black	Electrically and thermally conductive, good mechanical properties

Bakshi *et al.* summarised the improvements in mechanical properties attained by adding CN to various metal matrix systems. The processes utilised to make the composites had been scrutinised to achieve a uniform distribution of CN in the matrix. The structural and chemical stability of CN in various metal matrixes were discussed, as well as the significance of the CNT/metal contact [18]. Sharma *et al.* conducted a tribological study *via* FSP using the surface alteration of Al6061-SiC through the dressing of CN, Graphite and Graphene. The friction coefficient was decreased because of the Graphene and SiC reinforcements. The same effect was reported in the case of wear rate. The combination of CN and SiC reinforcements within the alloy affected the wear resistance significantly. The major reasons for the improvement in wear resistance were the layered structure of the graphene flakes. The large specific surface area and wrinkled morphology also participated in increasing wear resistance. The surface properties were greatly affected by the impact of plastic flow stresses found in graphene-layered nano-platelets. Fig. (3) shows the FSP technique to prepare the composite surface on the metal alloy.

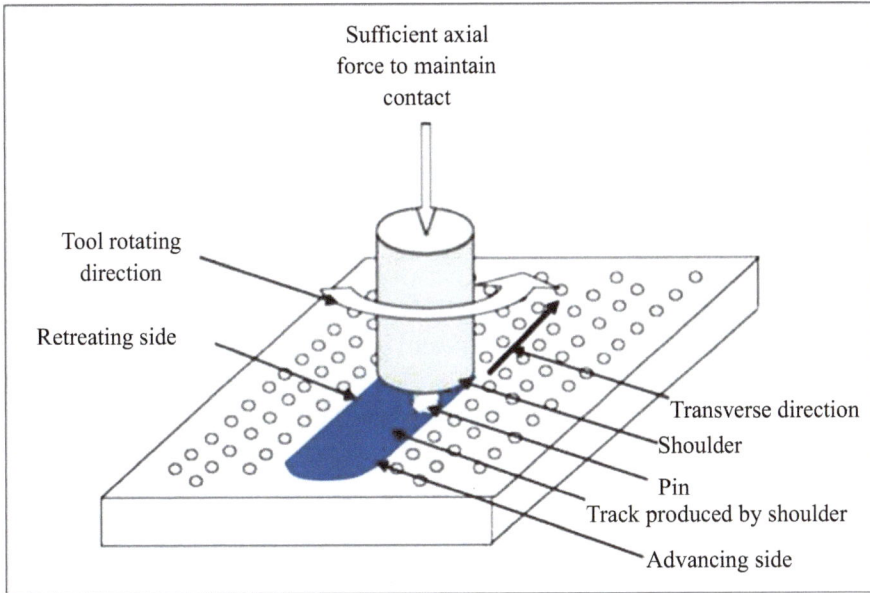

Fig. (3). Metal matrix composite preparation through friction stir processing [19].

Multi-pass friction stir processing (FSP) was used to disperse CN into an Al matrix with 4.5 vol. percent CN/2009Al composites. Fig. (**4**) shows the effect of CN tube addition in aluminium alloys. The maximum strength was achieved using three-pass FSP, which was attributable to a combination of CN cluster reduction, grain refining, and CN shortening. To characterise the change in CN tube length with FSP passes, a CN shortening model was presented by the authors [15].

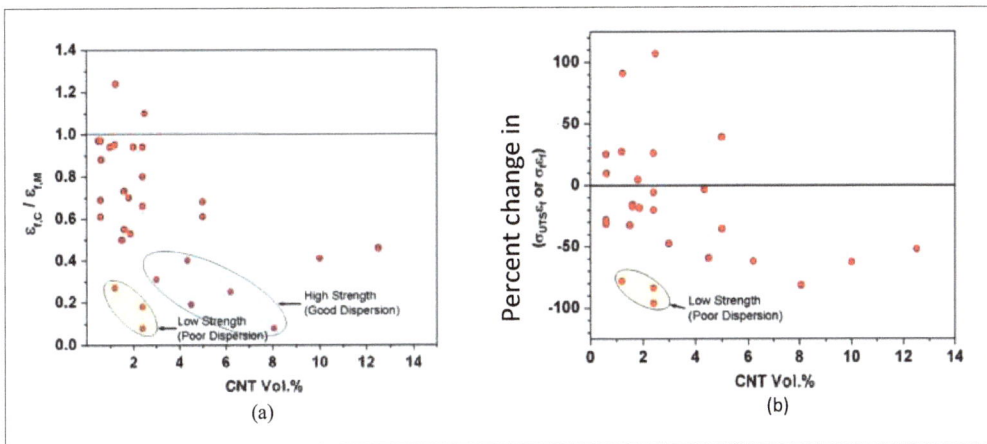

Fig. (4). CN effects on the ratio of failure (a) strain (b) stress of Al/CNT composite [18].

FACTORS RESPONSIBLE FOR WELD STRENGTH

The process parameters and the tool geometry affect the weld strength. Many researchers provided the correlations between process parameters and strengths.

Process Parameters

Two significant process parameters influence the weld properties of the aluminium-based composites, the V (traverse speed) with which the tool moves in the forwarding direction and R (rotational speed) with which the tool rotates around its axis. The material flow, as well as heat input, are affected by both the parameters correspondingly (V and R). The heat generated is directly proportional to R. It is inversely proportional to V. The appropriate choice among these parameters leads to quality weld. Uttam Acharya *et al.* investigated the effect of R on AA 6092/17.5 SiCp – T6 composites. They reported that the hardness of the joint is maximum at 2000 rpm. The impact strength was reduced at a higher R [20]. Tanvir Singh *et al.* recorded the optimal R and V as 2000 rpm and 70 mm/min, respectively, for quality weld [21, 22]. The effects of process parameters on welding AA6082/SiC/10P composite plates were studied by Bhushan *et al.* They have investigated the UT of FSW welds. The 1800 R, 100 V and α of 2 degrees was the best combination found to obtain quality weld of said composites. The heat generation at low R was not sufficient to strain the semisolid material. Therefore, the tunnel defects were observed and resulted in lower tensile strength. The heat input was drastically increased at high R, which resulted in macro-level defects, and low strength was reported. The higher heat input was given at lower V at the adjoining line as a longer contact time was provided. It affected decelerating the cooling rate, which results in grain coarsening [23]. Table **2** reveals the work done with different tool designs and process parameters on metal matrix composites.

Table 2. Work done with various tool design and process parameters on metal matrix composites.

Author and Year	Material	Tool Size and material	Parameter	Findings
Rajesh Kumar Bhushan (2019) [23]	AA6082/SiC/10P Size 100X50X6	Shoulder 15 mm, Cylindrical Pin 6 mm D and 5.7 mm L, AISI steel	R 1500, 1800, 2100, V 50,100, 150, α 1, 2, 3°	The UT of friction stir weld of composite is significantly dependent on the tool R, V and α.
Uttam Acharya (2020) [20]	AA6092/17.5 SiCp-T6, Size 150X60X6	Taper Pin 6:3 D. 5.7 mm L, 18 mm D shoulder, H13 steel	R 1000, 1500, 2000 1 mm/s V, α 2 °	The defect-free welds were formed. The hardness of the weld zone varied throughout the thickness

(Table 2) cont.....

Author and Year	Material	Tool Size and material	Parameter	Findings
Tanvir Singh (2019) [21]	AA6061-T6 Thickness 2.5 mm	Cylindrical shoulder with 15mm D, Pin 5 mm D, L 2.30 mm	R1800, 2000, and 2200 rpm, V 60, 70, and 80 mm/min, α 0 °	The presence of nanoparticles affected the ultimate strength due to grain refinement at the nugget zone.
Bodaghi (2017) [24]	AA5052 4 mm thick sheets with (SiC nanoparticles)	A shoulder (concave) with 14 mm D, Threaded pin, 4 mm D, L 3.7 mm, H13 steel	R 800, 1000, and 1250, V 0, 50, and 80 mm/ min	The grain size was reduced due to SiC nanoparticles which resulted in improvements in properties.
Birru (2019) [3]	AA6082-T6, Size 150x75x2	20 mm shoulder,7 mm pin L, hexagon profile of 4mm side, Tungsten carbide	R 710 and 900, V 40	Minimum of three passes were found to be required to achieve the highest of the composite.
Almomani (2013) [10]	AMC (with SiC and Graphene particles) plates Size 100X75 X7.5	Square, Hexagonal, Octagonal pin profiles, Tool size is not mentioned	R 630, 800, 1000,1250 V 35, 45,55, 65	The square pin profile provided enhanced corrosion resistance. An increase in R gave a negative effect, while V gave a positive effect on corrosion resistance.
Devanathan (2014) [25]	Aluminum alloy (5% SiC particles reinforcement) Size 100 X 50 X 6 mm	Taper pin 6:4, shoulder 19.5 mm thick, and pin L 5.7 mm	R 1200, 1500, 1800 V 20, 40, 60 F 6, 7, 8 kN	35% contribution of axial force followed by 25% of V and 12% of R in creating the quality weld were noted.
Nikoo (2016) [26]	AA6061-T6 alloy 6 mm thick, with Al2O3 in the groove	Shoulder 18 mm D, threaded pin M6.1, height 5.7 mm. H13 steel, 58 HRC	R 1600, V 25, 40, 45, 63	At 40 mm/min V most homogeneous dispersions of nanoparticles were observed by compromising on the hardness of the weld.
Khodabakshi (2018) [27]	AA5052, 70 x210 x5, MWCNT of 20-50 nm dia, 1-2 um L	Shoulder 18 mm Threaded pin M5, 1 4 mm, concave angle 2.5 degrees, depth of insertion 0.5 mm, H13 steel	R 1250, V 25	55% improved hardness and 110% improved tensile strength was obtained.
Huang (2018) [28]	Mg Alloy, with CNTs of 40~60 nm D and ~2 μm L	Shoulder 15 mm, Threaded pin D 3 mm and L 4.5 mm, A 1.5°	R 1000 V of 50	A remarkable improvement in YS171 MPa, UT330MPa, and E 15% was reported as compared to the base alloy.

Tool Design and its Position

Informing the quality weld, tool design plays a vital role. The tool consists of two distinct parts named shoulder and pin. Various tools are experimented with by the researchers to understand the joining mechanism information of welds, as mentioned in Table **2**. The pin profile and its diameter is responsible for stirring the metals in contact, while the shoulder keeps the metals in their place by providing vertical force. The material of the tool should be hard enough to withstand heavy shear stresses. H13 is the tool material used in FSW. Other tool profiles and materials are given in Table **2**. Plain, cylindrical, threaded patterns or flute-type designs of the pin are tried forgetting the flow behaviour of metals at the weld line. When the tool axis line shifts toward either of the plates, then it is considered to offset positioning [5]. This arrangement is found advantageous in the case of dissimilar metal welding. In many cases, zero tool offset (which means the tool is exactly located at the adjoining edges of the metals to be joined) is used to form the weld. It is observed that the proper tool positioning provides a continuous blending of the materials at the adjoining line.

FSW MACHINES

By using the specialized machine, it is possible to carry the FSW for long-distance. Fig. (**5**) shows some of the specialized machines where precise control of the forces would be possible. The downward force is controlled with the help of a sensor which is one of the essential factors for quality weld formation [9]. FSW was carried on a vertical milling machine or VMCs in most of cases. The controlling of forces in all directions is quite difficult.

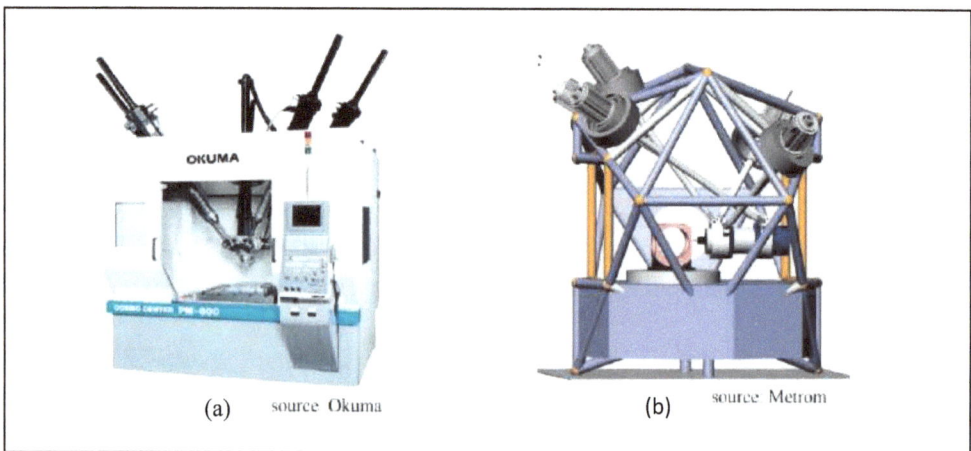

(a)   source Okuma   (b)   source Metrom

Fig. (5). (a) Hexapod (b) Pentapod machines [9].

FSW provided improved weld seam characteristics that were needed for dissimilar welding (for example, aluminium-copper) in the battery design of automobiles. Author Huang *et al.* used the numerical control machine for FSP [24].

MECHANICAL PROPERTIES OF WELD

An aluminium alloy reinforced with multiwalled CN was produced with the help of the FSP by Devanathan et.al. The effect of R and the penetration depth of the tool were studied on the homogeneity dispersion of nanotubes within the alloy [25]. Multi-pass FSP was used to uniformly distribute CN within the metal matrix. The uniform dispersions of nanoparticles resulted in the enhancement of mechanical properties. Fig. (6) shows the preparation of the specimen for testing from welded plates to understand the mechanical properties of the welds formed by FSW.

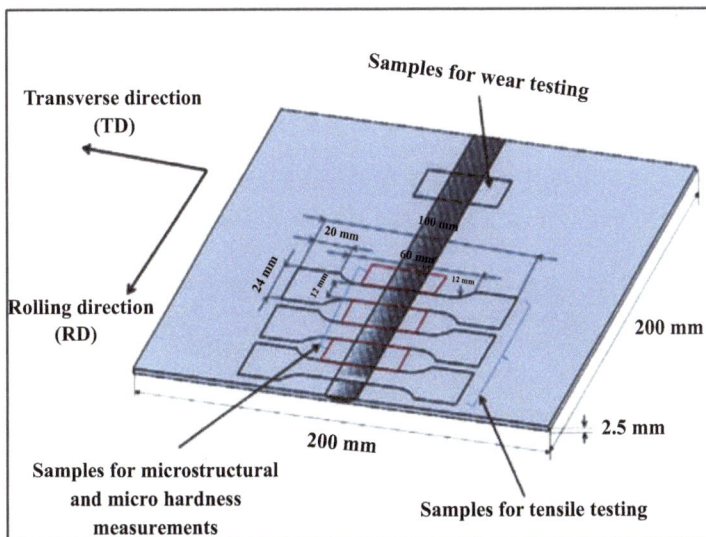

Fig. (6). Specimens for testing purposes from FSW plates [21].

The strengthening mechanisms responsible for improving the YS or UT were observed in grain refinement. The results indicated that the YS 171 MPA, UT 330 MPa, and E 15% of the composites, respectively, were 144%, 156%, and 87% higher than the received Mg-6Zn alloy. The phases were found and resistant to wear and corrosion. In comparison to the Al-Mg-based alloy, the microstructural change increased the hardness and tensile strength of the treated nanocomposite by 55% and 110%, respectively [26]. The fabrication route was effective in developing novel CN-reinforced metal matrix composites with superior mechanical properties [27]. The microstructural characteristics help in the better understanding of the welds formed by FSW.

Microstructural Characteristics

SEM and TEM microscopy testing was adopted to understand the grain development and recrystallization mechanisms within the composite material. By reactive FSP of an Al-Mg alloy containing about 2.5 vol. % embedded multi-walled CN, a fine-grained aluminium matrix hybrid nanocomposite was made-up, as shown in Fig. (**7**) [26]. It was demonstrated that 1400 rpm R, 50 mm/min V and 5 cumulative FSP passes gave rise to 510°C temperature, which was capable of producing a homogeneous distribution of reinforcement. The applied severe plastic deformation also changes the structure of precipitates and grains, which improves the indentation and tensile properties of the processed alloy [28]. The characterization images of the material revealed the strengthening mechanisms of the composites. These TEM images exposed the microstructural evolution and mechanical properties. The compact bonding of the singly dispersed CN with the matrix contributed to the grain refinement of the Mg-6Zn matrix [27]. The load transfer, grain refinement, and Orowan looping mechanisms were used to make the strong joints. Though FSW makes firmed joints, defects are found near the vicinity of the nugget zone. Inappropriate process parameters lead to defective welds.

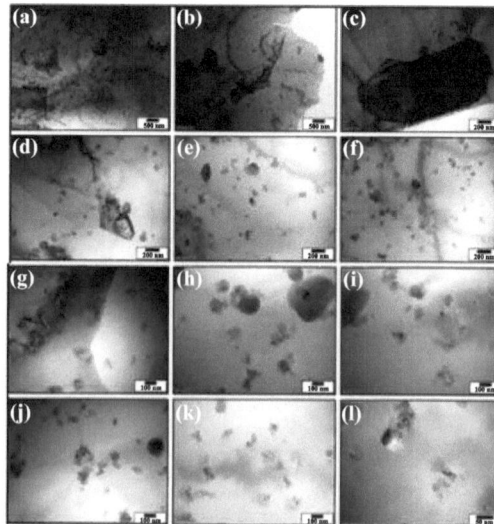

Fig. (7). TEM images showing the dispersion of nanoparticles within aluminium metal matrix [26].

Weld Defects

The defect rate is quite low for FSW (as compared to conventional welding methods). The various defects observed within FSW joints are shown in Fig. (**8**). The process parameters, tool profiles, tool offset, clamping devices and vertical forces are important factors that are significantly responsible for forming the weld. The higher temperature generated at the adjoining line due to friction between the tool and metal matrix composites leads to higher grain growth. At lower R, the tunnel defect may appear as a result of a deficiency of stream of welded material. The voids may form due to progressive wear of the pin profile, which is responsible for material flow at the abutting line of composite plates.

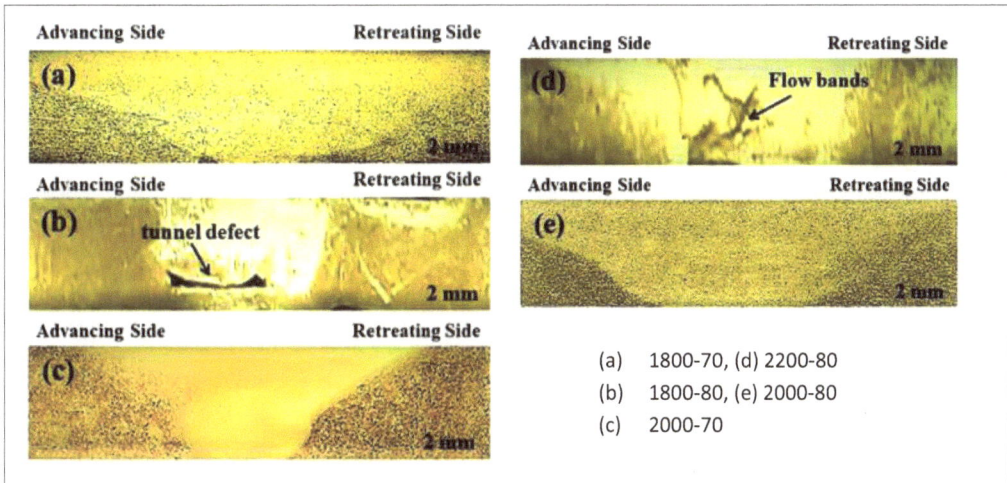

Fig. (8). Defects appeared at different combinations of processing parameters [21].

SCOPE FOR FURTHER RESEARCH

The scope of further research is mentioned after reviewing the research papers.

- The reinforcement within aluminium alloy steers the improvement in mechanical properties. Detailed analysis of CN dispersion within the metal matrix through the FSW is needed.
- The tool design is one of the essential factors in obtaining quality welds; therefore, the optimum profile concerning metal matrix composite thickness needs to be determined.
- The thin metal matrix sheets were used for experimentation. Comprehensive research is required for thick plates (> 3 mm).
- The right combination of process parameters ushers to uniform weld formation.

The appropriate method for the selection of the process parameters needs to be identified.

• The addition of reinforcing particles plays a vital role in enhancing material properties. The identification of reinforcing particle dispersion and strengthening mechanisms within the weld can be explored for a better understanding of the material behaviour.

CONCLUSION

In aerospace applications, aluminium metal matrix composite materials have the potential to prepare light structures. If nanoparticles like CN are reinforced in an aluminium-based metal matrix, they will have more prominent metal properties. The FSW techniques can be used to join advanced metal matrix composites. The scanning electron microscope, transmission electron microscope, electron backscattered diffraction or X-ray powder diffraction machines can be employed to characterize the weld zone for understanding the evolution phenomenon. The mechanical properties, including tensile strength, hardness, or impact strength, can be achieved by suitable experimental arrangements.

ABBREVIATION

FSW Friction Stir Welding

FSP Friction Stir Processing

Al Aluminum

AMC Aluminium Metal Composite

CN Carbon Nanotubes

R Rotational speed (RPM)

V Traverse speed (mm/min)

α Tilt angle (degree)

UT Ultimate Tensile Strength

YS Yield Strength

E Elongation (mm/mm)

D Diameter (mm)

L Length (mm)

REFERENCES

[1]   F.C. Campbell, *Manufacturing technology for aerospace structural materials*. Butterworth-Heinemann Publication, An Imprint of Elsevier Publications: New York, NY, USA, 2006.

[2]   P. Rambabu, N. Eswara Prasad, V.V. Kutumbarao, and R.J.H. Wanhill, Aluminium Alloys for AerospaceApplications.*In Aerospace Materials and Material Technologies* Indian Institute of Metals Series: India: Springer, 2017, pp. 29-52.

[3]   S.M. Birru, and A. Kumar, "Friction Stir Welding of AA6082 Thin Aluminium Alloy Reinforced with Al2O3 Nanoparticles", *Trans. Indian Ceram. Soc.,* vol. 78, no. 3, pp. 1-9, 2019.

[4]   T. Prater, "Friction stir welding of metal matrix composites for use in aerospace structures", *Acta Astronaut.,* vol. 93, pp. 366-373, 2014.

[5]   Virendra Pratap Singh, Surendra Kumar Patel, Alok Ranjan, and Basil Kuriachen, "Recent research progress in solid-state friction-stir welding of aluminium–magnesium alloys", *jmaterrestechnol* , pp. 6217-6256, 2020.

[6]   A.K. Sharma, R. Bhandari, and C. Pinca-Bretotean, "A systematic overview on fabrication aspects and methods of aluminum metal matrix composites", *Mater. Today Proc.,* pp. 1-6, 2021.

[7]   L. Ceschini, I. Boromei, G. Minak, A. Morri, and F. Tarterini, "Effect of friction stir welding on microstructure, tensile and fatigue properties of the AA7005/10 vol.%Al2O3p composite", *Compos. Sci. Technol.,* vol. 67, pp. 605-615, 2007.

[8]   A. Heidarzadeh, A. Radi, and G.G. Yapici, "Formation of nano-sized compounds during friction stir welding of Cu-Zn alloys: effect of tool composition", *J. Mater. Res. Technol.,* pp. 15874-15879, 2020.

[9]   A. Grimm, S. Schulze, A. Silva, G. Göbel, J. Standfuss, B. Brenner, E. Beyer, and U. Füssel, "Friction stir welding of light metals for industrial applications", *Mater. Today Proc.,* vol. 2, pp. S169-S178, 2015.

[10]   M. Almomani, A.M. Hassan, T. Qasim, and A. Ghaithan, "Effect of process parameters on corrosion rate of friction stir welded aluminium SiC–Gr hybrid composites", *Corros. Eng. Sci. Technol.,* vol. 48, no. 5, pp. 346-353, 2013.

[11]   D. L. Chen and J. 1 Zhan. Friction Stir Welding – From Basics to Applications. Woodhead Publishing Limited, Kaw, K. Second Edition, 2006. Mechanics of Composite Materials. Taylor and Franscis, 2010.

[12]   R. Rajan, P. Kah, B. Mvola, and J. Martikainen, "Trends in aluminium alloy development and their joining methods", *Rev. Adv. Mater. Sci.,* vol. 44, no. 4, pp. 383-397, 2016.

[13]   O.S. Salih, H. Ou, W. Sun, and D.G. Mc. Cartney, "A review of friction stir welding of aluminium matrix composites", *Mater. Des.,* pp. 61-71, 2015.

[14]   N. Kumar, A. Das, and S.B. Prasad, "An analysis of friction stir welding (FSW) of metal matrix composites (MMCs)", *Mater. Today Proc.,* p. 7, 2020.

[15]   Z.Y. Liu, B.L. Xiao, W.G. Wang, and Z.Y. Ma, "Analysis of carbon nanotube shortening and composite strengthening in carbon nanotube/aluminum composites fabricated by multi-pass friction stir processing", *Carbon,* vol. 69, pp. 264-274, 2014.

[16]   F. Khodabakhshi, A.P. Gerlich, and P. Švec, "Reactive friction-stir processing of an Al-Mg alloy with introducing multi-walled carbon nano-tubes (MW-CNTs)", *Mater. Charact.,* 2017.

[17]   P.P. Awate, and S.B. Barve, "Formation and characterization of aluminium metal matrix nanocomposites", *Technology,* vol. 64, no. 557, pp. 933-942, 2015.

[18]   S R Bakshi, D Lahiri, and A Agarwal, "Carbon nanotube reinforced metal matrix composites: A review", *International Materials Reviews,* vol. 1, pp. 41-64, 2010.

[19]   Y. Huang, J. Li, L. Wan, X. Meng, and Y. Xie, "Strengthening and toughening mechanisms of CNTs/Mg-6Zn composites via friction stir processing", *Mater. Sci. Eng. A,* vol. 732, pp. 205-211, 2018.

[20]   U. Acharya, B.S. Roy, and S.C. Saha, "Effect of tool rotational speed on the particle distribution in friction stir welding of AA6092/17.5 SiCp-T6 composite plates and its consequences on the mechanical property of the joint", *Defence Technology,* vol. 16, pp. 381-391, 2020.

[21]   S.K. Tanvir Singh, D.K. Tiwari, and S.K. Shukla, "Friction-stir welding of AA6061-T6: The effects of Al2O3 nanoparticles adition", *Results in Materials,* vol. 1, pp. 1-12, 2019.

[22]   S. Raja, M. R. Muhamad, M. F. Jamaludin, and F. Yusof, "A review on nanomaterials reinforcement in friction stir welding", *J. Mat. Res. and Techno.,* pp. 16459-16487, 2020.

[23]   R.K. Bhushan, and D. Sharma, "Optimization of FSW parameters for maximum UTS of AA6082/SiC/10P composites", *Adv. Compos. Lett.,* vol. 28, pp. 1-7, 2019.

[24]   M. Bodaghi, and K. Dehghani, "Friction stir welding of AA5052: the effects of SiC nano-particles addition", *Int. J. Adv. Manuf. Technol.,* vol. 88, no. 9-12, pp. 2651-2660, 2017.

[25]   C. Devanathan, and A. Suresh Babu, "Friction Stir Welding of Metal Matrix Composite using Coated tool", *Procedia Materials Science,* vol. 6, pp. 1470-1475, 2014.

[26]   M. Nikoo, H.A. Farahmand, N. Parvin, and H. Yousefpour Naghibi, "The influence of heat treatment on microstructure and wear properties of friction stir welded AA6061-T6/Al2O3 nanocomposite joint at four different travelling speed.""", *J. Manuf. Process.,* vol. 22, pp. 90-98, 2016.

[27]   F. Khodabakhshi, M. Nosko, and A.P. Gerlich, "Influence of CNTs decomposition during reactive friction-stir processing of an Al-Mg alloy on the correlation between microstructural characteristics and microtextural components", *J. Microsc.,* vol. 271, pp. 1-19, 2018.

[28]   Y. Huang, X. Meng, Y. Xie, L. Wan, and Lv. Zongliang, "Friction stir welding/processing of polymers and polymer matrix composites", *Compos., Part A Appl. Sci. Manuf.,* pp. 1-84, 2017.

CHAPTER 3

Binder Jetting: A Review on Process Parameters and Challenges

Kriti Srivastava[1,*] and **Yogesh Kumar**[1]

[1] *Department of Mechanical Engineering, National Institute of Technology, Patna-800005, Bihar, India*

Abstract: Binder jetting (BJ) is a 3D printing technology in which objects are manufactured from ceramics, metals, polymers, and composites. Binder jetting process incorporates various types of technologies, such as printing, deposition of powder, complex combination of the binder with powder, and post-processing of sintered part. BJ has high productivity with the utilization of a wide variety of powders. In BJ, the binder is combined with powder of materials that bond together to create an object in a layer-wise fashion that is generally modeled on CAD. In order to obtain desired product accuracy, the main challenges are balancing proper process parameters with manufacturing time, such as characteristics of powders (distribution of particle size, packing density and flowability of powders, green strength), characteristic of binders, *etc.* This paper gives a brief review of technologies, materials, defects and challenges of the binder jetting additive manufacturing process and their future trends.

Keywords: Additive manufacturing, Binder jetting, 3D printing.

INTRODUCTION

AM is a uniquely designed production technology that involves layering materials to make 3-D components right from CAD designs. The most noteworthy profit of AM over other manufacturing is the capability to cope with complex shapes and material complications those subtractive manufacturing technologies cannot create [1].

It enables the development of printed components along with the models. One of the primary advantages of this approach to product production is the capacity to make nearly any profile that would not be possible to the machine. Other profits comprise reduced time and cost, amplified human engagement, and, as an outcome, a rapid product development cycle [2]. Each design and development

---

* **Corresponding author Kriti Srivastava:** Mechanical Engineering Department, National Institute of Technology, Patna, Bihar, India; E-mail: kritis.phd20.me@nitp.ac.in

Amar Patnaik, Albano Cavaleiro, Malay Kumar Banerjee, Ernst Kozeschnik & Vikas Kukshal (Eds.)

procedure utilizing a 3D printing machine necessitates the operator performing a specific set of tasks. The accessibility of such a task sequence is emphasized by convenient 3D printing devices. These desk machines are distinguished by their low price, ease of usage, and capability to be used in an industrial and academic environment. Each stage in this system is expected to have a restricted number of alternatives and require the least effort. Nevertheless, this implies that there are restricted options, like restricted selection of materials and many other parameters with which to do experimentation. Bigger, further flexible machines are extra competent at being tailored to encounter a variety of user needs, but they are also more complex to handle.

The additive manufacturing technology has the following process chain, as shown in Fig. **(1)** [3].

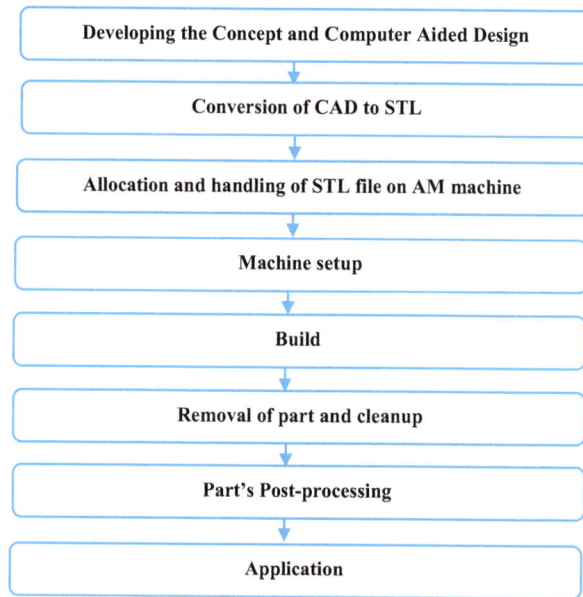

Developing the Concept and Computer Aided Design

Conversion of CAD to STL

Allocation and handling of STL file on AM machine

Machine setup

Build

Removal of part and cleanup

Part's Post-processing

Application

Fig. (1). Process chain of AM [3].

Classification of AM

According to the American Society for Testing and Materials (ASTM)/ International Organization for Standardization (ISO), AM is classified into seven categories, as shown in Fig. **(2)** [4]. All the major types of AM processes are as follows:

1. Powder Bed Fusion

2. Binder Jetting

3. Material Extrusion

4. Metal Jetting

5. Vat Polymerization

6. Sheet Lamination

7. Director Energy Deposition

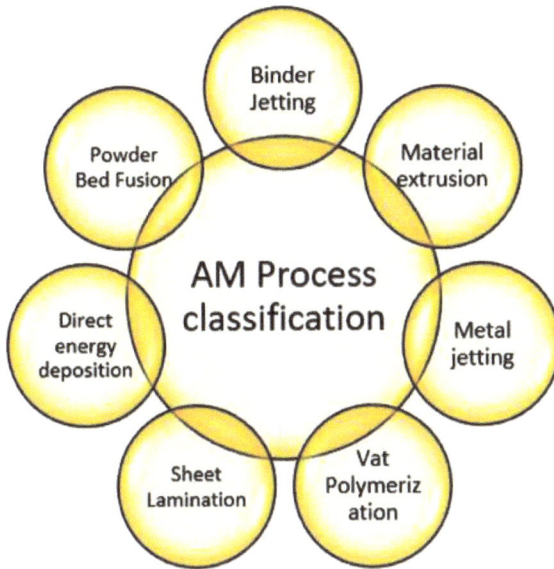

Fig. (2). Classification of AM technology [4].

Table **1** shows a comparison between different AM processes, and Table **2** shows a comparison of binder jetting with Selective Laser Sintering (SLS), Stereolithography (SLA), and Laminated Object Manufacturing (LOM).

Table 1. Comparison between the various AM processes on the basis of some features [5 - 12]

Process	Technologies	Form of feed material
Binder Jetting	Binder and powder 3D Printer	Powders and binder
Powder Bed Fusion	Electron Beam Melting, Selective laser sintering,	Powders
Metal Extrusion	FFF	Rods or wire
Sheet Lamination	Sheet forming	Sheet
Direct Energy De position	Laser Cladding	Wire

(Table 1) cont.....

Process	Technologies	Form of feed material
Vat Polymerization	Stereolithography	Photosensitive resin
Material Jetting	Polyjet	Molten wax as filler or Liquid photosensitive resin

Table 2. Comparison of BJ with SLS, SLA, BJ, LOM [13, 14]

Technology	Dimensional Accuracy	Advantages	Disadvantages
SLS	<0.05-0.1 mm	Support structures are not required. Varieties of materials are available such as ceramics, metals, and polymers. Curing after the process is not required.	Surface finish obtained is poor. Not suitable for the office atmosphere. Heating and cooling time for the printers are more.
SLA	<0.05 mm	Fast Stable Free of mask Aspect ratio is high Suitable For Office Environment.	Fragile Less cost-effective Less materials available for making filament.
BJ	0.2-0.3 mm	Cost is relatively low More choices of materials No support required	Strength is poor. Rough appearance. Requirement of post-processing.
LOM	0.15 mm	Support materials are not required. Large parts are manufactured quickly. Inexpensive for producing large parts.	Surface finish is poor. Accuracy is low. Difficult in making complex geometries and internal structures.

The criteria for selection of a suitable AM process, as shown in Fig. (**3**) depend on the availability of technology, materials available for that process, post-processing methods required, properties of the final product, and surface quality of the product produced [15].

The AM method of manufacturing is adjustable, adaptable, and highly configurable, making it ideal for a wide range of industrial applications. The four Ms that are considered important in additive manufacturing are shown in Fig. (**4**). AM's major focus has been on customizing low-volume, highly valued items that can be produced fast. Materials are extremely important in AM, particularly since the way AM manages materials differs significantly from traditional production. Certain varieties, shapes, and phases of materials are connected to machines and certain AM methods [16].

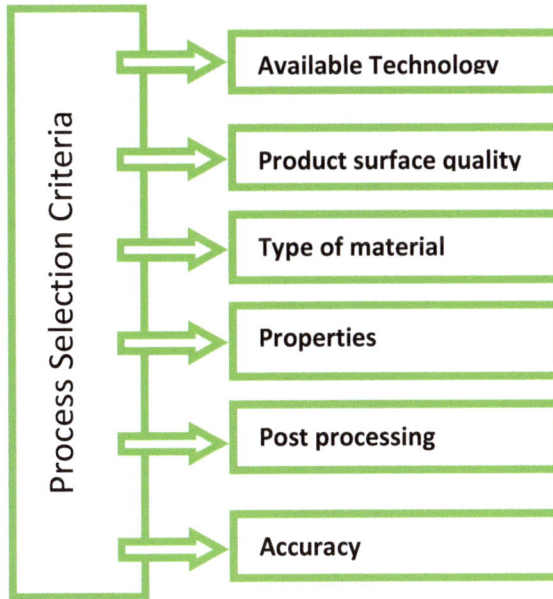

Fig. (3). AM Process selection criteria [15].

There were numerous facts to back up additive manufacturing's potential in these zones, including (i) Customized healthcare products to improve population health, wellness, and supremacy of life (ii) A simplified distribution chain to improve demand fulfilment durability and attentiveness, and (iii) A reduced environmental footprint for manufacturing adaptability [17].

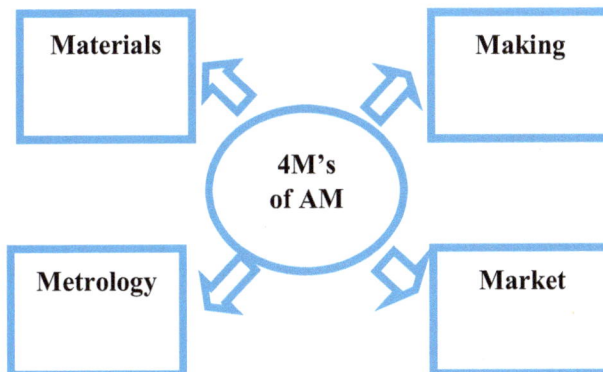

Fig. (4). Four M's of AM [16]

BINDER JETTING

Binder Jetting was developed by Sachs et al. in 1993 and is also called powder bed and inkjet head 3D printing [18]. The aim is to fetch 2D printing into the 3D model. It operates by injecting a liquid binder with one or more nozzles onto the top part of a bed of powder, cementing the powder particles together. Furthermore, nozzle tracks the predetermined path till the application of a slim coating of powder. Ultimately, layers are set to create a three-dimensional object. Binder jetted parts' post-processing is generally more difficult than other additive manufacturing processes, specifically for metals, because it results in the shrinkage of parts after the removal of the binder [19].

In commercialized BJT setups, material jetting processes are used to efficiently apply binding materials within a build platform. Before the binder jetting system traverses, a rake or wiper is used to ensure that the jetted binding agent combines with the powder that is freshly deposited. The powder and binding agent mixture can then be cured either thermally or with ambient air or by curing with UV [20].

Basic Principle

Powdered particles coupled with a liquid binder are utilized to create the component layer by layer. The powder roller forms the powder bed by spreading a slight layer of powder particles on the powder bed from the powder pool.

The printer then shoots a binder at the locations designated by the layer outline of the 3-Dimensional prototypical data subsequently, the particles of powder in those places are bound with particles adjacent to them.

After the deposition of one layer is finished, the build platform is adjusted downwards to a predetermined height, and a coating of powder is applied to the previously completed film. The repetition of process is done till the complete portion is completed. After that, the final part, known as the green part, is segregated from the free powder. Sintering, curing, debinding, and optional compaction are the post-processing stages. The basic phases of binder jetting additive manufacturing are depicted in Fig. (**5**). Following the preparation for raw material and printing procedures, the green portion will be given heat to a certain temperature (typically about 201°C). After that, the binder will be activated thermally, toughening the green sections by cross-linking, polymerization, evaporation of solvent, or other mechanisms.

The debinding procedure next removes the binder, leaving a brown-colored part in which the binder can be thermally dissolved or burned away in the air by raising the temperature of the component to between 600-800 degrees Celsius and sustaining the temperature for a period of time. To assure that the gaseous byproducts escape the part, the heating rate is controlled. Sintering is accomplished by heating the component to a high enough temperature (typically above 800°C) that is lesser than the material's melting point, than staying and cooling it. By the diffusion of mass across the boundaries of the particles, the sintering phase increases part density.

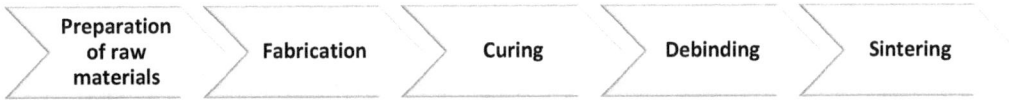

Fig. (5). Steps of binder jetting AM [21].

Densification techniques, such as infiltration, isostatic pressing, *etc.* could be utilized to enhance the density and hence mechanical qualities of a material [21]. Table 3 shows the salient features of binder jetting technology. The advantages and disadvantages of binder jetting are shown in Table. 4.

Table 3. Salient features of Binder Jetting Technology [16, 22, 23]

Features	Values
Specific energy density	0.026 Jcm^{-3}
Fabrication speed	1.27-1.9 cmh^{-1}
Resolution	1900 elements mm^{-3}
Dimension of BJ 3D printer	x≤4000, y≤2000, z≤1000
Company	3D System, ExOne,Voxelijet
Materials	Metal, ceramic, stainless steel, glass, polymer and sand. Other materials like tungsten carbide, ceramic beads, chromite, silica sand, zircon, bonded tungsten, soda lime glass *etc.* are available for industrial use.

Table 4. Benefits and Limitations of Binder Jetting Technology [24, 25]

Advantages	Disadvantages
Cost is relatively low.	Strength of the final product is poor.
More choices of materials.	The appearance of the product is rough.
Specific energy for imaging is low.	To improve the strength post processing is required.
Build volume is large	Final product has limited mechanical properties.
No support required	

POWDER AND BINDER

Powder Selection

The powder selection is one of the key components of binder jetting. Binder jetting's powder selection necessitates a thorough grasp of the entire production process. The powder utilized should be simple to process and ensure the system's constancy and sturdiness, along with a uniform bed of powder and part features. One of the other characteristics of powder particles should be that they must help in facilitating the process of sintering. Now the issue is that if the powder is made processable, then there will be a lack of sintering ability, and the same is the case with the reverse [26].

As a result, the deposition of powder is a critical factor in producing parts that are both reliable and fast. Powders could flow in an analogous way to fluids, but their performance is far more complicated. As forces between particles vary depending on particle size, geometry, chemical structure, and dampness, several techniques may be required to reliably generate thick, defect-free layers. The deposition method is determined by the flow characteristics of powders. Usually, large binder jetting powders are treated dry, while finer particles have been successfully deposited and distributed by fine-tuning the process conditions. It's possible that additional powder treatments are required. Small powders, on the other hand, can be agglomerated. To enhance packing density and uniformity, the powder is frequently disseminated in a fluid for the smallest size ranges [27].

Green Strength

Green Strength is the powder's strength before the compaction. The force of bonding between inter-particle metallic interactions during the process of compaction is primarily responsible for a powder compact's green strength [28]. Several factors can affect the green strength, such as the contact radius of tiny particles, the powder spreading method, and point of contact between the particles of the powder.

Flowability

The capability of powder to flow is critical for AM techniques that are based on layers. This feature allows for thin powder layers and, as a result, higher resolution printing. For successful 3D printing, an optimal flow ability is essential; however, a value greater than this affects the powder bed's stability, while a value lower than this diminishes the printed part's resolution. Flow ability is influenced by particle size and density. There will be a reduction in flowability if the agglomerated powder particles are of irregular shapes. Powder morphology also plays a key role [29].

Packing Density

The powder packing density might have a big impact on the final bulk density of 3D printed parts. For powders, three main sorts of densities that can be assessed are apparent density, skeletal density and tap density. The apparent density of a powder is useful in determining material properties that powder producers and users can use to determine the quality and good amount of uniformity. Scott volumeter, Hall flowmeter, Arnold meter, *etc.*, can be used to determine apparent density [30]. In order to have the maximum density of layer, good packing powder is necessary [31].

Selection of Binder

Choosing an efficient binder is the solution for achieving BJ. The following points should be kept in mind while doing so. They are:

1. The binder should be printable.

2. Binders should be able to withstand the shear stresses introduced by printing.

3. A low viscous binder would be excellent, allowing the spray of separate droplet particles to develop and then break free from the nozzles quickly.

4. Other requirements include better interaction with powder, clean burnout properties, a long-life cycle, and a low ecological [32].

In common, the bonding methods for a primary particle are categorized as:

- In-Bed binding, and
- In-Liquid

In liquid, jetting liquid carries the entire binding agent. In the case of bed binding, the entire binding agent is carried by the jetting liquid in liquid form. Binder in the dry form that is immersed in a powder bed interacts with the sample liquid to be printed. Hydration initiates the formation of inter-particle connections. In some cases, the fusion between the powder particles is facilitated by the solvent binder, such as in the case of polymers and also practiced with magnesium [33].

As the solvent effectively increases surface diffusion—a sintering process that is non-densifying as described in [34]. This minimizes impurities but could also limit the densification of sintering.

Several researchers have looked into the use of nanoparticles in binder. The objective is to fill the spaces between the bigger powders with jetted nanoparticles to increase sinterability and shrinkage [35]. Common types of binder are shown in Table **5**.

Table 5. Types of binders [36].

Liquid binders	Liquid binders are of 3 types. a) Binders that activate the reaction between the powders but can't bind to powder. b) Reactive binders c) Binder that possess binding effect
Solid binders	They are used in combination with liquid binders that activate the reaction between the powders.

Behavior of the binder drops is generally governed by three numbers that are, Weber numbers (We), Reynolds number (Re) and Ohnesorge Number (Oh) [37],

$$WeberNumber = \frac{InertialForce}{SurfaceTension}$$

$$We = \frac{\rho L V^2}{\sigma}$$

$$Reynolds\ Number = \frac{InertialForce}{ViscousForce}$$

$$Re = \frac{\rho V L}{\mu}$$

Where ρ is density, L is the characteristic length, μ is the dynamic viscosity, σ is the surface tension, V is velocity.

$$Ohnesorgenumber = \frac{\sqrt{Webernumber}}{Reynoldsnumber}$$

$$Oh = \frac{\mu}{\sqrt{\sigma L \rho}}$$

DEFECTS AND CHALLENGES

Binder jetting enables creation of complicated shapes, but the mechanical qualities of the sintered preform may be a difficulty. Zachary et al [38] performed the binder jetting of ferrous and have used bronze as an infiltered material to boost the strength of sintered parts. The infiltration largely influenced the transverse rupture strength (4 times more) by reducing the stress concentration spots at inter-particle necks, according to an examination of the surfaces of the infiltrated and sintered materials that are fractured.

Binder jetting technique has a commercial system that is fairly intricate.

Improvement of green density is required. Because of the bonding force between the binders and powders, the strength of green bodies generated by binder jetting is currently relatively low. Using a modified silicate binder in the process of printing, Shang et al. examined the influence of these parameters on the quality of the surface and relative density of samples produced from binder jetting. Green density can be improved by raising the saturation of binder within a particular range and reducing layer height. Excessive binder saturation lowers the green density, which can lead to the shifting of layers and geometric distortion. Future research may also concentrate on producing a high-loading binder to increase the strength and green density of ceramic pieces [39].

Shrinkage-This defect is observed after sintering which is not easy to control.

Coffee stain defect- This defect is caused by the dried droplets that have left the printer head, which may lead to non-uniform deposits on the initial droplets. This deposition is called a coffee stain [29].

CONCLUSION AND FUTURE SCOPE

Ceramic Powder needs an advanced milling process for better binder jetting technology. This will break the limitations and make it available for using this technology in other fields. High-strength binders are needed to increase the bonding force between the binders and powders. After mixing, the initiators and powders must crosslink the reaction either the atmospheric temperature (25°C or slight heating is much more useful. Very less number of simulations and experimental validations are reported. In this regard, the software simulations must be performed with adopted parameters and also m validated with experimental results. Post parameters, it is an independent process of AM. Densification of the process is needed to improve the properties of ceramics [36].

Droplet spreading and absorption dynamics may be influenced by the mobility of droplets impinging on the surface of a powder bed. For accurate dimensional products, the role of printing speed and its impact is also considered as an individual parameter to study the binder jetting process [40]. The role of Nano-particle densifiers and a printing liquid should be focused [41]. High temperatures of sintering result in more advanced sintering, which improves mechanical characteristics. For further examination, a research emphasis on mechanical characterization is required [42]. The influence of anisotropic infiltration may be researched in greater depth in the future, including how more complicated geometries are affected. The infiltration approach described here might be used to other materials for additive manufacturing with a higher base strength in future research. Sand or metal powders, or plaster with reinforcing elements like fibers, might be used in the BJ process. Polymer material extrusion and polymer powder bed fusion are two further methods that might be beneficial [43]. Powder having less flowability, on the other hand, necessitates a variety of methods to promote the spreading of powder, permeability of binder, and quality of part [44]. Sintering is a crucial procedure for increasing the density of the part and strength by removing the binder and hardening the part [45].

REFERENCES

[1]   H.A. Hegab, "Design for additive manufacturing of composite materials and potential alloys: A review", In: *Manufacturing Review.* vol. 3. EDP Sciences, 2016.

[2]   S. Ashley, *Rapid prototyping systems.* vol. 113. , 2008, p. 119892.

[3]   I. Gibson, D.W. Rosen, and B. Stucker, "Additive manufacturing technologies: Rapid prototyping to direct digital manufacturing", *Additive Manufacturing Technologies: Rapid Prototyping to Direct Digital Manufacturing.*, pp. 1-459, 2010.

[4]   M.D. Monzón, Z. Ortega, A. Martínez, and F. Ortega, "Standardization in additive manufacturing: Activities carried out by international organizations and projects", *Int. J. Adv. Manuf. Technol.,* vol. 76, no. 5–8, pp. 1111-1121, 2015.

[5]   A. Mostafaei, J. Toman, E.L. Stevens, E.T. Hughes, Y.L. Krimer, and M. Chmielus, "Microstructural evolution and mechanical properties of differently heat-treated binder jet printed samples from gas- and water-atomized alloy 625 powders", *Acta Mater.,* vol. 124, pp. 280-289, 2017.
[http://dx.doi.org/10.1016/j.actamat.2016.11.021]

[6]   Q. Yang, P. Zhang, L. Cheng, Z. Min, M. Chyu, and A.C. To, "Finite element modeling and validation of thermomechanical behavior of Ti-6Al-4V in directed energy deposition additive manufacturing", *Addit. Manuf.,* vol. 12, pp. 169-177, 2016.
[http://dx.doi.org/10.1016/j.addma.2016.06.012]

[7]   P. Ma, K. Prashanth, S. Scudino, Y. Jia, H. Wang, and C. Zou, "Influence of annealing on mechanical properties of al-20si processed by selective laser melting", *Metals,* vol. 4, no. 1, pp. 28-36, 2014.

[8]   M. Annoni, H. Giberti, and M. Strano, "Feasibility study of an extrusion-based direct metal additive manufacturing technique", *Procedia Manuf.,* vol. 5, pp. 916-927, 2016.
[http://dx.doi.org/10.1016/j.promfg.2016.08.079]

[9]   W. Gao, Y. Zhang, D. Ramanujan, K. Ramani, Y. Chen, and C.B. Williams, "The status, challenges, and future of additive manufacturing in engineering", *CAD Comput Aided Des,* vol. 69, pp. 65-89, 2015.
[http://dx.doi.org/10.1016/j.cad.2015.04.001]

[10]   S.M. Thompson, L. Bian, N. Shamsaei, and A. Yadollahi, "An overview of direct laser deposition for additive manufacturing; Part I: Transport phenomena, modeling and diagnostics", *Addit. Manuf.,* vol. 8, pp. 36-62, 2015.
[http://dx.doi.org/10.1016/j.addma.2015.07.001]

[11]   A. Peng, and Z. Wang, "Researches into influence of process parameters on FDM parts precision", *Appl. Mech. Mater.,* vol. 34–35, pp. 338-343, 2010.

[12]   K.G. Prashanth, and J. Eckert, "Formation of metastable cellular microstructures in selective laser melted alloys", *J. Alloys Compd.,* vol. 707, pp. 27-34, 2017.
[http://dx.doi.org/10.1016/j.jallcom.2016.12.209]

[13]   V. Manoharan, S.M. Chou, S. Forrester, G.B. Chai, and P.W. Kong, "Application of additive manufacturing techniques in sports footwear: This paper suggests a five-point scoring technique to evaluate the performance of four AM techniques, namely, stereolithography (SLA), PolyJet (PJ), selective laser sintering (SLS) and three-dimensional printing (3DP), in four important aspects of accuracy, surface finish, range of materials supported and building time for prototyping sports footwear", *Virtual Phys. Prototyp.,* vol. 8, no. 4, pp. 249-252, 2013.

[14]   S Singh, S Ramakrishna, and F. Berto 3D, *Printing of polymer composites: A short review.,* vol. 2, no. 2, pp. 0-1, 2020.*Mater Des Process Commun,* vol. 2, no. 2, pp. 0-1, 2020.

[15]   P.K. Gokuldoss, S. Kolla, and J. Eckert, "Additive manufacturing processes: Selective laser melting, electron beam melting and binder jetting-selection guidelines", *Materials,* vol. 10, no. 6, 2017.

[16]   S.A.M. Tofail, E.P. Koumoulos, A. Bandyopadhyay, S. Bose, L. O'Donoghue, and C. Charitidis, "Additive manufacturing: Scientific and technological challenges, market uptake and opportunities", *Mater. Today,* vol. 21, no. 1, pp. 22-37, 2018.
[http://dx.doi.org/10.1016/j.mattod.2017.07.001]

[17]   S.H. Huang, P. Liu, A. Mokasdar, and L. Hou, "Additive manufacturing and its societal impact: A literature review", *Int. J. Adv. Manuf. Technol.,* vol. 67, no. 5–8, pp. 1191-1203, 2013.

[18]   Haggerty S, Michael J, Williams PA, Lawrence PE. United States Patent (19). 1993;(19).

[19]   Y. Zhang, W. Jarosinski, Y.G. Jung, and J. Zhang, "Additive manufacturing processes and equipment", In: *Additive Manufacturing: Materials, Processes, Quantifications and Applications.,* 2018, pp. 39-51.

[20]   Y.L. Tee, P. Tran, M. Leary, P. Pille, and M. Brandt III, "Printing of polymer composites with material jetting: Mechanical and fractographic analysis", *Addit. Manuf.,* vol. 36, no. August, p. 101558, 2020.
[http://dx.doi.org/10.1016/j.addma.2020.101558]

[21]   W. Du, "Imece2017-70344 Binder Jetting Additive Manufacturing of Ceramics", *A Literature,* p. 1p. 12, 2017.

[22]   S.P. Michaels, E.M. Sachs, and M.J. Cima, *Metal Parts Generation by Three-Dimensional Printing.,* 1992, pp. 244-250. Available from: http://sffsymposium.engr.utexas.edu/Manuscripts/1992/1992-28-Michaels.pdf%5Cnhttp://sffsymposium.engr.utexas.edu/1992TOC

[23]   S.C. Daminabo, S. Goel, S.A. Grammatikos, H.Y. Nezhad, and V.K. Thakur, "Fused deposition modeling-based additive manufacturing (3D printing): Techniques for polymer material systems", *Mater. Today Chem.,* vol. 16, p. 100248, 2020.
[http://dx.doi.org/10.1016/j.mtchem.2020.100248]

[24]   J.Y. Lee, J. An, and C.K. Chua, "Fundamentals and applications of 3D printing for novel materials", *Appl. Mater. Today,* vol. 7, pp. 120-133, 2017.
[http://dx.doi.org/10.1016/j.apmt.2017.02.004]

[25]   M. Gatto, and R.A. Harris, "Non-destructive analysis (NDA) of external and internal structures in 3DP", *Rapid Prototyping J.,* vol. 17, no. 2, pp. 128-137, 2011.

[26]   K.J Seluga, "Three dimensional printing by vector printing of fine metal powders." PhD diss., Massachusetts Institute of Technology, 2001.

[27]   M. Ziaee, E.M. Tridas, and N.B. Crane, "Binder-jet printing of fine stainless steel powder with varied final density", *JOM,* vol. 69, no. 3, pp. 592-596, 2017.

[28]   I.H. Moon, and K.H. Kim, "Relationship between compacting pressure, green density, and green strength of copper powder compacts", *Powder Metall.,* vol. 27, no. 2, pp. 80-84, 1984.

[29]   F. Dini, S.A. Ghaffari, J. Jafar, R. Hamidreza, and S. Marjan, "A review of binder jet process parameters; powder, binder, printing and sintering condition", *Met. Powder Rep.,* vol. 75, no. 2, pp. 95-100, 2020.
[http://dx.doi.org/10.1016/j.mprp.2019.05.001]

[30]   CK Chua, CH Wong, and WY Yeong, "Standards, quality control, and measurement sciences in 3D printing and additive manufacturing", 2017 Jun 3. 2017.

[31]   M. Ziaee, and N.B. Crane, "Binder jetting: A review of process, materials, and methods", *Addit. Manuf.,* vol. 28, pp. 781-801, 2019.
[http://dx.doi.org/10.1016/j.addma.2019.05.031]

[32]   A. Lores, N. Azurmendi, I. Agote, and E. Zuza, "A review on recent developments in binder jetting metal additive manufacturing: materials and process characteristics", *Powder Metall.,* vol. 62, no. 5, pp. 267-296, 2019.
[http://dx.doi.org/10.1080/00325899.2019.1669299]

[33]   B. Utela, D. Storti, R. Anderson, and M. Ganter, "A review of process development steps for new material systems in three dimensional printing (3DP)", *J. Manuf. Process.,* vol. 10, no. 2, pp. 96-104, 2008.
[http://dx.doi.org/10.1016/j.jmapro.2009.03.002]

[34]   M. Salehi, S. Maleksaeedi, S.M.L. Nai, G.K. Meenashisundaram, M.H. Goh, and M. Gupta, "A paradigm shift towards compositionally zero-sum binderless 3D printing of magnesium alloys via capillary-mediated bridging", *Acta Mater.,* vol. 165, pp. 294-306, 2019.
[http://dx.doi.org/10.1016/j.actamat.2018.11.061]

[35]   J.G. Bai, K.D. Creehan, and H.A. Kuhn, "Inkjet printable nanosilver suspensions for enhanced sintering quality in rapid manufacturing", *Nanotechnology,* vol. 18, no. 18, 2007.

[36]   X. Lv, F. Ye, L. Cheng, S. Fan, and Y. Liu, "Binder jetting of ceramics: Powders, binders, printing parameters, equipment, and post-treatment", *Ceram. Int.,* vol. 45, no. 10, pp. 12609-12624, 2019.
[http://dx.doi.org/10.1016/j.ceramint.2019.04.012]

[37]   B. Derby, "Inkjet printing of functional and structural materials: Fluid property requirements, feature stability, and resolution", *Annu. Rev. Mater. Res.,* vol. 40, pp. 395-414, 2010.

[38]   Z.C. Cordero, D.H. Siddel, W.H. Peter, and A.M. Elliott, "Strengthening of ferrous binder jet 3D printed components through bronze infiltration", *Addit. Manuf.,* vol. 15, pp. 87-92, 2017.
[http://dx.doi.org/10.1016/j.addma.2017.03.011]

[39]   S.J. Huang, C.S. Ye, H.P. Zhao, and Z.T. Fan, "Parameters optimization of binder jetting process using modified silicate as a binder", *Mater. Manuf. Process.,* vol. 35, no. 2, pp. 214-220, 2020.
[http://dx.doi.org/10.1080/10426914.2019.1675890]

[40]   H. Miyanaji, N. Momenzadeh, and L. Yang, "Effect of printing speed on quality of printed parts in Binder Jetting Process", *Addit. Manuf.,* vol. 20, pp. 1-10, 2018.
[http://dx.doi.org/10.1016/j.addma.2017.12.008]

[41]   P. Kunchala, and K. Kappagantula, "3D printing high density ceramics using binder jetting with nanoparticle densifiers", *Mater. Des.,* vol. 155, pp. 443-450, 2018.
[http://dx.doi.org/10.1016/j.matdes.2018.06.009]

[42]   I. Rishmawi, M. Salarian, and M. Vlasea, "Tailoring green and sintered density of pure iron parts using binder jetting additive manufacturing", *Addit. Manuf.,* vol. 24, pp. 508-520, 2018.
[http://dx.doi.org/10.1016/j.addma.2018.10.015]

[43]   T.J. Ayres, S.R. Sama, S.B. Joshi, and G.P. Manogharan, "Influence of resin infiltrants on mechanical and thermal performance in plaster binder jetting additive manufacturing", *Addit. Manuf.,* vol. 30, p. 100885, 2019.
[http://dx.doi.org/10.1016/j.addma.2019.100885]

[44]   H. Miyanaji, M. Orth, J.M. Akbar, and L. Yang, "Process development for green part printing using binder jetting additive manufacturing", *Front. Mech. Eng.,* vol. 13, no. 4, pp. 504-512, 2018.

[45]   Y. Wang, and Y.F. Zhao, "Investigation of sintering shrinkage in binder jetting additive manufacturing process", *Procedia Manuf.,* vol. 10, pp. 779-790, 2017.
[http://dx.doi.org/10.1016/j.promfg.2017.07.077]

CHAPTER 4

Spot Welding of Dissimilar Materials Al6061/ Ss304 with an Interfacial Coating of Graphene Nano Platelets

Velavali Sudharshan[1,*] and **Jinu Paul**[1]

[1] *Department of Mechanical Engineering, National Institute of Technology Calicut, Calicut-673601, India*

Abstract: Spot welding of dissimilar materials Al6061 and SS304 with an interfacial coating of Graphene Nano platelets (GNPs) by using the Resistance spot welding (RSW) technique is described in this chapter. A thin layer of GNPs (a few micrometers thick) is incorporated as an interlayer between Al6061/SS304 sheets (thickness 1 mm) and spot welded in lap configuration. RSW parameters play a vital role in this joining process. It was observed the strength of the lap joint increases with welding time and current. The method of welding held up the metal ejection, which occurs at high welding conditions. The influence of process parameters (welding current, time) on the mechanical and microstructural properties of the spot-welded joint was evaluated in detail. A tensile test was carried out in lap-shear mode to evaluate the strength of the joint and to determine the peak load. The fusion zone and the weld nugget were characterized by Scanning Electron microscopy and hardness analysis. SEM analysis observes that Fe and Al matrix well looped around the Graphene particles leads to the formation of inter-metallic compounds of Fe-GNP, Al- GNP, and Fe-Al-GNP at the joint interface. The results are compared with the properties of joints without and with GNP interlayer. It was found that the GNP interlayer enhances the properties of the spot-welded joint and increases joint strength. Possible strengthening mechanisms include enhanced grain refinement and dislocation pile-ups in the presence of the GNPs.

Keywords: Dissimilar welding, Graphene nanoplatelets, Interfacial coating, Lap joints, Resistance spot welding.

INTRODUCTION

Resistance Spot Welding (RSW) is one of the most commonly used welding methods. It is used in a broad range of industries, prominently for joining thin

* **Corresponding author Velavali Sudharshan:** Department of Mechanical Engineering, National Institute of Technology Calicut, Calicut- 673601, India; Email: sudharshan_m190618me@nitc.ac.in

Amar Patnaik, Albano Cavaleiro, Malay Kumar Banerjee, Ernst Kozeschnik & Vikas Kukshal (Eds.)

sheets for light vehicle parts. The RSW is normally performed by overlapping sheet metals in a lap configuration. RSW is essentially used to join pieces, usually up to 4 mm in thickness. The intensity of the RSW joint depends on the nugget size of the weld zone. RSW nugget zone or weld zone diameter ranges from 2 mm to 13 mm depending on the electrode's tip [1 - 8].

In Fig. (**1**), the principle of resistance spot welding is illustrated. In stage-1, metal sheets with a good surface finish are placed on the bottom fixed electrode in overlapping positions. In stage-2, pressure is applied by the movable upper electrode, and now both electrodes hold overlapped workpieces and high current supplied through the electrodes. In stage-3, the resistance at the interface generates heat at the interface, causing the melting of material up to the plastic melting stage. In stage-4, after the formation of the molten metal nugget between overlapped workpieces, the current supply is stopped, but electrodes are not free, and it still holds the overlapped workpieces. In stage-5, a squeezing force is applied to electrodes leading to the joining of the plastic molten metal at the interface. The excess heat from the joint is removed by the water cooling facility provided on the electrode. In stage 6, the electrode is made free by removing the forces, and the finished welded lap joint is taken out.

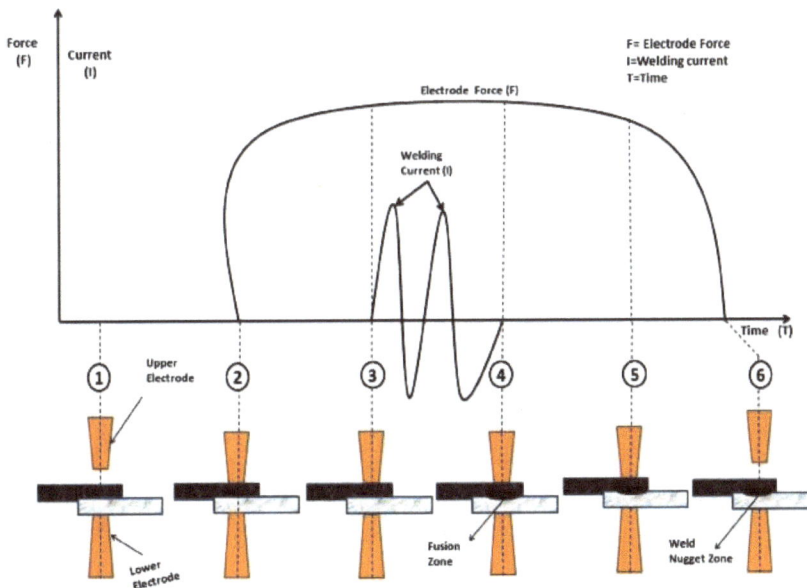

Fig. (1). Lap joint spot-welding processes.

In Resistance spot welding, the joining at the interface is attained by simultaneously applying current and force. The overlapped workpieces are in between the electrodes. The overlapped work pieces compressed together between the electrodes leads to melting at the interface of the contact surface. Thus, while the current stopped, the molten nugget region of workpieces solidified to form a joint. RSW has also been demonstrated for the joining of dissimilar materials like magnesium/aluminium [2, 4, 9], low carbide steel /aluminium [3], etc.

One of the major problems associated with the resistance spot welding process is the weldability of highly conductive materials like Aluminium. Since Aluminium is highly conductive, the resistance at the interface may not be sufficient to generate the required amount of heat. Further, the high thermal conductivity of Aluminium easily transfers the heat from the nugget zone, which prevents its melting.

In order to address this issue, a commonly used method is the usage of interlayers. It was reported that interfacing a layer of suitable material at the interface of the joint leads to increase the strength of the joint [2 - 7, 9, 10]. The interfacial element or the interlayer could be Ni [2], Tin [9], CNT [5, 7], Lead, Zinc [4], etc. Graphene, a single-layer honeycomb lattice [6, 10 - 13] possessing a lighter weight, high conductivity, strength, and flexibility, has also been reported as an interlayer. Baidehish Sahoo [6] modified Aluminium surface with graphene using a resistance heat pressing technique shown in Fig. (**2**). This study describes that introducing graphene leads to surface modification on the topography of Al6063 and forms an internal nanometallic compound on the top surface resulting in ~5 times enhancement in surface hardness in optimum conditions. Also, approximately 200% improvement in young's modulus and 60% improvement in the tensile strength as compared to the bare sample was reported. This result in significant improvement in the surface mechanical properties of Al6063.

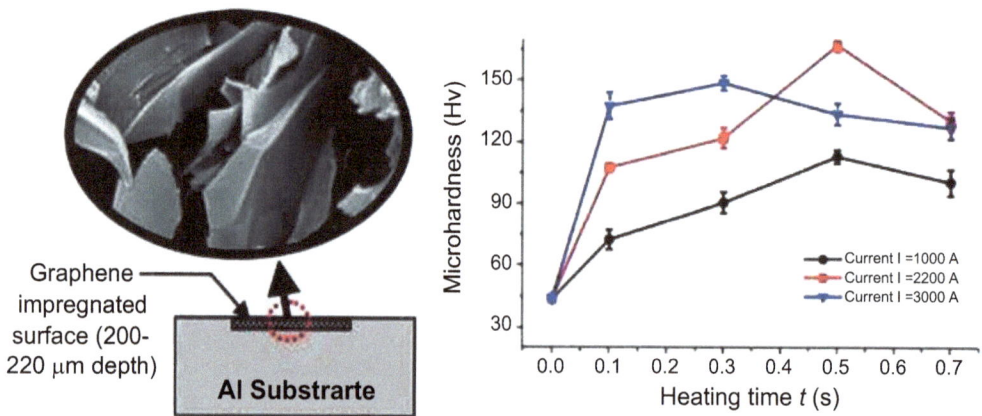

Fig. (2). Graphene impregnated (left), Hardness vs. heating time(right) [6].

Surface modification can also be done with single-channel reinforcement filling (SCRF) and multiple microchannel reinforcement filling (MMCRF) with GNP and Al6061 to form nanocomposite at the surface of the material by using friction stir processing [14]. This observes the effect of MMCRF and SCRF. After the process, the SEM shows that GNP uniformly distributes and is well embedded with the Al matrix. Process temperature is high at MMCRF (18%) than the SCRF. In SCRF, GNP observes as an Al matrix leads to nano composite fabrication at SCRF, while the MMCRF side with of process zone increases and forms high grain refinement than at SCRF. The interfacial strength of the nanocomposite is high at MMCRF when compared with SCRF. Resistance spot welding of dissimilar AZ31/AA5754 lap joint, as shown in Fig. (**3**), was also demonstrated in a recent study. The study held with an interfacial nickel interlayer in between the dissimilar materials. By adding nickel interlayer, while processing, it reacts with Al and Mg matrix, which leads to forming an intermetallic compound like Al-Ni and Mg-Ni. Enhancement of mechanical properties is observed in the nickel interlayer coated sample when compared with the bare sample. Fractured sample analysis showed that more defect occurs at low weld conditions; thus, deficiency reduces lap joint strength. It also [10] reported the resistance spot welding similar material interfacing graphene to AISI 1008 and observed that the effect of graphene in the joint is promising. The addition of graphene leads to enhancement in the mechanical properties and their strength. The results showed that the hardness increased due to the entrapment of graphene. The joint strength depends upon the weld current and time. SEM observed that GNP are well embedded in the Fe matrix in the fusion zone. In microstructural analysis, fine grains were observed in FZ, coarser grains were observed in HAZ, and elongated columnar grain was observed in BM. Carbon particles are also observed at the fracture surface. In RSW, most of the steel fracture occurs as shear dimple fracture. Most of the week, joint failure occurs at low welding conditions.

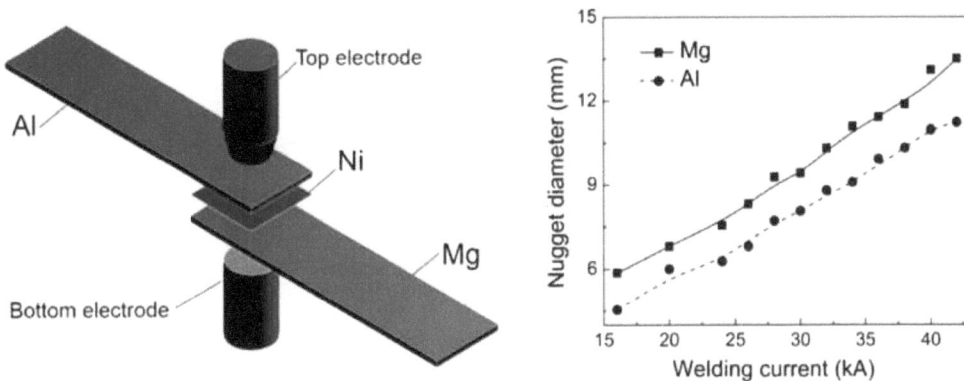

Fig. (3). Welding setup (left), Nugget diameter vs welding current [2].

Further, welding dissimilar materials like Aluminium and steel is always challenging due to the difference in their thermaland physical properties. Herein this study, we report the resistance spot welding of dissimilar materials SS304/Al6061. The main challenge is that SS304/Al6061 has different thermal and mechanical properties. To address this issue, an interlayer of Graphene Nano Platelets (GNPs) has been used. PVA has been used as a binder for GNPs, which reduces the conductivity and spreads the GNP particles uniformly on the joint while processing.

MATERIALS AND METHODOLOGY

Materials

SS304 is an austenite stainless steel, also known as chromium-nickel austenitic alloy, which contains Cr (18 -20%), and nickel (8-10.5%) leading to high corrosion resistance. Al6061 is a high conductivity material, to weld the aluminium materials by using an RSW machine need a current flow of more than 20 kVA with a high load to the joining of the aluminium sheets. SS304/Al6061 samples with dimensions 90 mm x 19 mm x 1 mm with 25- gauge length have been used for welding. The mating surfaces are first cleaned using ethanol.

GNPs are multilayer carbon atoms packed into a honeycomb lattice structure. GNP thickness (5-10 nm), interlayer 5-10 layers, an average of lateral dimension (X&Y) 10 µm, specific surface area 60 - 200* m^2/g, bulk density 0.45 g/cm^3, physical form like fluffy light powder was used.

Polyvinyl alcohol (PVA) is a polymer material that involves nanoparticles of surface materials that should resist engagement on the water surface. PVA is used in a wide range of specialties as glues and films because of its excellent biocompatibility and thermal & biochemical characteristics, opposition to heat fluctuation, and non-toxicity. PVA swells when it retains water due to the exhibit of hydrogels (Formula [-CH2CHOH-]n). It exhibits excellent mechanical properties. PVA enhances metal matrix composition due to the uniform distribution of nanoparticles on the surface to enhance the mechanical properties of the sample, and biomedical tools to fulfill the elevated need for those substances in pharmaceutical applications.

Methodology

In order to prepare the sample, GNPs in powder form with appropriate amounts of distilled water and PVA is sonicated for 1 hour to get a well-dispersed solution of

GNPs. The solution is then drop casted onto the sample surface (SS304/Al6061) to get the GNP coating (0.02 to 0.03 mm thickness). Coating of GNPs is done on one of the mating surfaces of the sample. The specimen placed in a vacuum oven for 3 hours at 65°C for drying. After the drying process, a thin uniform layer coating of approximately a few micrometers thickness is obtained on the sample surface. To do the welding, a ZT SQ -25 kVA -Rocker arm type spot welding machine is used. Resistance spot welding is carried out with different current time combinations 3 kA (0.3, 0.5, 0.7), 4 kA (0.3, 0.5, 0.7) with a constant load for both coated (C) and uncoated (UC) specimens. The preparation of the GNP coating and welding process is shown schematically in Fig. (4).

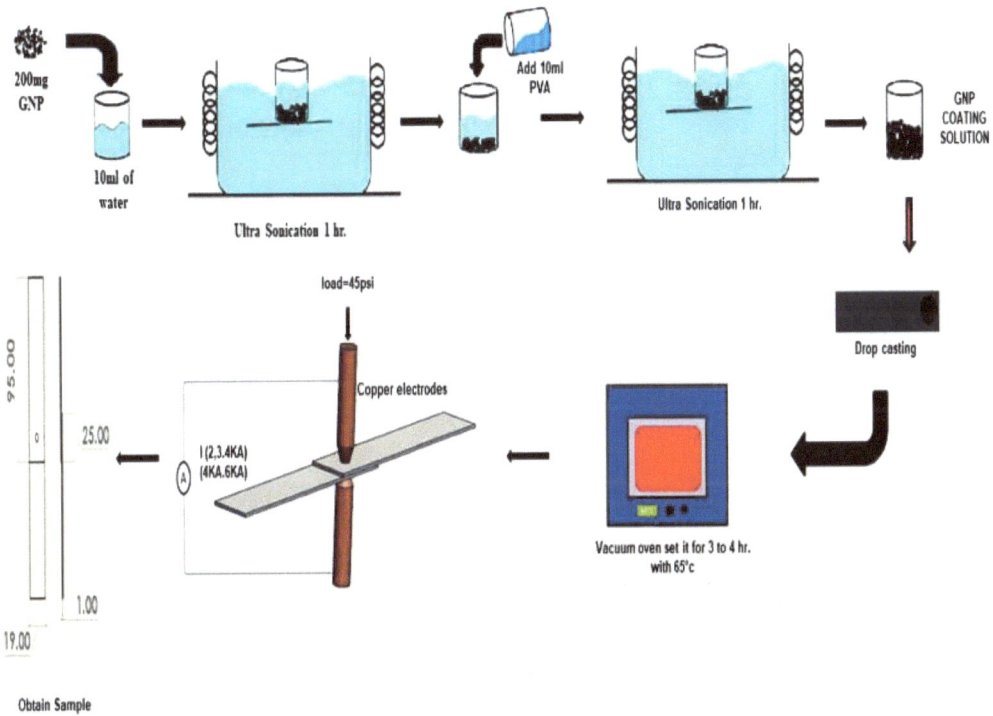

Fig. (4). Schematic showing the sample preparation procedure.

For the microstructure study of the samples, the etched samples are placed on the platform under the objective lens of the optical microscope. The objective lens is changed to suit various magnifications, say, 100x, 200x, 50x, etc., and the platform is adjusted and focused to get the distinct microstructure of the specimen. The microstructure is displayed on the screen, which shows various phases, precipitates, and grains present on the surface of the sample. Scanning Electron microscopy with EDAX has also been employed for the microstructural study.

Lap shear strength test was employed in AG-XPLUS SHIMADZU (Fig. **5a**). A peal of the test was conducted, and ultimate lap shear strength was taken at maximum weld strength at which lap joint failure or crack propagation occurs. The load has been given 10 mm/min by griping on both sides. As per the weld schedule in the plan, lap shear strength was conducted on each sample.

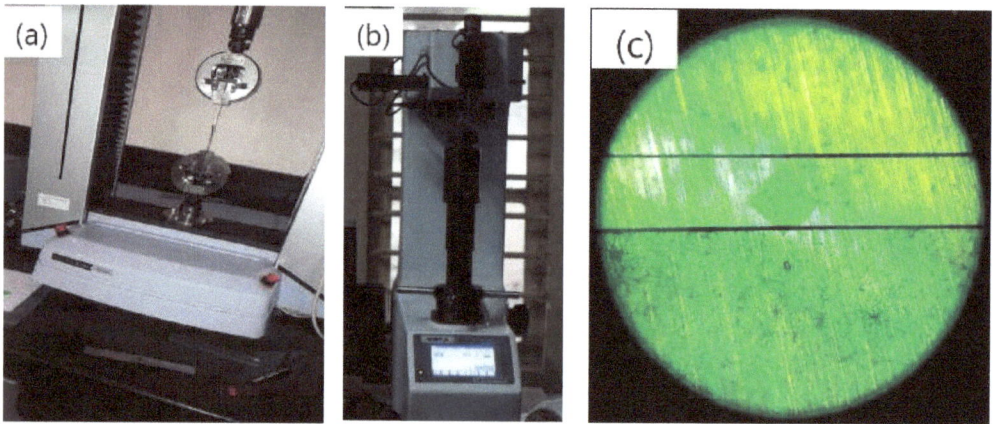

Fig. (5). (a) Lap shear test setup **(b)** Vickers Hardness Testing set up **(c)** Rhombus-shaped indent formed during Vickers Hardness Test.

For the Hardness Test, the specimen is placed on the specimen holder of the Vickers Hardness Testing Machine (Fig. **5b**). The load is set to 1 kgf, which is preferable for Fe and Al alloy samples. Then the Vickers Hardness Tester is started. The diamond indent puts an indent on the sample for 10 seconds, after which a rhombus-shaped indent is visible on the sample surface, as shown in Fig. **(5c)**. Then the diagonals are measured using the index lines on the eyepieces of the microscope. By measuring the diagonals, we can calculate the Vickers Hardness Number by using the equation.

RESULTS AND DISCUSSION

Microstructure Analysis

Microstructure analysis was carried out on the fracture surface of the SS304/Al6061 weld joint. After the lap shear strength test of the joint, fracture surfaces are examined in an Olympus microscope. A variation in the grain structure between the fusion zone (FZ), thermal mismatched zone (TMA), heat-affected zone (HAZ), and base metal (BM) was observed. Fig. (**6a**) shows the microstructural details of samples welded at 3 kA-0.5 s. For the internal fracture steel side, as shown in Fig. (**6a1**), shear deformation can be observed. Elongated grains were observed in the base metal shown in Fig. (**6a2**), and coarse grains are observed in the heat-affected zone shown in Fig. (**6a3**). Dynamic grain refinement in the FZ may lead to the formation of a lath martensitic structure. The formation of martensite occurs due to the rapid cooling of these materials in a critical range of 430 °C to 760 °C.

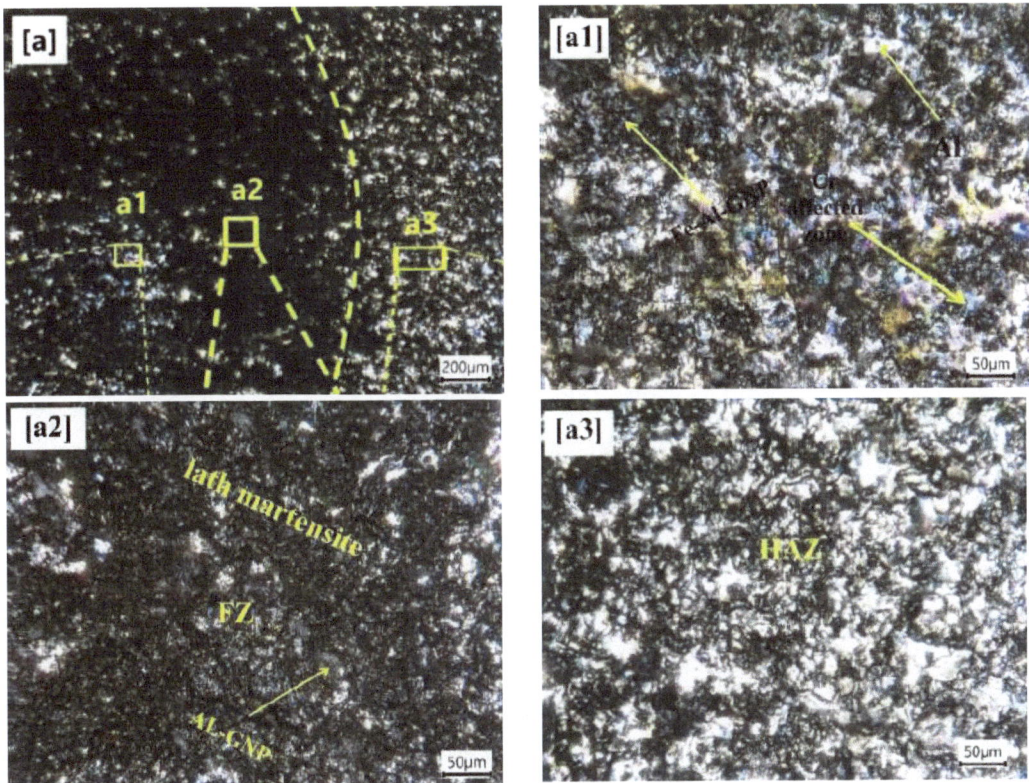

Fig. (6a). Microstructural details of samples welded at 3 kA-0.5 s (a) SS304 side.

The fracture surface of the aluminium side is shown in Fig. (**6b**). During the lap shear test, the Al side undergoes shear dimple deformation leading to internal cracks due to the thermal mismatch of metals. Elongated grains were observed in base metal. The plastically deformed recrystallized grain in the heat affect zone contains Fe looped around GNP, and Al is shown in Fig. (**6b2**). Dynamic recrystallized or refined grains are observed in the thermal mismatch zone in the fusion zone. The shear dimple fracture zone contains Fe looped around shown in Fig. **(6b1)**.

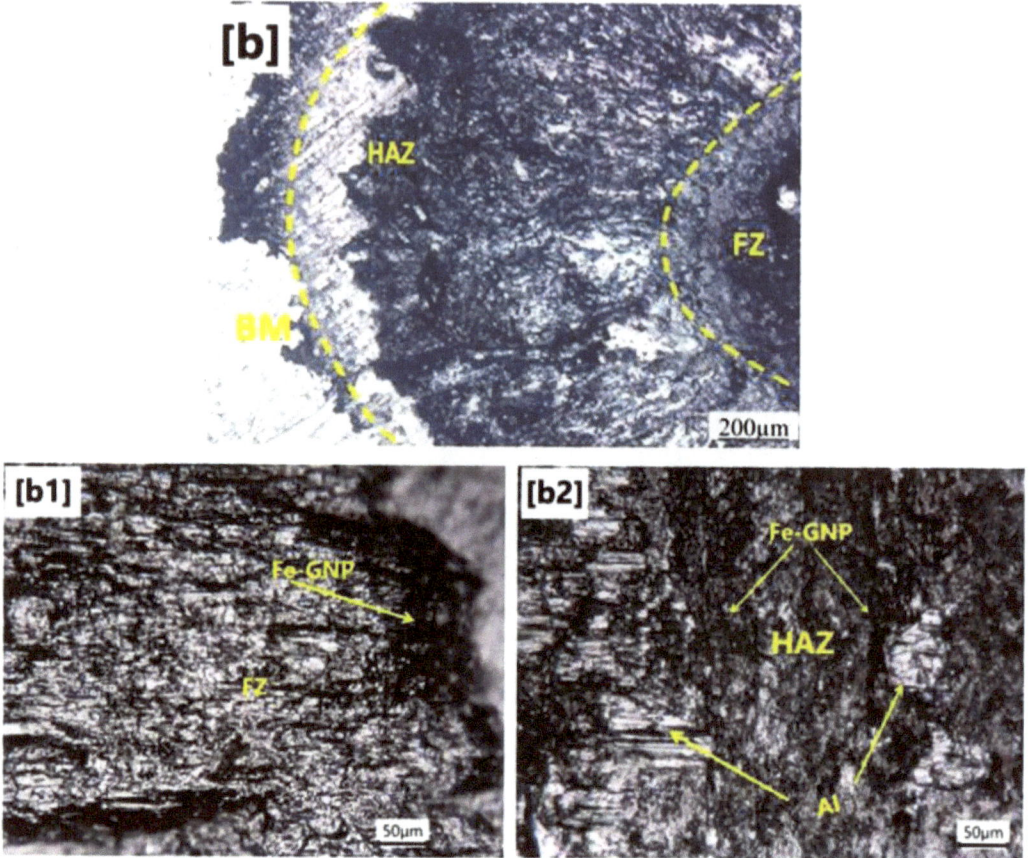

Fig. (6b). Microstructural details of samples welded at 3 kA-0.5 s (b) Al6061 side.

Scanning Electron Microscopy (SEM) Analysis

SEM images of the fractured surfaces are shown in Fig. **(7)**. In SEM analysis, the fracture surface of aluminium and steel shows that the intermetallic compound matrix of Fe loop around the GNP. The formation of shear dimples and internal cracks due to thermal mismatch between materials can be observed on the Al side, as shown in Fig. **(7a)**. Area EDS on the Fe side is shown in Fig. **(7c)**. Fe-58%, Al-6.5%, Carbon-14%, Oxygen-9.6% was obtained. The presence of oxygen indicates that oxidation of Al or Fe is taking place during the process. The presence of Oxygen could also be due to remnants of PVA present in the matrix.

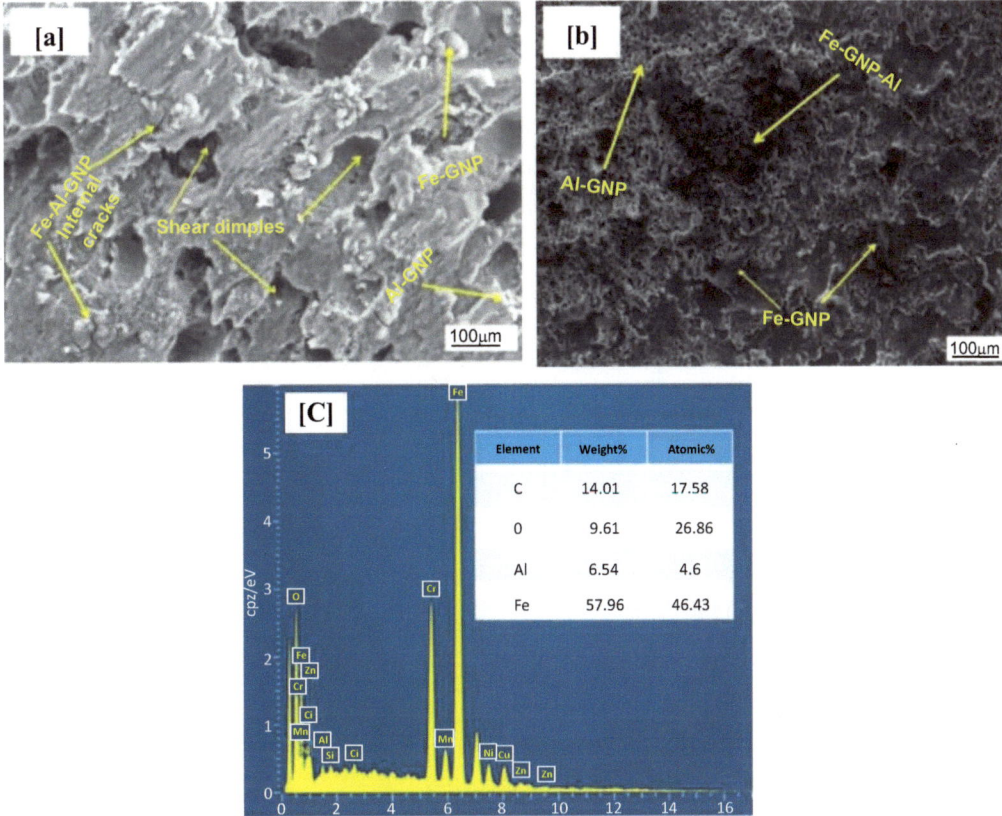

Fig. (7). SEM analysis of fractured samples welded at 3 kA-0.5 s (a) Al6061 side and (b) SS304 side (c) EDS analysis at SS304 side.

Lap Shear Strength Analysis

Lap shear strength analysis Al6061/SS304 with and without GNP coating at different welding currents and time conditions was carried out. Results are summarized in Fig. (**8**). In this Fig., C represents the GNP-coated sample, and UC represents samples without GNP coating. As per Fig. (**8a1**), the maximum load-bearing capacity for the uncoated sample is approximately 1260 N. This occurs at a current time combination of 3 kA-0.5 s. However, for GNP-coated samples, under the same welding conditions, a higher load-bearing capacity is observed, as illustrated in Fig. (**8a2**). With GNP coating, a maximum load-bearing capacity of 1978 N is obtained for the 3 kA-0.3 s. At a higher welding condition of 4 kA, a decrease in the maximum load-bearing capacity for uncoated samples can be observed in Fig. (**8b1**). However, the GNP-coated samples maintain their higher load-bearing capacity at 4 kA, which is evident in Fig. (**8b2**) . The peak load at this current is approximately 1950 N, which occurs at a time of 0.5 s. It could be noted, for coated samples, from Figs. (**8a2** and **8b2**) , that peak load occurs at a welding time of 0.5 s for both the welding currents 3 kA and 4 kA.

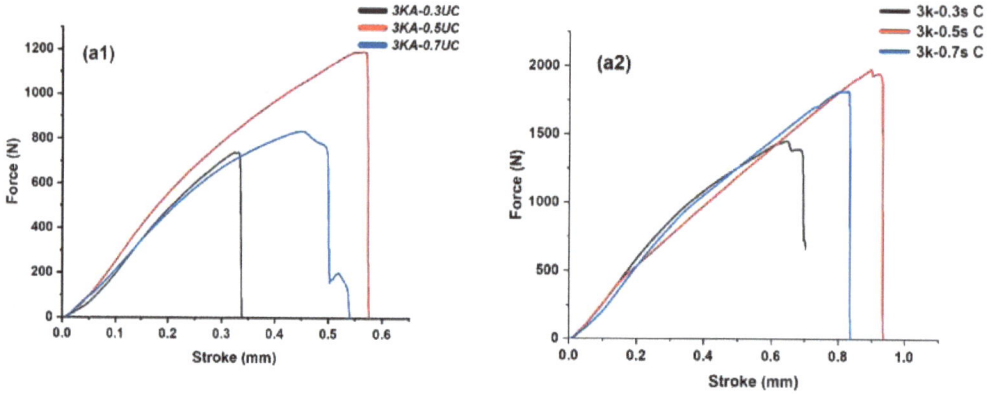

Fig. (8a1, a2). Lap shear strength analysis Al6061/SS304 with and without GNP coating at a welding current of 3 kA and at different time conditions. (C-GNP coated sample; UC- without GNP coating).

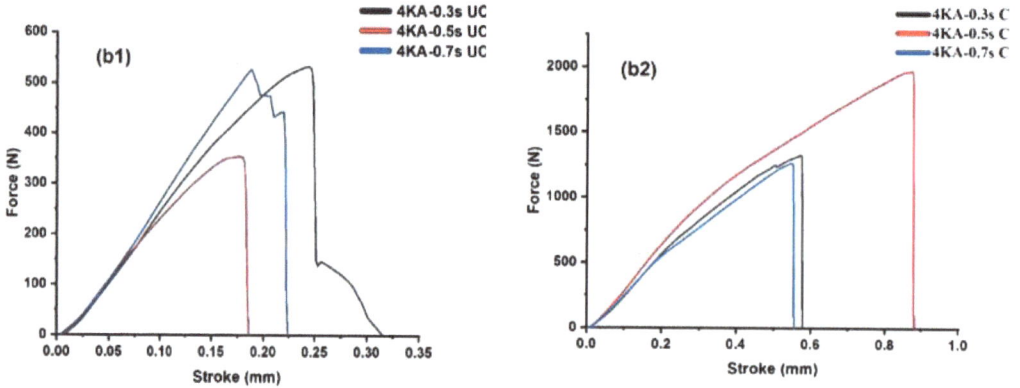

Fig. (8b1, b2). Lap shear strength analysis Al6061/SS304 with and without GNP coating at a welding current of 4 kA and at different time conditions. (C-GNP coated sample; UC- without GNP coating).

The lap joint's strength is usually dependent on nugget size, which depends on welding parameters (current, time, load). The variation in the nugget diameter (N-D) at various welding parameters for GNP-coated samples and uncoated samples is summarized in Fig. (**8c**). Nugget diameter (N-D) is proportional to current and time; an increase in current and time leads to an improvement in the nugget diameter size, as shown in Fig. (**8c**). Nugget diameter (N-D) size also depends on the conductivity of materials. Nugget diameter (N-D) is high on the steel side than the Al side because steel has low conductivity when compared with Al. Further, the low conductivity of steel reacts during the process resulting weld zone area higher than the electrode tip diameter.

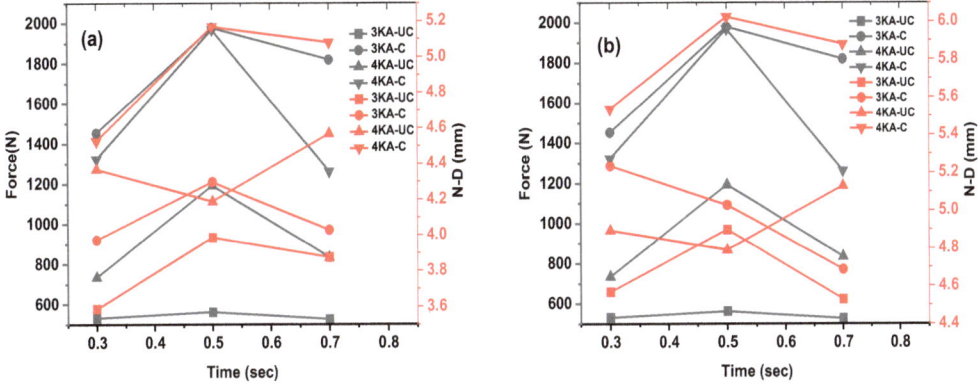

Fig. (8c). Variation in maximum force and nugget diameter (N-D) at **(a)** Aluminium side and **(b)** steel side for various welding conditions of GNP coated (C) and uncoated (UC) samples.

Further, (Fig. **8c**) shows that the nugget size is high at GNP coated sample compared to the uncoated sample. This could be because the PVA-bonded GNPs may reduce the conductivity of the interlayer coating. In the Al side, the highest Nugget diameter (N-D) size obtains as 4.57 mm in the uncoated sample (4 kA-0.5 s) and 5.17 mm in the coated sample (4 kA-0.3 s). The lowest Nugget diameter (N-D) size was obtained as 3.59 mm in the 3 kA-0.3s uncoated samples and 3.97 mm in 3 kA-0.3 s in the coated sample. Thus, the nugget size remains always high at GNP-coated aluminium than the uncoated aluminium sample. On the steel side, the highest nugget diameter (N-D) size obtains as 5.13 mm in the 4 kA-0.7 s uncoated sample and 6.02 mm in the 4 kA-0.5 s coated sample. The lowest size obtained was 4.53 mm in the 3 kA-0.7 s uncoated sample and 4.69 mm in the 3 kA-0.7 s coated sample. Overall, there is approximately 9% enhancement in the nugget size on the aluminium side and ~ 13% enhancement in the steel side due to the incorporation of the GNP coating interlayer.

Analysis of Microhardness

Hardness was investigated along the weld zone of BM, HAZ, and FZ with 10 s dwell time. Results are summarized in Fig. (**9**). Hardness gradually increments from the left side zone of the BM zone to the FZ zone and decrements towards the right side of the FZ to Base metal. It could be noticed that the RSW thus does not affect the hardness of the BM. An enhancement of ~20% on the steel side and ~ 55% on the Al side is observed for GNP-coated samples as compared to uncoated samples. 20% enhancement in the steel side due to the fusion of carbon (GNP) to steel leads to the formation of chromium carbide, thus adding the depth of hardenability with improving resistance to abrasion and wear. ~55% enhancement in Al due to the formation of an intermetallic compound of MgSi-GNP improves the ductility and hardness of the joint.

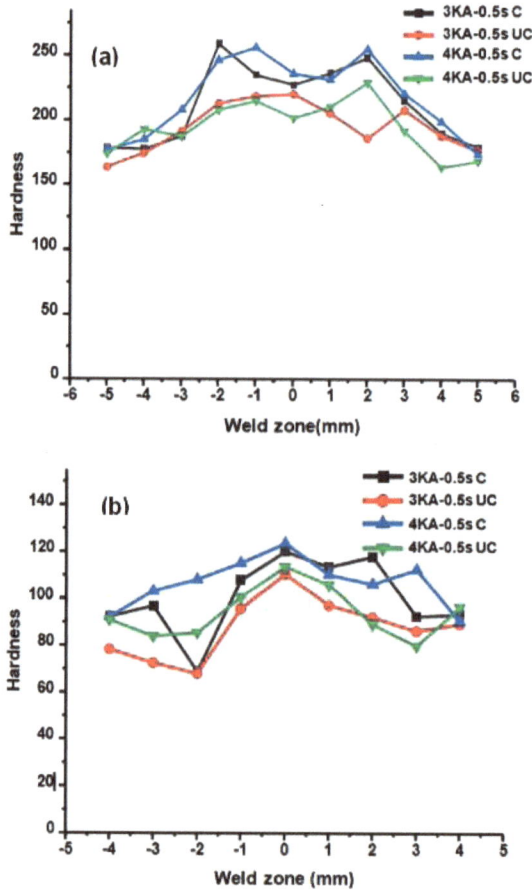

Fig. (9). Variation in microhardness at various welding conditions for GNP-coated and uncoated samples. (a) Steel side (b) Al side.

CONCLUSION

Resistance spot welding of dissimilar materials SS304/A6061 was carried out with an interlayer coating of Graphene nanoplatelets. Optimization of the welding parameters (current and time) was carried out to evaluate the best processing conditions. Overall, the GNP interlayer enhances the performance of the lap joint. The following conclusions may be drawn from the present study.

- The processing parameters (welding current & time) perform an essential role in the performance of the welded joint.
- An increase of ~71% in the lap shear strength was observed by comparing GNP-coated samples with uncoated samples at their respective best welding conditions.
- An enhancement of ~20% on the steel side, and ~ 55% on the Al side is observed for GNP-coated samples as compared to uncoated samples.
- Microstructural analysis of fractured surfaces revealed the uniform distribution of GNPs in the fusion zone. The presence of GNPs in the FZ shall enhance the strength of the joint.
- It was observed that the presence of a GNP interlayer increases the weld nugget size. There is approximately 9% enhancement in the nugget size on the Al6061 side and ~ 13% enhancement on the SS304 side due to the incorporation of the GNP coating interlayer.

CONSENT FOR PUBLICATION

The authors give consent for the publication of identifiable details, which can include photographs, graphs, etc., within the text materials to the Bentham Science Publishers.

CONFLICT OF INTEREST

The authors declare that they have no known competing financial interests or personal relationships that could have appeared to influence the work reported in this paper.

ACKNOWLEDGEMENTS

The authors are thankful to HOD, MED, NIT Calicut, Director and M/s Carborandom Universal Limited Kochi for the support extended to carry out the various phases of the work.

REFERENCES

[1]   N. Charde, "Effects of electrode deformation of RSW on SS304 weld geometry, journal of mechanical engineering and sciences", *J. Mech. Eng. Sci.,* vol. 3, no. August, pp. 261-270, 2012.

[2]   M. Sun, S.T. Niknejad, G. Zhang, M.K. Lee, L. Wu, and Y. Zhou, "RSW AZ31/AA5754 using a nickel interlayer Mater", *Des.,* vol. 87, pp. 905-913, 2015.

[3]   N. Farman bar, S.M. Mousavizade, and H.R. Ezatpour, "FSSW a simple novel method to produce dissimilar joints of galvanized steel/aluminum sheets", *Mater. Res. Express,* vol. 6, p. 026575, 2019.

[4]   Y. Zhang, Z. Luo, Y. Li, Z.M. Liu, and Z.Y. Huang, "Microstructure characterization and tensile properties of Mg/Al dissimilar joints RSW with Zn interlayer", *Mater. Des.,* vol. 75, pp. 166-173, 2015.

[5]   M. Hashempour, A. Vicenzo, F. Zhao, and M. Bestetti, "Direct growth of MWCNTs on 316 stainless

steel by chemical vapor deposition", *Carbon,* vol. 63, pp. 330-347, 2018.

[6]   B. Sahoo, J. Joseph, A. Sharma, and J. Paul, "Al Surface modification by grapheme", *Mater. Des.,* vol. 116, pp. 51-64, 2017.

[7]   T. Das, S.K. Panda, and J. Paul, "RSW AISI-1008 Steel Lap Joints Using MCNT as an Interlayer", *J. Mater. Eng. Perform.,* 2021.

[8]   J. Hirsch, "Al in innovative light-weight car design", *Mater. Trans.,* vol. 52, no. 5, pp. 818-824, 2011.

[9]   M. Sun, S.B. Behravesh, L. Wu, Y. Zhou, and H. Jahed, "Al 5052 /Mg AZ31 RSW with Sn Interlayer Fatigue Fact", *Eng. Mater. Struct,* vol. 40, pp. 1048-1058, 2017.

[10]   T. Das, B. Sahoo, P. Kumar, and J. Paul, "GNP interlayer on RSW of AISI-1008 steel joints", *Mater. Res. Express,* vol. 6, no. 8, 2019.

[11]   U.K. Sur, "Graphene", *Int. J. Electrochem.,* pp. 1-12, 2012.

[12]   B. Sahoo, S.D. Girhe, and J. Paul, "Influence of process parameters and temperature on the solid-state fabrication of multilayered graphene-aluminium surface nano composites", *J. Manuf. Process.,* vol. 34, pp. 486-494, 2018.

[13]   A. Sharm, V.M. Sharma, A. Gugaliya, P. Rai, S.K. Pal, and J. Pal, "Friction stir lap welding of AA6061 aluminium alloy with a graphene interlayer", *Materials and Manufacturing Processes,* vol. 35, no. 3, pp. 258-269.

[14]   A. Sharma, V.M. Sharma, B. Sahoo, S.K. Pal, and J. Paul, "Effect of multiple micro channel reinforcement filling strategy on Al6061-graphene Nano composite fabricated through friction stir processing", *J. Manuf. Process.,* vol. 37, pp. 53-70, 2018.

CHAPTER 5

Optimization of Laser Welding Parameters of Aluminium Alloy 2024 using Particle Swarm Optimization Technique

Aparna Duggirala[1,*], Upama Dey[1], Souradip Paul[1], Bappa Acherjee[2] and Souren Mitra[3]

[1] *School of Laser Science and Engineering, Jadavpur University, Kolkata, 700032, India*

[2] *Department of Production and Industrial Engineering, Birla Institute of Technology: Mesra, Ranchi, 835215, India*

[3] *Department of Production Engineering, Jadavpur University, Kolkata, 700032, India*

Abstract: Laser welding is a viable method of joining aluminium alloys. The input parameters employed in the welding process have a significant impact on the weld quality. There are several parameters that influence weld quality, however, describing their relationship with weld seam characteristics is challenging. This study uses the Taguchi approach and particle swarm optimization (PSO) techniques for improving the weld quality in an Al 2024 lap joint to achieve a consistent and reliable joint. The experiments are performed on a laser welding machine following an L9 orthogonal array experimental design with peak power, scanning speed, and frequency as input parameters. Here, breaking load, bond width and throat length are considered as the responses. Experimentally a maximum breaking load of 1233 N and a minimum bond width of 398.81 μm is achieved. The throat length ranged from 340.72 μm to 983.94 μm. Regression analysis is used to establish the relationship between the input and the responses. The regression equations are utilized as the objective function in an optimization problem. The crowding distance PSO is used to acquire the global optima. Finally, the optimal process parameters for achieving the desired goals are presented.

Keywords: Aluminium alloy 2024, Design of Experiments, Laser welding, Particle Swarm Optimization, Taguchi method.

INTRODUCTION

There is an increase in demand for fuel-efficient automobiles, low-cost flights, and low-cost goods. As a result, there is an ongoing need for research into novel

* **Corresponding author Aparna Duggirala:** School of Laser Science and Engineering, Jadavpur University, Kolkata, 700032, India; E-mail: aparna.sudhakiran@gmail.com

Amar Patnaik, Albano Cavaleiro, Malay Kumar Banerjee, Ernst Kozeschnik & Vikas Kukshal (Eds.)

materials that have a higher strength-to-density ratio while staying economically viable. Aluminium alloys continue to be critical for structural components due to their availability, simplicity of fabrication, and low cost. 2xxx (Al-Cu) series alloys with a high damage tolerance are frequently employed in fatigue-critical applications [1]. Due to their corrosion resistance, ease of manufacture, and high specific strength, alloys of the 5xxx (Al-Mg) and 6xxx (Al-Mg-Si) series are in great demand. The 7xxx (Al-Zn) series alloys are the strongest of all aluminium alloys and are utilised in high-stress aerospace equipment [1]. Despite their increased specific strength, the usage of 7xxx series alloys is restricted owing to the reduced dependability of welded components that occur due to flaws such as fusion zone softening, cracking, and porosity [2]. Certain aluminium alloys are difficult to weld using traditional welding procedures. As a result of its heat source, which is very intense, laser welding is favoured for welding alloys of aluminium. Among its benefits are its rapid rate of manufacturing, high energy density, and low deformation [3, 4]. Numerous process factors have an effect on the quality of the welded component. Various combinations of input parameters result in excellent joints, but it is critical to pick the most effective combination [5]. Kovacocy *et al.* [6] investigate the role of beam traverse speed, shielding gas, and laser power in making reliable welds. The authors determine the effectiveness of various laser welding settings on the production of flaws, impaired microstructure, and fusion zone fracture [7 - 9]. To obtain high-quality welds, experimental design optimises and establishes mathematical correlations between the input parameters and their related outputs. Even an emaciated layer may be adequate to supply the joints with the necessary strength [10]. In many cases, a thicker welded layer leads to increased power consumption and also in faults. Additionally, the geometry of the weld pool, which is influenced by heat conditions, affects the development of weld grains significantly. As a result, selecting the appropriate settings is critical to obtaining the greatest outcomes. Additionally, thanks to developments in numerical modelling approaches, the solidification of a weld may be explored more efficiently and precisely [11]. The Taguchi approach is frequently used in offline mode to optimise quality features. Multi-objective particle swarm optimization (MOPSO) is a relatively new and commonly accepted computer technique that employs a nature-inspired approach that is simple to implement and efficient in terms of time. Numerous references [12 - 14] describe a range of statistical strategies for developing and improving welding process parameters. This article discusses numerous optimization strategies and their use in manufacturing processes.

The effect of the laser beam's maximum power, frequency, and scanning speed on the weld width, load-bearing capacity, and throat length is investigated in this study utilising aluminium alloy 2024. Due to the alloy's physical and thermal qualities, conventional welding is challenging. As a result, the input parameters are optimised to produce a narrow weld width while keeping a high load-bearing capacity and nominal throat length. The trials are developed using the Taguchi approach; moreover, MOPSO is used to determine the global optimal process parameters that result in dependable lap joints.

EXPERIMENTAL SETUP

Table **1** contains the chemical composition of aluminium alloy 2024 as acquired from the material source. The specimen is 75 mm long, 20 mm wide and 2 mm thick. The experiment is conducted utilizing a 300W pulsed laser welding system with a maximum obtainable power of 300W and a wavelength of 1.064 m. When focused, the laser beam has a diameter of 0.6 m. Argon is employed as the shielding gas, with its supply concentric with the output of the beam. Fig. (**1**) illustrates the setup used for lap welding the components.

Table 1. Material composition of AA 2024.

Material	Cu	Mg	Mn	Si	Zn	Ti	Cr	Others	Al
Composition	4.9	1.8	0.9	0.5	0.25	0.15	0.10	0.05	Remainder

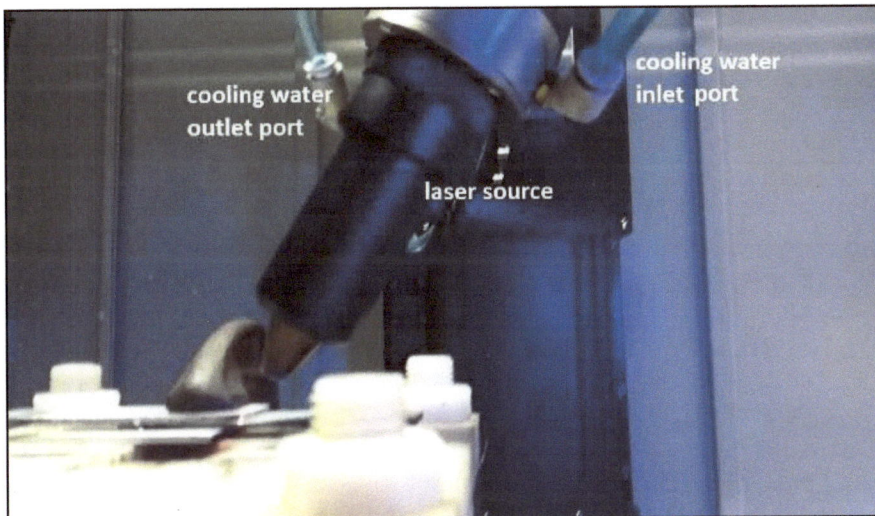

Fig. 1. Laser welding setup.

EXPERIMENTAL DESIGN

To achieve the targeted breaking load capacity, nominal throat length, and reduced weld width, the investigation's parameters include laser peak power, pulse frequency and scanning speed. Table **2** lists the process variables that were employed in the experiment, along with their associated units and values. In order to assess the impact of altering the input processing factors on the defined goals, experiments are carried out. Table **3** summarizes the findings of the tests in terms of reaction.

Table 2. Process parameters used for experiments with symbols, units, and levels.

Sl. No	Process Parameter	Symbol	Unit	Values	
				Minimum value	Maximum value
1.	Laser peak power	P	kW	4.43	14
2.	Scanning speed	V	mm/s	1	2
3.	Pulse frequency	F	Hz	6	8

Table 3. Results of lap welding experiments

Sl. No	Laser peak power (P) kW	Scanning Speed (V) mm/sec	Pulse frequency (f) Hz	Breaking load (L) N	Weld width (W) μm	Throat length (t) (μm)
1	4.43	1	6	188	398.81	340.72
2	4.43	1.5	7	520	434.79	350.95
3	4.43	2	8	727	474.37	427.40
4	4.79	1	7	1233	713.01	731.53
5	4.79	1.5	8	594	954.76	983.94
6	4.79	2	6	445	496.56	369.99
7	14	1	8	230	854.45	772.51
8	14	1.5	6	330	507.23	385.39
9	14	2	7	712	552.69	417.32

RESULTS AND DISCUSSIONS

Main Effects Plots for the Responses

Breaking load, which is the main parameter to assess the load-bearing capacity of the weldments, is discussed here. Welded joints are intended to withstand greater loads. As seen in Fig. (**2a**), increasing peak power and frequency enhances load-bearing capacity up to a point. Increases in scanning speed first boosted the breaking load capacity, but eventually reduced it. By increasing the laser's peak power, the energy available to counteract the alloy's greater reflectivity is increased. Frequency and scanning speed both contribute to the integrity of the weld. Due to the increased thermal conductivity of the material at higher energies, heat conducts away into the substance, thereby reducing the load-bearing capability. Increased scanning speed results in less evaporation inside the weld zone, resulting in a more refined grain structure and hence a larger breaking load. The maximum breaking load is determined by the laser's maximum power of 4.79 kW, a frequency of 7Hz, and a scanning speed of 2 mm/s.

Weld width is an objective that indicates the bonding between the two mating components. The purpose is to achieve the smallest feasible weld width to minimize damage to the parent material and eventual joint weakness. Fig. (2b) illustrates how weld width increases with frequency. By adjusting the peak power, the weld width first increases and then decreases. Due to the faster speed of the laser beam, the weld width was lowered. To achieve the smallest possible weld width, the laser's peak power and frequency are kept low while the scanning speed is increased. Minimum power and frequency reduce the amount of energy absorbed by the material, hence reducing the weld width. To accomplish the smallest possible weld width, the maximum laser power of 4.43 kW, a frequency of 6Hz, and a scanning speed of 2 mm/s are required.

Throat length is similar to that of the depth of penetration for the lap joint. For a high-quality weld, nominal throat length is preferable. Fig. (2c) illustrates how the throat length increased first with increasing laser peak power and later on decreased. With a shift in frequency and a decrease in scanning speed, a reduction in throat length is noticed. The nominal throat length is achieved with the use of the given parameters: a maximum laser beam power of 4.79 kW, a frequency of 7 Hz, and a scanning speed of 1.5 mm/s for the laser beam.

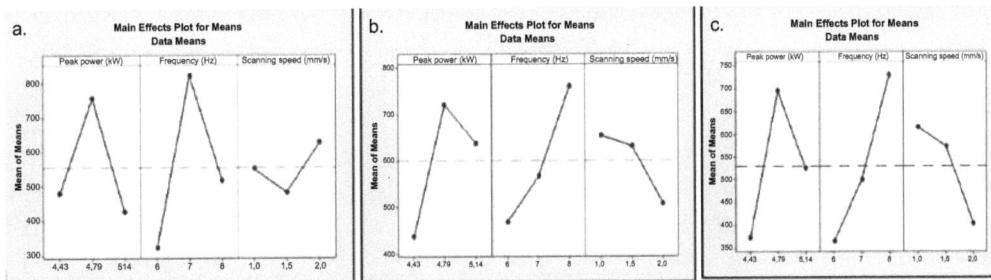

Fig. (2). Main effect plots for (a) breaking load, (b) weld width, (c) throat length.

Prediction of Objectives using Regression Equations

The present study used a regression model, shown in Eq. 1 – 3, to determine the relationship between the objective function and the input parameters. The following are the regression equations for the three responses:

$$Bond\ width\ (W) = -27709 + 13248P - 411.7V - 1182f - 1452P * P - \tag{1}$$
$$261.9\ V * V + 67.28\ f * f + 225.7\ P * V + 84.96P * f$$

$$Breaking\ load(L) = -100293 + 29581P - 6555V + 10030f - 2433\ P * \tag{2}$$
$$P + 1156V * V - 351.6\ f * f + 583.7\ P * V - 1036\ P * f$$

$$Throat\ length\ (t) = -40597 + 18472 * P - 1820 * V - 766.2 * f - \tag{3}$$
$$1945P * P - 218.9\ V * V + 87.84\ F * F + 469.99\ P * V - 50.21\ P * f$$

Particle Swarm Optimization

PSO is a class of evolutionary optimization methods that are inspired by the social behaviour of birds. It is a strategy that is often used to solve non-linear issues. To reduce or maximize the objective function, people in a crowd settle toward a possible solution based on their own knowledge and also on their interactions with other group members. Multi-objective PSO maintains two archives: one for each individual's best position and another for the best global solution. By updating the PSO governing equations, each individual attempts to replicate the global best and personal best solutions accessible in the archives. Crowding distance PSO uses mutation operations to ensure that the solutions saved in the archive retain variety. Convergence toward evenly distributed, non-dominated solutions is faster in the crowding distance variation of PSO [15].

Computational Procedure

The computations were carried out in MATLAB (R2020b version), and a function was applied to it to optimise the formulations of goal functions such as breaking load (BL), weld width (WW), and throat length (TL). A programme is constructed in such a manner that optimization is possible. In the written programme, the number of particles for the optimization approach was set to 500. Additionally, the repository was set to 500 generations, the maximum number of generations to 25, the inertia coefficient to 0.4, the personal and swarm confidence factors to 1, and the uniform mutation percentage to 0.5. In a simplified flowchart, Fig. **(3)** depicts the whole computation technique.

This research considers three input parameters (P, V, and f). Each particle's location is assessed using the objective function. The objective is to discover the ideal combination of design variables (P, V, f) for optimal welding, while still adhering to prescribed limitations.

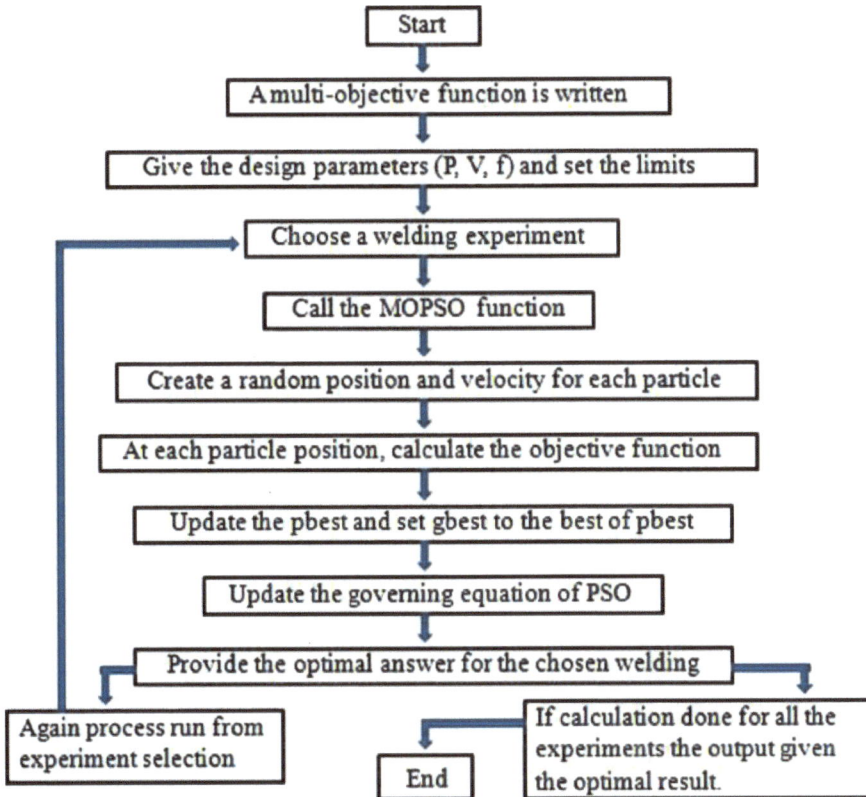

Fig. (3). Computation flowchart of the MOPSO optimization program.

Three design parameters and three goals are used in this study. The number of feasible solutions and the beginning population is determined by the process parameters shown in Table **4**. The derived Pareto fronts are shown in Fig. (**4**). The sparse distribution of points in the Pareto front between weld width and breaking load (Fig. **4a**) implies that there are not many possible combinations of input parameters accessible in the specified range at smaller weld width. The Pareto front indicates that for weld widths between 750 and 850 m, the breaking load stayed approximately 1200N. The population size rose progressively as weld width increased. At around 930 m of weld width, a kink is seen where the downhill tendency becomes abrupt and practically horizontal. The aim of increasing load-bearing capacity while maintaining a small weld width is advantageous for values less than 850 m. The trade-off was achieved by combining a peak power of 4.5kW, scanning speed of 1 mm/s, and frequency of 7.7Hz. The Pareto is densely crowded throughout, with the exception of the very beginning, to achieve the goals of increased breaking load and nominal throat length (Fig. **4b**). At shorter throat lengths, the breaking load is much more than

1250N. However, with the load-bearing capacity limited to 1235.4566N, a throat length of around 950m is acceptable for the trade-off. The combined input parameters are 4.6kW, 1mm/s scanning speed, and 7.8Hz frequency. The aims of decreasing the weld width and increasing the notional breaking force resulted in a perfect Pareto front with a weld width of around 1000m and a throat length of 980m (Fig. **4c**). In this scenario, the input parameters are as follows: peak power of 4.5kW, scanning speed of 1 mm/s, and frequency of 7.7 Hz.

Table 4. Possible combination of input parameters to achieve the three objectives simultaneously.

Laser peak power (P) kW	Scanning speed (V) mm/sec	Pulse frequency (f) Hz	Breaking load (L) N	Weld width (W) μm	Throat length (t) (μm)
4.76592	1.00	8	1288.7738	804.2346	720.2362
4.83411	1.18	7.9	1286.0135	804.8781	742.6109
4.89346	1.32	8	1283.5751	755.5948	756.546
4.83702	1.18	8	1281.4512	814.3238	759.3211
4.83298	1.18	8	1279.5283	818.3566	762.234
4.81431	1.13	7.9	1279.1967	820.283	766.6287
4.77711	1.00	8	1278.6654	824.1352	769.5866
4.78344	1.05	8	1277.4614	824.4594	772.2644
4.81192	1.11	8	1275.7409	768.6978	774.2179
4.88052	1.29	8	1271.2074	771.9461	783.2636
4.86393	1.26	8	1268.8676	834.279	784.8615
4.83546	1.19	7.9	1268.3812	836.2903	786.442
4.84515	1.21	8	1266.0925	837.6715	791.6305
4.87664	1.28	8	1264.3104	788.5579	794.4583
4.77938	1.02	8	1263.1263	762.6037	796.1367
4.77907	1.00	8	1254.9726	792.8561	797.3268
4.8166	1.12	8	1254.0429	797.9726	805.1718
4.88863	1.30	7.9	1252.0241	795.0297	810.2021
4.79494	1.07	8	1248.3234	793.8361	813.0504
4.79059	1.05	7.9	1242.771	771.3065	816.8496
4.83785	1.21	8	1242.6681	818.3338	820.4964
4.8285	1.16	8	1238.2192	737.368	824.655
4.7734	1.00	8	1234.6073	851.0254	826.2917
4.89073	1.31	8	1234.1782	773.8963	829.4557
4.89396	1.32	8	1232.5921	794.3748	830.1142

(Table 4) cont.....

Laser peak power (P) kW	Scanning speed (V) mm/sec	Pulse frequency (f) Hz	Breaking load (L) N	Weld width (W) µm	Throat length (t) (µm)
4.8006	1.09	8	1228.3176	759.1478	835.0541
4.87269	1.26	7.9	1224.2886	857.5032	838.2663
4.83092	1.18	8	1223.1278	828.6797	839.6798
4.88964	1.31	7.9	1217.1412	791.3809	843.3337
4.84319	1.22	7.9	1217.0502	823.3522	845.3899
4.84763	1.24	7.9	1215.7829	822.0996	846.265
4.84563	1.21	7.9	1214.6956	847.9116	847.1927
4.78293	1.02	7.9	1211.7809	845.5059	849.8457
4.88406	1.30	8	1207.9304	810.4031	852.3995

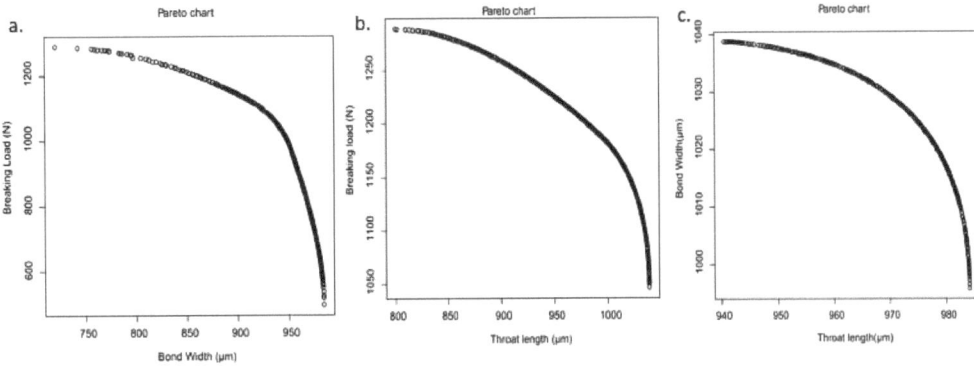

Fig. (4). Pareto charts showing the interaction between **(a)** weld width and breaking load **(b)** throat length ad breaking load **(c)** throat length and weld width.

Due to the desired increased load-bearing capability, the following are the highest values of the goals that may be achieved within the given range of input parameters: 1200N breaking load, 850m weld width, and 900m throat length. Peak power of 4.5kW, scanning speed of 1 mm/s, and frequency of 7.7Hz are the optimal input settings. The combination of input parameters shown in Table 4 is one possibility for achieving the intended goals. The load-bearing capability is maintained at a minimum of 1200N in this case. It is concluded that the values obtained are acceptable at greater scanning speeds and laser output powers. The objective measurements did not demonstrate any substantial difference in pulse frequency.

CONCLUSION

The purpose of this work is to increase the load-bearing capability of the Al 2024 welded junction. MOPSO is used to determine the optimal global input parameters and load-bearing capability.

The specified testing yielded a maximum breaking load of 1233 N. However, the weld width corresponding to this breaking load value is not the smallest that the system can achieve. To achieve the minimal weld width and greatest load bearing capacity while maintaining the nominal throat length, a trade-off between the goals is made, with the highest and lowest values being sacrificed. While some combinations of the input parameters result in a load-bearing capacity of up to 1288.7738N, this capacity is sacrificed in order to achieve the minimal weld width. The PSO algorithm returns the global optimal values, and the values of the input parameters and associated objective function are returned. 1207N breaking load, 868m weld width, and 900m throat length. The optimal set of input parameters is as follows: To accomplish three goals concurrently, the laser's peak power is 4.8kW, the laser beam's scanning speed is 1.3 mm/s, and the pulse frequency is 8Hz.

REFERENCES

[1]   A. Gloria, R. Montanari, M. Richetta, and A. Varone, "Alloys for Aeronautic Applications: State of the Art and Perspectives", *Metals,* vol. 9, p. 662, 2019.

[2]   S.B. Adisa, I. Loginova, A. Khalil, and A. Solonin, "Effect of laser welding process parameters and filler metals on the weldability and the mechanical properties of AA7020 aluminium alloy", *J. Manuf. Mater. Process,* vol. 2, no. 2, p. 33, 2018.

[3]   K. Chen, Z.Y. Wang, R.S. Xiao, and T.C. Zuo, "The powder welding technique of Al alloy with high power Slab CO_2 laser", *Laser Journal,* vol. 21, no. 5, pp. 45-48, 2000.

[4]   J. Zhou, and H.L. Tsai, "Developments in pulsed and continuous wave laser welding technologies". *Woodhead Publ. ser. Electron.,* S. Katayama, Ed., Woodhead Publishing, 2013, pp. 103-138.

[5]   K.Y. Benyounis, and A.G. Olabi, "Optimization of different welding processes using statistical and numerical approaches: A reference guide", *Adv. Eng. Softw.,* vol. 39, pp. 483-496, 2007.

[6]   P. Kovacocy, B. Simekova, I. Kovarikova, E. Hodulova, M. Damankova, J. Ptacinova, and P. Jurci, "Investigation of the microstructure and mechanical characteristics of disk laser-welded Ti-6Al-4V alloy joints", *J. Mater. Eng. Perform.,* vol. 29, pp. 593-606, 2020.

[7]   N. Eisenreich, M. Aeckerle, C. Bantel, A. Heider1, and A. Olowinsky, "Influence of laser parameters on tensile shear strength of copper welds", *J. Laser Appl.,* p. ICALEO2018, 2019.

[8]   N. Radek, J. Pietraszek, J. Bronček, and P. Fabian, Properties of Steel Welded with CO_2 Laser.*Current Methods of Construction Design. Lect. Notes Mech. Eng.,* Š. Medvecký, S. Hrček, R. Kohár, F. Brumerčík, V. Konstantová, Eds., Springer: Cham, 2019.

[9]   X. Li, F. Li, X. Hua, and M. Wang, "Analysis of back-weld spatters in laser welding of CP-Ti", *J. Manuf. Process.,* pp. 48-54, 2020.

[10] S. Paul, A. Duggirala, and S. Mitra, "Study of laser beam welding of AA 2024 using taguchi methodology", *IEEE India Council International Subsections Conference (INDISCON)* 2020, pp.248-

253.

[11] Q. Gao, C. Jin, and Z. Yang, "Morphology and texture characterization of grains in laser welding of aluminum alloys", *Weld. World,* vol. 65, pp. 475-483, 2021.

[12] S.B. Halim, S. Bannour, K. Abderrazak, W. Kriaa, and M. Autric, "Numerical analysis of intermetallic compounds formed during laser welding of Aluminum-Magnesium dissimilar couple", *Therm. Sci. Eng. Prog.,* vol. 22, p. 100838, 2021.

[13] M.A. Ahmad, A.K. Sheikh, and K. Nazir, "Design of experiment based statistical approaches to optimize submerged arc welding process parameters", *ISA Trans.,* vol. 94, pp. 307-315, 2019.

[14] S. M. Karazi, M.M. Malayer, and K.Y. Benyounis, "Statistical and numerical approaches for modeling and optimizing laser micromachining process", In: *Reference Module in Materials Science and Materials Engineering,* 2019.

[15] C. Raquel, and P.C. Naval Jr, "An effective use of crowding distance in multiobjective particle swarm optimization", *Genetic and Evolutionary Computation Conference, GECCO 2005, Proceedings.* Washington DC, USA, 2005.

CHAPTER 6

A Review on Theories and Discharge Mechanisms in Electro-Chemical Discharge Machining

Mahaveer Prasad Sharma[1,*]**, Pankaj Kumar Gupta**[1] **and Gaurav Kumar**[2]

[1] *Department of Mechanical Engineering, Malaviya National Institute of Technology Jaipur, Rajasthan, India*

[2] *Department of Mechanical Engineering, NIT Uttarakhand, Srinagar-246174, India*

Abstract: Electro-chemical discharge machining (ECDM) is a hybrid machining process that can machine conductive and non-conductive materials at the micro level. It caters to the benefits of two well-established constituent processes, namely, electro-chemical machining (ECM) and electric discharge machining (EDM). The technology is quite established. However, the control of discharges in ECDM still needs further research. In this view, the present study reviews the various theories and mechanisms of discharge in ECDM given by researchers. The study also comprises an introduction to the ECDM technique, its various names given by different researchers, applications, and historical developments.

Keywords: Discharge mechanism, Electro-chemical discharge machining, Electro-chemical machining, Electric discharge machining, Hybrid machining.

INTRODUCTION

In today's world of miniaturization, glass and ceramics are extensively being used in micro-systems manufacturing. Machining of such materials by traditional machining methods experiences considerable machinability challenges due to their inherent properties, such as hardness and brittleness. Excessive tool wear, surface cracks, and damages are some of the challenges. Several non-traditional micro-machining technologies like abrasive-based machining, ultrasonic machining, chemical etching and laser-based machining are suitable for machining these materials. However, these technologies have some limitations regarding surface quality, productivity, dimensional accuracy, aspect ratio, and sometimes even hazardous [1, 2].

---

* **Corresponding author Mahaveer Prasad Sharma:** Department of Mechanical Engineering, Malaviya National Institute of Technology Jaipur, Rajasthan, India; E-mail: mahaveer.gpc@gmail.com

Amar Patnaik, Albano Cavaleiro, Malay Kumar Banerjee, Ernst Kozeschnik & Vikas Kukshal (Eds.)

Electric discharge machining (EDM) and electro-chemical machining (ECM) are the two non-conventional machining processes assisted by electrical energy. Though these two processes are well-established, these two processes are suitable for electrically conductive materials only. The absence of electrical conductivity in the workpiece limits the use of these two processes [3]. Electro-chemical discharge machining (ECDM) has emerged as a hybrid machining process to overcome these limitations, which exploits the benefits of ECM and EDM processes [2, 4]. ECDM is an economical and promising technique for non-conductive materials which cannot be easily machined by other techniques. However, ECDM is not restricted to electrically non-conductive material. Researchers have used ECDM for various materials such as glass (*i.e.*, soda-lime glass and borosilicate glass), ceramics (*i.e.*, alumina and silicon nitride), quartz, copper, and some composites (*i.e.*, Carbon fiber and Kevlar epoxy composites). Electro-chemical arc machining (ECAM) was capable of giving higher machining rates than ECM and EDM alone [5]. Researchers studied and found ECDM performs better regarding material removal, dimensional integrity and surface quality during ECDM than ECM and EDM alone [6].

On being an economical and potential micro-machining process, the commercial use of this technology is still limited because of challenges in the control of discharges. The current limitations in the universal commercial acceptance of this technology open up opportunities for further research in this field. In view of attracting researchers to explore the process capabilities, this article presents a comprehensive review of theories and discharge mechanisms reported by researchers. A brief introduction of the ECDM process, its historical development in chronological order, and process names are given by various researchers, which also make up part of this article.

HISTORICAL DEVELOPMENTS IN ECDM

The first development of spark-assisted chemical engraving (SACE) occurred in Japan in the late 1950s for application in diamond die workshops. Further developments in this field are tabulated below Table (**1**) [7].

Table 1. Historical developments in ECDM [7].

Year	Chronological Development in ECDM	Researcher
1968	First reported	Kurafuji and Suda
1973	First characterization	Cook *et al.*
1985	First variant "wire electro-chemical discharge machining"	Tsuchiya *et al.*
1990	First application in the field of MEMS published	Esashi *et al.*

(Table 1) cont.....

Year	Chronological Development in ECDM	Researcher
1996	First theoretical model of spark generation	Ghosh *et al.*
1999	First model of discharge phenomenon	Jain *et al.*
2002-04	Study in light of electrochemistry	Fascio *et al.*
2006	Use of pulsed power	Kim *et al.*
2009	Machining structures (<100µm)	Cao *et al.*
2014	First commercial machine	By Posalux SA, the Swiss company

Different researchers have given different names to the ECDM process, such as "Electro Chemical Arc Machining" ECAM [5, 8], "Electro Erosion Dissolution Machining" EEDM [6], "Electro Chemical Discharge Machining" ECDM [9], "Electro Chemical Spark Machining" ECSM [10], "Spark Assisted Chemical Engraving" SACE [11, 12] and "Spark Assisted Etching" SAE [13].

ELECTRO-CHEMICAL DISCHARGE MACHINING (ECDM) SYSTEM

An electro-chemical cell forms the primary basis of an ECDM system, as shown in Fig. (**1**). An electro-chemical cell consists of two electrodes with grossly different sizes (about a factor of 100). An electric spark is generated at the electrode (tool)-workpiece interface above a critical value of voltage. This phenomenon is called the electro-chemical spark/discharge. The workpiece material gets removed in the close vicinity of the discharge zone primarily by the thermal effect of discharge energy.

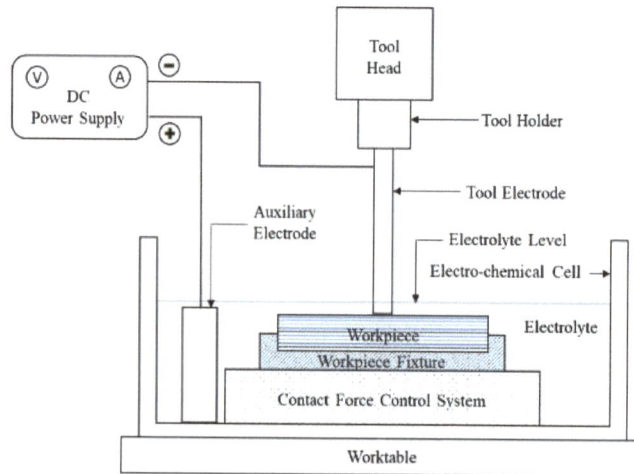

Fig. (1). Schematic of electro-chemical discharge machining (ECDM) system.

The discharge mechanism in ECDM principally depends on the principles of both constituting processes. Initially, the electrolysis process of ECM causes the generation of hydrogen gas on the cathode. These hydrogen bubbles accumulate at the cathode tool surface, causing an increase in resistance to current flow between the tool and electrolyte, which results in ohmic heating of the electrolyte in the close proximity of the cathode, further resulting in the vaporization of electrolyte and formation of vapor bubbles [11]. These vapor and hydrogen bubbles coalesce with increasing voltage, forming larger bubbles. At a specific voltage (critical voltage), the merging of bubbles and the gas film formation surrounding the immersed tool surface occurs, restricting the contact of the tool surface with the electrolyte. This gas film serves the purpose of dielectric and poses resistance to cause a large potential difference between electrodes. Due to this large resistance, the current drops suddenly and raises current density, leading to the generation of an intense electric field (10V/μm) in the gas film, causing the ionizing of hydrogen gas. The movement of free ions and electrons leads to the development of a plasma channel, which causes discharge between the electrolyte solution and tool electrode during ECDM. The temperature in the discharge zone ranges between 8000 – 10000 K [3]. Workpiece material kept close to the cathode tool gets machined primarily by melting, vaporization and thermal erosion due to localized heating of work material and simultaneous chemical etching [1, 14].

THEORIES AND MECHANISMS OF DISCHARGE IN ECDM

Kurafuji and Suda reported ECDM first in the year 1968. A hole of 310 μm was drilled in glass using the process by them. However, the reason behind the discharge was not clarified. In the year of 1984, Crichton and McGeough used streak photography and reported that bubble formation and growth were responsible for spark generation. The cause behind the discharge was still unclarified [1].

The first theoretical model was reported in 1996 by Ghosh *et al.* Formation of spark was described based on the switching theory [15]. According to this theory, bubble density increases (which constricts the current path at the tool electrolyte interface) due to ohmic heating. Dense bubbles lead to gas film formation, resulting in the breaking of bubble bridges that restrict electrolyte solution and tool contact. Subsequently, the current flowing in the circuit dips to zero. This phenomenon was considered analogous to switching off electrical circuits. Isolation caused by blanketing of the cathode tool with hydrogen gas bubbles and water vapor is responsible for electric discharge (due to switching-off action) between electrode and electrolyte. Basak and Ghosh [16, 17] presented theoretical models for estimating critical electrical parameters (*i.e.*, critical voltage and current) and MRR based on the switching phenomenon. Machining of electrically conductive materials using the electro-chemical discharge phenomenon was termed ECAM, whereas it was termed ECDM for non-conductive materials. MRR was predicted for varying input parameters (concentration, additional inductance and applied voltage).

Owing to the limitations of the switching theory, Jain *et al*. [10] proposed another theory known as "valve theory". The switching theory was based on Paschen's curve, valid for Townsend type of discharge (starts when pressure is a few mm of Hg and current is a few µA). In comparison, Jain *et al*. [10] reported similarities of discharge in ECSM with arc discharge (viz. pressure is atmospheric, a current is up to 1.5 A, and similar V-I characteristics) based on V-I characteristics of a discharge tube. In valve theory, bubbles were modeled as arc discharge valves, and the breakdown of these valves (bubbles) in the presence of an intense electric field (~10V/µm) results in arc discharge. The author made some assumptions for theoretical analysis, such as the frequency of the spark was assumed to be constant at 2kHz. Spark time was taken as 50 µs. The inductive effect was assumed to be insignificant. They computed the energy of spark and the approximate diameter of hydrogen gas bubble based on the proposed valve theory. The approximate diameter of the H_2 bubble was calculated as 25 to 55 µm by treating it as gap between two distant electrodes arranged in parallel. In conjunction with the finite element method, the proposed valve theory was used for computing material removal rate (MRR). Simulation temperature contours were used for estimating overcut, limited depth of penetration, and material removal per spark. Results of material removal rate were compared with experimental results of Basak (1992) graphically.

Kulkarni *et al*. [18] carried out experimentation and reported that the discharge phenomenon depends on the tool electrode immersion depth in the electrolyte. During experimentation, a crucial observation was that the discharge does not occur when the cathode is immersed beyond a certain depth in the electrolyte. The ECDM operation was carried out for a small time span to study the single discharge effect. According to time-varying current graphs, current spikes correspond to the discharges. Different materials showed different values of parameters (*i.e.*, peak current, pulse width, time span between two current pulses and discharge frequency). The zone affected by discharge appeared as a circular zone with a shiny appearance, which seemed to be a quick solidification of the melted material. A discharge mechanism was proposed based on time-varying current. According to this mechanism, resistance goes on building up with the increasing number of bubbles (decreasing experimental current). High dynamic resistance of gas film results in negligible current through the ECDM circuit.

Consequently, an intense electric field across the gas film causes an arc discharge, which is observed as a large current spike in time-varying current graphs [18].

Fascio *et al.* [19] reported that electrolyte properties and temperature influences the critical voltage and electrolyte concentration. Based on percolation theory, critical conditions were predicted, which were expressed as functions depending on electrolyte properties and tool geometry. It was reported that electrolyte temperature strongly influenced the conductivity of the solution. The voltage-current characteristic, along with photographs, was explained by dividing it into five regions.

- Thermodynamic and overpotential region: Electrolysis and current flow are negligible in this region.
- Ohmic heating region: Current varies almost linear to voltage.
- Limiting current region: Current is almost constant up to critical current value (limiting current value).
- Instability region: Mean current value drops rapidly due to the formation of gas film surrounding the immersed tool surface.
- Discharge region: Discharge occurs within this region. Isolation at the electrode-electrolyte interface results in discharge.

The coalescence of small bubbles results in bigger bubbles, which subsequently grow and help in forming a gas film surrounding the tool surface. It was reported that the mean current was very low (~mA) in the arc region. The current at that moment was characterized by a series of short-span pulses of a few micro-seconds due to electric discharges. By counting these pulses, it was concluded that the discharges follow a Poisson distribution. The author estimated machining depth by the application of numerical simulation and compared it with experimental results.

The capability of ECDM has been proved by its diverse applications. Microfabrication on glass is done due to its vast use in MEMS, electronics, microfluidic devices, and 3D structures. Micro drilling on ceramic materials [20, 21], machining of some composites [22], and industries using localized modification with high accuracy and efficiency, such as electronics, aviation and medical (surgery). Non-machining applications of ECDM are micro-welding of thermocouple wire and engraving [23, 24]. ECDM is also used for the fabrication of miniature machine tools, micromachining in aeronautics, and electrical and mechanical engineering. Many researchers have performed different types of operations using the ECDM process. Some of them are tabulated here (Table **2**):

Table 2. Type of operations performed using the ECDM process [14].

Operation	Researcher
Micro-channel	T. F. Didar *et al.* [25],
Die sinking	A. B. E. Khairy and J. A. McGeough [6],
Deep hole by ECS trepanning	V. K. Jain and S. K. Chak [26],
Machining of a cylindrical rod	K. Furutani and H. Maeda [27],
Dressing of micro-grinding tools	M. Schöpf *et al.* [28],
Slicing of glass rods	W. Y. Peng and Y. S. Liao [29],
Engraving	V. Fascio *et al.* [11],
Complex and intricate 3D microstructures	X. D. Cao [30],

CONCLUSION

A review of various theories and mechanisms of discharge in the ECDM process has been presented in this paper, based on the research carried out in ECDM by different researchers. This article briefly introduces the ECDM process more simply, along with its development over time. After the review study carried out in this article, the following crucial aspects are reported below:

- Valve theory emerged out of the enfeeblement of switching theory.
- The discharge in ECDM is analogous to arc discharge.
- The discharge occurs at or beyond the application of critical potential difference between two electrodes. Thus, only electrolysis occurs below critical voltage.
- The density of bubble formation increases with voltage applied because of ohmic heating. At critical parameters, these bubbles coalesce and lead to the film formation of H_2 gas and electrolyte vapor.
- The high electric field across the gas film causes plasma channel formation across the gas film, which is responsible for the discharge.
- A series of small pulses, which correspond to arc discharges, characterize the current at any instant.

The future scope of the work requires research for the establishment of a standard discharge phenomenon. It will improve the reproducibility of the machining operation.

REFERENCES

[1]   N. Kumar, N. Mandal, and A.K. Das, "Micro-machining through electrochemical discharge processes: A review", *Mater. Manuf. Process.,* vol. 35, no. 4, pp. 363-404, 2020.

[2]   P.K. Gupta, A. Dvivedi, and P. Kumar, "Effect of pulse duration on quality characteristics of blind

hole drilled in glass by ECDM", *Mater. Manuf. Process.*, vol. 31, no. 13, pp. 1740-1748, 2016.

[3]   A. Ghosh, "Electrochemical discharge machining: Principle and possibilities. Sadhana - Acad Proc", *Eng. Sci.*, vol. 22, no. pt 3, pp. 435-447, 1997.

[4]   V.K. Jain, S.K. Choudhury, and K.M. Ramesh, "On the machining of alumina and glass", *Int. J. Mach. Tools Manuf.*, vol. 42, no. 11, pp. 1269-1276, 2002.

[5]   JA McGeough, ABM Khayry, W Munro, and JR Crookall, "Theoretical and experimental investigation of the relative effects of spark erosion and electrochemical dissolution in electrochemical arc machining", *CIRP Ann - Manuf Technol*, vol. 32, no. 1, pp. 113-118, 1983.

[6]   ABE Khairy, and JA McGeough, "Die-sinking by electroerosion-dissolution machining", In: *CIRP Ann - Manuf Technol*, vol. 39. , 1990, no. 1, pp. 191-5.

[7]   R. Wüthrich, and J.D.A. Ziki, *Micromachining using electrochemical discharge phenomenon : Fundamentals and application of spark assisted chemical engraving. Second edi.* Elsevier: Oxford, 2014.

[8]   H. Krötz, R. Roth, and K. Wegener, "Experimental investigation and simulation of heat flux into metallic surfaces due to single discharges in micro-electrochemical arc machining (micro-ECAM)", *Int. J. Adv. Manuf. Technol.*, vol. 68, no. 5–8, pp. 1267-1275, 2013.

[9]   B. Bhattacharyya, B.N. Doloi, and S.K. Sorkhel, "Experimental investigations into electrochemical discharge machining (ECDM) of non-conductive ceramic materials", *J. Mater. Process. Technol.*, vol. 95, no. 1–3, pp. 145-154, 1999.

[10]   V.K. Jain, P.M. Dixit, and P.M. Pandey, "On the analysis of the electrochemical spark machining process", *Int. J. Mach. Tools Manuf.*, vol. 39, no. 1, pp. 165-186, 1999.

[11]   V. Fascio, H.H. Langen, H. Bleuler, and C. Comninellis, "Investigations of the spark assisted chemical engraving", *Electrochem. Commun.*, vol. 5, no. 3, pp. 203-207, 2003.

[12]   J.D.A. Ziki, and R. Wüthrich, "Tool wear and tool thermal expansion during micro-machining by spark assisted chemical engraving", *Int. J. Adv. Manuf. Technol.*, vol. 61, no. 5–8, pp. 481-486, 2012.

[13]   A. Daridon, J. Lichtenberg, E. Verpoorte, N.F. De Rooij, V. Fascio, and R. Wütrich, "Multi-layer microfluidic glass chips for microanalytical applications", *Fresenius J. Anal. Chem.*, vol. 371, no. 2, pp. 261-269, 2001.

[14]   T. Singh, and A. Dvivedi, "Developments in electrochemical discharge machining: A review on electrochemical discharge machining, process variants and their hybrid methods", *Int. J. Mach. Tools Manuf.*, vol. 105, pp. 1-13, 2016.

[15]   I. Basak, *"Electrochemical discharge machining: Mechanism and a scheme for enhancing material removal capacity"*, PhD thesis. IIT, Kanpur, 1992.

[16]   I. Basak, and A. Ghosh, "Mechanism of spark generation during electrochemical discharge machining: A theoretical model and experimental verification", *J. Mater. Process. Technol.*, vol. 62, no. 1–3, pp. 46-53, 1996.

[17]   I. Basak, and A. Ghosh, "Mechanism of material removal in electrochemical discharge machining: A theoretical model and experimental verification", *J. Mater. Process. Technol.*, vol. 71, no. 3, pp. 350-359, 1997.

[18]   A. Kulkarni, R. Sharan, and G.K. Lal, "An experimental study of discharge mechanism in electrochemical discharge machining", *Int. J. Mach. Tools Manuf.*, vol. 42, no. 10, pp. 1121-1127, 2002.

[19]   V Fascio, R Wüthrich, and H Bleuler, "Spark assisted chemical engraving in the light of electrochemistry", *Electrochim Acta.*, vol. 49, no. 22-23 SPEC. ISS., pp. 3997-4003, 2004.

[20]   B.R. Sarkar, B. Doloi, and B. Bhattacharyya, "Parametric analysis on electrochemical discharge machining of silicon nitride ceramics", *Int. J. Adv. Manuf. Technol.*, vol. 28, no. 9, pp. 873-881, 2006.

[21]   A. Behroozfar, and M.R. Razfar, "Experimental study of the tool wear during the electrochemical discharge machining", *Mater. Manuf. Process.,* vol. 31, no. 5, pp. 574-580, 2016.

[22]   V.V. Nesarikar, V.K. Jain, and S.K. Choudhury, "Traveling wire electrochemical spark machining of thick sheets of Kevlar-Epoxy composites", *Proceedings of the sixteenth AIMTDR conference* year. 1334, pp. 672-677.

[23]   A. Ghosh, M.K. Muju, S. Parija, and A. Kanjrathinkal, "Microwelding using electrochemical discharge", *Int. J. Mach. Tools Manuf.,* vol. 37, no. 9, pp. 1303-1312, 1997.

[24]   D.K. Mishra, A.K. Verma, J. Arab, D. Marla, and P. Dixit, "Numerical and experimental investigations into microchannel formation in glass substrate using electrochemical discharge machining", *J. Micromech. Microeng.,* vol. 29, no. 7, p. 075004, 2019.

[25]   T.F. Didar, A. Dolatabadi, and R. Wüthrich, "Characterization and modeling of 2D-glass micro-machining by spark-assisted chemical engraving (SACE) with constant velocity", *J. Micromech. Microeng.,* vol. 18, no. 6, 2008.

[26]   V.K. Jain, and S.K. Chak, "Electrochemical spark trepanning of alumina and quartz", *Mach. Sci. Technol.,* vol. 4, no. 2, pp. 277-290, 2000.

[27]   K. Furutani, and H. Maeda, "Machining a glass rod with a lathe-type electro-chemical discharge machine", *J. Micromech. Microeng.,* vol. 18, no. 6, 2008.

[28]   M Schöpf, I Beltrami, M Boccadoro, D Kramer, and B. Schumacher, "ECDM *(electro chemical discharge machining),* a new method for trueing and dressing of metal-bonded diamond grinding tools", *CIRP Ann - Manuf Technol,* vol. 50, no. 1, pp. 125-218, 2001.

[29]   W.Y. Peng, and Y.S. Liao, "Study of electrochemical discharge machining technology for slicing non-conductive brittle materials", *J. Mater. Process. Technol.,* vol. 149, no. 1–3, pp. 363-369, 2004.

[30]   X.D. Cao, B.H. Kim, and C.N. Chu, "Micro-structuring of glass with features less than 100 μm by electrochemical discharge machining", *Precis. Eng.,* vol. 33, no. 4, pp. 459-465, 2009.

CHAPTER 7

Investigations on Magnetic Field Assisted Electrochemical Discharge Machining Process

Botcha Appalanaidu[1,*], Rajendra Kumar Arya[2] and Akshay Dvivedi[1]

[1] *Department of Mechanical and Industrial Engineering, Indian Institute of Technology, Roorkee, Uttarakhand, India*

[2] *Department of Mechanical Engineering, Indian Institute of Technology, Mumbai, Maharastra, India*

Abstract: Electrochemical discharge machining (ECDM) process is an arising unconventional machining process for the micromachining of non-conducting materials. During the ECDM process, surface damages, machining continuity at higher depths and hole over cut (HOC) are the main issues during drilling. Previous researchers reported that gas film thickness, debris evacuation and electrolyte replenishment are the prime reasons for the lack of surface quality and lower hole depth. The present investigation has employed a magnetic field during the machining process, and they found a positive effect on the above-mentioned issues. Lorentz force was produced during the machining process, and created a circular motion of the electrolyte around the tool electrode. This phenomenon helped to control the gas film thickness, debris flushing, and electrolyte replenishment at the tool end. In the present work, the authors used a 1300 Gauss Fe-based ceramic permeant ring magnet. Magnetic field strength for both south and north poles was measured using a digital Gauss meter. A high-speed image-capturing camera was used to understand the bubble generation, gas film formation, and debris evacuation during the machining process. The authors applied both north and south-pole magnetic fields for the investigation of the machining process and compared the results with the conventional ECDM process. Better results in surface quality, hole depth, and HOC were achieved with the south pole magnetic field compared to the traditional ECDM process.

Keywords: Borosilicate glass, ECDM, Gas film, Lorentz force, Magneto hydrodynamic effect, NaOH.

---

* **Corresponding author Botcha Appalanaidu:** Department of Mechanical Engineering, GITAM University, Visakhapatnam, India; Email: botchaappalanaidu1@gmail.com.

Amar Patnaik, Albano Cavaleiro, Malay Kumar Banerjee, Ernst Kozeschnik & Vikas Kukshal (Eds.)

INTRODUCTION

The electrochemical discharge machining (ECDM) process is a hybrid micromachining process and gained researcher's attention in recent times [1]. The eminent feature of the ECDM process is that all kinds of work materials can be machined [1, 2].

However, in recent times, the ECDM process gained popularity for machining hard-to-machine materials such as borosilicate, soda lime glass, quartz, zirconia, *etc* [1, 3]. The experimental facility of an ECDM process has two electrodes (tool and counter electrode), a power supply unit, an electrolyte medium (in general alkaline solution), and a feeding unit [2]. The tool electrode (highly conducting material) is placed in a tool holder and immersed in the electrolyte bath nearly about 2 mm [3]. The counter electrode (highly conducting material) is fully dipped in an electrolyte medium. A potential difference was created using a DC power supply. Hydrogen bubbles are generated at the cathode, and oxygen bubbles are at the anode. Due to the lower surface area, bubbles are densely concentrated at the tool electrode and coalesce with each other to form a gas layer on the tool. As the potential difference increases beyond the critical voltage, the film gets ionized, and the peak voltage increases. At this condition, the film breaks by producing thermal discharges around the tool. When the work material is in the vicinity of the discharges, the thermal discharges strikes to the work material. As result of this, material gets melted and evaporated in micro level. These thermal discharges also increase the temperature in the machining zone. As a result, high-temperature chemical etching also takes place on the work material. Eventually, along with material melting and vaporization, high-temperature chemical etching also involved in the material removal mechanism [4].

Being thermal energy is the main source in the material removal mechanism during the ECDM process, the scope for thermal damage at the machined surface is high. The same was reported in the past literature with micro-cracks over the hole periphery, hole entrance damages, hole over cut (HOC), and micro-cracks propagation into macro cracks, and sometimes this leads to breakage of work material [2]. There are several reasons to state the above deteriorations. During the machining process, a depth of around 250-μm is considered as the discharge regime. In the discharge regime, the thermal discharges are responsible for the major material removal process. As the machining progress, the tool electrode enters into the hydrodynamic regime. In the hydrodynamic region, evacuation of debris from the machining area (beneath the tool electrode and between tool side walls and the machined hole) is difficult. Therefore, the tool electrode is surrounded by molten material. In addition, due to the lack of fresh electrolyte availability, it is difficult to continue the machining process further. Meanwhile, the electrolyte evaporates in the machining zone and leaves salt. This slat accumulates as sludge at the tool end. Both the phenomenon of lack of availability of the electrolyte and sludge/debris formation around the tool tip hinders further machining. On the other hand, an abundance of electrolytes at the hole entrance continued the machining process and channeled the major energy to the hole entrance region. As a result, hole depth was reduced, and HOC and hole taper increased [5].

In order to minimize the above-mentioned phenomenon, several works have been reported in the past. Sabahi *et al.* added two different surfactants, such as C-TAB and SDS, to minimize the surface tension of the gas bubbles. This phenomenon produced a thin gas layer over the tool. The thin gas layer generated low potential sparks in the machining area. Low-intensity sparks reduced the thermal damage over the hole entrance. However, this technique failed to provide the solution for the evacuation of sludge and debris and the addition of fresh electrolytes into the machining area [6]. Zheng *et al.* applied tool rotation motion during the machining process, as depicted in Fig. (**1a**). Tool rotation produced the centrifugal motion of gas bubbles. As a result, larger bubbles were moved from the machining region, and smaller bubbles participated in the film formation. Electrolyte rotation aided to evacuate sludge from the hole bottom and electrolyte availability in the machining zone [7]. Rathore and Dvivedi provided the ultrasonic motion to the tool, as shown in Fig. (**1b**). The vibration motion of the too electrode separated the larger bubbles from the machining zone. This phenomenon helped to develop a thin film over the tool. Due to rapid up-an--down motion of the tool helped debris to evacuate from the machining zone [8]. Han *et al.* applied the vibration to the electrolyte medium during the machining, as shown in Fig. (**1c**). This phenomenon improved the spark discharge efficiency by modifying the gas film. As a result, hole depth was improved, but HOC and machined geometry as still major issues. Singh *et al.* used a textured tool during machining to increase nucleation sites for better spark discharges by lowering film thickness, as shown in Fig. (**1d**). Though this process enhanced the machined geometry quality, it still had issues with sludge removal [9]. Tool shape modification [10], supplying electrolytes through tool electrodes [5], and application of an external magnetic field in the machining zone [11] also tried to achieve better results. All the above-mentioned processes solved only one of the major issues (flushing, replenishment, increment in hole depth, reduction in HOC and hole damages), and some process variants needed external attachments, which include mechanical forces and alignment issues. Therefore, a novel method is needed in this field to enhance hole quality and hole depth without affecting existing experimental facility.

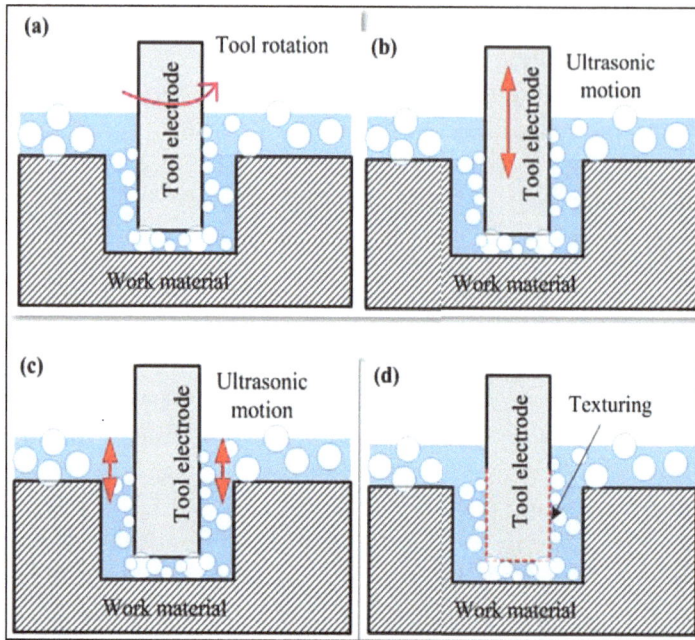

Fig. (1). Schematic images of **(a)** Rotation tool-assisted ECDM process, **(b)** Ultrasonic tool assisted ECDM, **(c)** Ultrasonic electrolyte-assisted ECDM process, and **(d)** Textured tool ECDM process.

During the ECDM process, the application of the magnetic field does not need any kind of external energy assistance or attachments. A simple magnet was used to produce a constant magnetic field in the machining zone. Cheng *et al.* used an Nd-Fe-B-based ring magnet and reported that the magnetic field created a circular motion of electrolytes, improving the machining efficiency [12]. Xu *et al.* employed N52 neodymium disc magnet for their experiments and reported that feature quality was improved and machining time was reduced during magnetic-assisted ECDM process [13].

In the present study, extensive experiments were performed to analyse the impact of the applied magnetic field on hole surface finish and machining speed using Fe based ring magnet. The present article also discussed the comparison results of conventional machining, upward magnetic flied, and downward magnetic field for machining quality. In the current study, a special arrangement was used to measure the machining rate during the machining process. Authors reported that the application of a magnetic field improved the machined hole quality and hole depth. Magnetic field assistance has reduced the machining time significantly. Out of the three methods, the upward magnetic field (South Pole facing tool electrode) gave better results regarding low machining time, higher hole depth, and lower HOC.

MATERIALS AND METHODOLOGY

Experimental Facility

The electrochemical discharge machining process experimental facility mainly consists of an electrolyte bath, two electrodes, a power supply and a feeding facility. Experiments were performed on a self-developed experimental facility. The same is depicted in a schematic way. (Fig. **2**). In the present study, a NaOH solution with 20% Wt./Vol. was employed as an electrolyte medium and placed in a rectangular acrylic vessel and mounted on the X-Y axis. A work hold device along with work material was immersed in an electrolyte medium, and a 2mm electrolyte level was maintained over the work material. A 500 μm diameter cylindrical SS303 was used as a tool electrode and placed in a tool holder, that was mounted on the z-axis. Gupta and Dvivedi reported that pulsed DC power supply hadbetter results than direct DC power [14]. Therefore, a programmable pulsed power supply was used in the present investigations. $50 \times 50 \times 15$ mm^3 graphite block was used as a counter electrode. A ring-shaped Fe-based ceramic magnet with 1300 G magnetic field strength was used as a constant external magnetic field source provider. Experimental conditions during the machining process are incorporated in Table **1**.

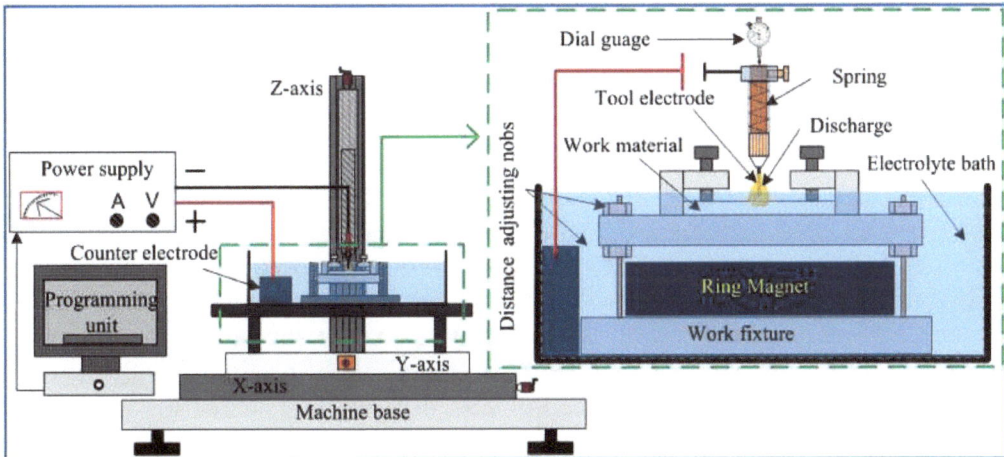

Fig. (2). Schematic view of magnetic field-assisted ECDM facility.

Table 1. Parameters used in experimental investigation.

	Constant Parameters	
1	T_{off}	1 ms
2	T_{on}	3 ms

(Table 1) cont.....

3	Applied voltage	50 V
4	Counter electrode	Graphite ($50 \times 50 \times 10$ mm^3)
5	Tool electrode	SS-304, 440 μm diameter
6	Machining time	1 min
7	Electrolyte	20% wt./vol. NaOH
8	Electrolyte level above work material	2 mm
9	Work material	Borosilicate glass
10	Feeding	Pressurized spring tool feed

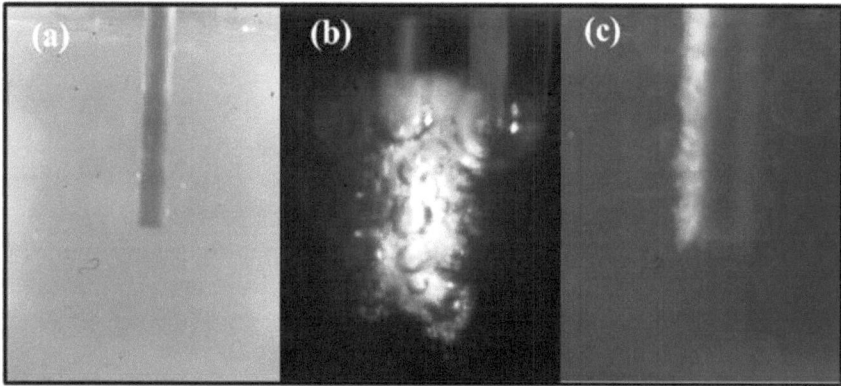

Fig. (3). High-speed camera images of **(a)** tool electrode, **(b)** bubble generation in ECDM, **(c)** in MFAECDM.

Material Removal Mechanism

As mentioned in the earlier sections, the major factor for removing material during the ECDM process is thermal energy generated from discharges. By controlling the amount of gas film thickness, the thermal energy can be controlled. To handle the above-mentioned issues, in the present research work, a constant external magnetic field was used. The basic principle involved in the - assisted ECDM process is that when a moving charged particle is placed in the magnetic field, a normal force perpendicular to the velocity direction is induced. This force is called Lorentz force. As a result, the charged particles (ions) tend to rotate around the tool electrode. It was reported that electrolyte is nothing but a pool of ions. As a result, the electrolyte in the vicinity of the tool also rotates. Due to this phenomenon, gas bubbles were expelled from the machining zone. As a result, the film thickness was reduced. Thin gas film in ECDM produced low potential and high-frequency sparks, which are in favour of the fine machining process [15]. The electrolyte rotation helped to evacuate the molten material from the machining zone. Electrolyte rotation also increased the availability of the electrolyte at the tool electrode bottom in the hydrodynamic region. Due to this phenomenon, the machining process further extended to higher depths and reduced HOC. Continuous replenishment of electrolytes maintains a constant temperature in the machining zone. Therefore, micro-cracks generation at the hole entrance periphery was reduced. Electrolyte rotation also minimized the re-solidification of molten material over the machined hole entrance periphery. As a result, the hole entrance thickness was reduced. High-speed camera images of the bubble generation process for both conventional and magnetic field-assisted ECDM (MFA-ECDM) are depicted in Fig. (3).

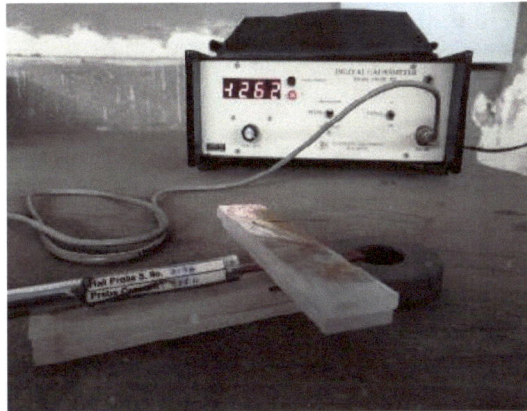

Fig. (4). Measurement of magnetic field using Gauss meter.

Measurements

A Digital gauss meter (SES Instruments PVT. LTD, India. Model no. DGM 202) was used to define the poles (North & South) and the strength of the ceramic-based magnetic field. The reason to choose ceramic magnet is that ceramic magnets are chemically inactive, which is an essential factor in the ECDM process. As shown in Fig. (4), the sensor probe with the North Pole indicator facing magnet surface is placed over the ring magnet. A digital reading was recorded in the gauss meter and can be observed from Fig. 4. If the reading is positive, then it is the North Pole, otherwise, it is the South Pole. From the measurement of magnetic field strength, it was noticed that the field strength from the South Pole is more compared to the field strength from the North Pole. The hole diameter was measured using a stereo microscope, and hole quality was estimated using the same. Hole depth was measured using a dial gauge having 1 μm least count. A special arrangement was made in the tool-holding device to measure instantaneous hole depth during the machining process. The same was plotted in a graph. The same graph was used to measure and compare machining speed for conventional and magnetic field-assisted ECDM processes.

RESULTS AND DISCUSSION

Effect of Magnetic Field on Machined Hole Quality

During the conventional ECDM process, as discussed earlier, the bubble generation rate is high. These bubbles coalesce each other leading to an increase in their size. This phenomenon increased the film thickness over the tool. The film thickness has produced high potential and low-frequency discharges. These high-intensity discharges strike the hole periphery with a rapid force and remove material. Due to the high intensity of discharges, crack generation also takes place over the hole periphery, as shown in Fig. (**5a1**) . As the machining time increases, improper debris evacuation and unavailability of the fresh electrolyte hampers further tool electrode penetration. Therefore, the hole bottom profile quality deteriorated, as shown in Fig. (**5a2**). Meanwhile, a high abundance of electrolytes at the hole entrance has the continuous discharge phenomenon and channelizes the energy to remove the material from the hole entrance. As a result, HOC was increased, and hole depth was reduced. Due to a lack of electrolyte circulation, re-solidification of molten material over the hole periphery has increased the edge thickness, as shown in Fig. (**5a1**).

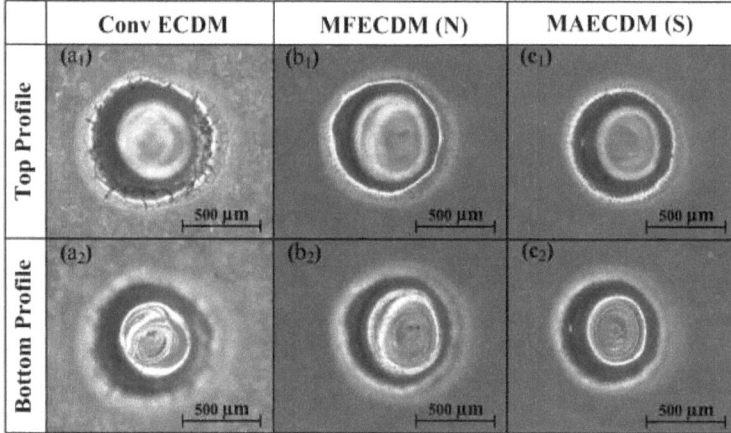

Fig. (5). Top and bottom profile of (a₁ & a₂) conventional ECDM, (b₁ & b₂) MFAECDM with north pole filed, (c₁ & c₂) MFAECDM with south pole filed.

Application of magnetic field in ECDM process induces electrolyte circulation. Due to this phenomenon, the generated bubbles rotate around the tool electrode and are ejected from the machining area by centrifugal force. As a result, a thin film was generated on the tool electrode. This film produced low potential and high-frequency discharges. These low-intensity discharges removed the material at a lower rate compared to the conventional ECDM process in the discharge regime. As the machining time increased, due to the rotation of the electrolyte medium, the removal of machined debris from the machining area increased significantly. Meanwhile, electrolyte replenishment also helped the availability of fresh electrolytes beneath the tool electrode, which increased further machining. This phenomenon increased the channelization of discharge energy at the tool bottom. As a result, hole depth and bottom profile quality improved, as shown in Fig. (**5b2**). Electrolyte circulation also diminished the re-solidification of molten material over the hole entrance and produced a smooth profile, as shown in Fig. (**5b1**) . As reported earlier, the magnetic field strength for the South pole is larger than that of the North pole. Due to this phenomenon, the generated Lorentz force was more, and this has increased the electrolyte circulation. As a result, the film thickness was reduced further, and debris evacuation and electrolyte replenishment improved. Eventually, hole depth was increased, and HOC was reduced marginally compared to the North pole magnetic field-assisted ECDM process, as shown in Fig. (**5c1**) and (**5c2**).

Effect of Magnetic Field on Machined Hole Depth and Machining Speed

During the conventional ECDM process, the machining speed or material removal rate is higher in the discharge regime. As the tool electrode proceeded to further depths or in a hydrodynamic regime, the machining speed was exponentially reduced. This phenomenon is clearly observed in Fig. (**6**). After 330 µm, the machining speed is nearly zero. The main reason behind this phenomenon is the lack of removal of debris from the machining area and insufficient electrolyte availability at the tool electrode end. These two factors hamper further tool penetration.

Unlike in the conventional machining process, during the magnetic field-assisted ECDM process, the electrolyte circulation helps to remove debris from the machining area even at higher depths. As a result, machining still progresses in a hydrodynamic regime. This phenomenon can be observed in Fig. (**7**) and Fig. (**8**). Electrolyte replenishment also boosted the machining rate and increased the hole depth by nearly 40% (North Pole) and 51% (South Pole) than the conventional machining process. Having said that application of the South Pole magnetic field had a little edge over the North Pole magnetic field, evidenced by (Fig. **7**) and (Fig. **8**).

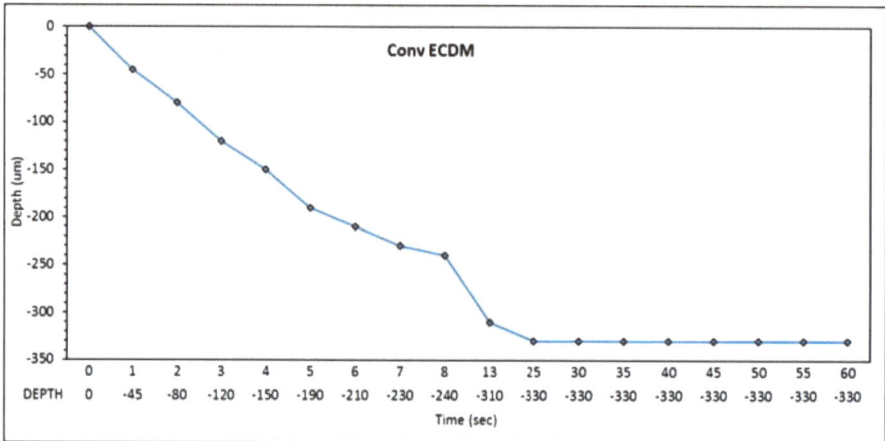

Fig. (6). Measurement of machining speed during the conventional ECDM process.

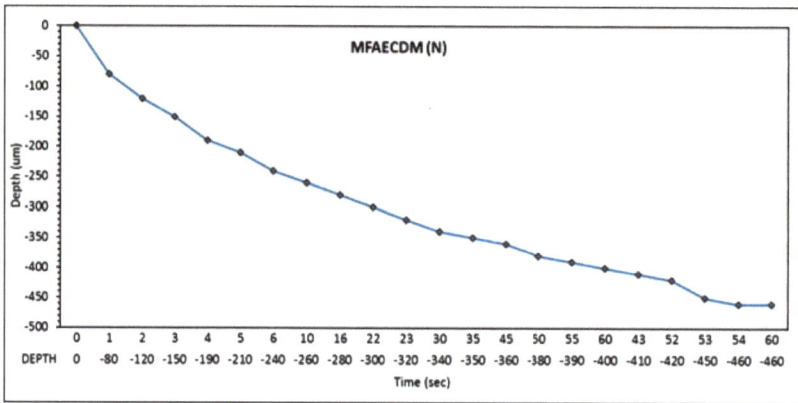

Fig. (7). Measurement of machining speed during MFAECDM (North Pole) process.

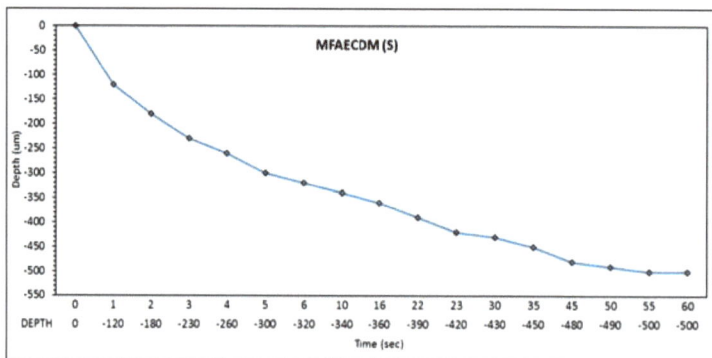

Fig. (8). Measurement of machining speed during the MFAECDM (South Pole) process.

CONCLUSION

- During the conventional machining process, the machining rate was high in the discharge regime. However, in the hydrodynamic regime, a lack of electrolyte replenishment and inefficient debris evacuation resulted in a low machining rate.
- Reduction in machining rate in hydrodynamic regime resulted in lower hole depth, high HOC, and thermal damages over the machined profile.
- Application of magnetic field in machining process induced Lorentz forces in the machining zone near tool region. This created an electrolyte circular motion and reduced the film thickness by ejecting gas bubbles from the machining region.
- Electrolyte circulation and thin film enhanced the discharge quality, debris evacuation, and electrolyte replenishment, even at higher depths.
- Electrolyte rotation improved the hole depth up to 51% and reduced HOC by 36%. The South Pole magnetic field has provided better results compared to the North Pole magnetic field.

REFERENCES

[1]   N. Kumar, N. Mandal, and A.K. Das, "Micro-machining through electrochemical discharge processes: A review", *Mater. Manuf. Process.,* vol. 35, pp. 363-404, 2020.

[2]   B. Appalanaidu, and A. Dvivedi, "On controlling of gas film shape in electrochemical discharge machining process for fabrication of elliptical holes", *Mater. Manuf. Process.,* vol. 36, pp. 558-571, 2020.

[3]   P.K. Gupta, "Effect of electrolyte level during electro chemical discharge machining of glass", *J. Electrochem. Soc.,* vol. 165, pp. E279-E281, 2018.

[4]   Wüthrich, "Physical principles and miniaturization of spark assisted chemical engraving (SACE)", *J. Micromech. Microeng.,* vol. 15, pp. 268-275, 2005.

[5]   R.K. Arya, and A. Dvivedi, "Investigations on quantification and replenishment of vaporized electrolyte during deep micro-holes drilling using pressurized flow-ECDM process", *J. Mater. Process. Technol.,* vol. 266, pp. 217-229, 2019.

[6]   N. Sabahi, M.R. Razfar, and M. Hajian, "Experimental investigation of surfactant-mixed electrolyte into electrochemical discharge machining (ECDM) process", *J. Mater. Process. Technol.,* vol. 250, pp. 190-202, 2017.

[7]   Z.P. Zheng, H.S. Su, F.Y. Huang, and B.H. Yan, "The tool geometrical shape and pulse-off time of pulse voltage effects in a Pyrex glass electrochemical discharge microdrilling process", *J. Micromech. Microeng.,* vol. 17, pp. 265-272, 2007.

[8]   R.S. Rathore, and A. Dvivedi, "Sonication of tool electrode for utilizing high discharge energy during ECDM", *Mater. Manuf. Process.,* vol. 35, no. 04, pp. 415-429, 2020.

[9]   T. Singh, A. Dvivedi, and R.K. Arya, "Fabrication of micro-slits using W-ECDM process with textured wire surface: An experimental investigation on kerf overcut reduction and straightness improvement", *Precis. Eng.,* vol. 59, pp. 211-223, 2019.

[10] L. Paul, and I. Jose, "Micro machining in ECDM process with tool modification", *Mater. Today Proc.,* vol. 5, pp. 11875-11881, 2018.

[11] M. Hajian, M.R. Razfar, and S. Movahed, "An experimental study on the effect of magnetic field orientations and electrolyte concentrations on ECDM milling performance of glass", *Precis. Eng.,* vol.

45, pp. 322-331, 2016.

[12]   C.P. Cheng, K.L. Wu, C.C. Mai, Y.S. Hsu, and B.H. Yan, "Magnetic field-assisted electrochemical discharge machining", *J. Micromech. Microeng.*, vol. 20, p. 07519, 2010.

[13]   Y. Xu, J. Chen, B.Y. Liu, and J. Ni, "Experimental investigation of magnetohydrodynamic effect in electrochemical discharge machining", *Int. J. machanical Sci.*, vol. 143, pp. 86-96, 2018.

[14]   P.K. Gupta, A. Dvivedi, and P. Kumar, "Effect of Pulse Duration on Quality Characteristics of Blind Hole Drilled in Glass by ECDM Effect of Pulse Duration on Quality Characteristics of Blind Hole Drilled in Glass by ECDM", *Mater. Manuf. Process.*, vol. 31, pp. 1740-1748, 2016.

[15]   J. Bindu Madhavi, and S.S. Hiremath, "Machining of micro-holes on borosilicate glass using micro-electro chemical discharge machining (μ-ECDM) and parametric optimisation", *Adv. Mater. Process. Technol.*, vol. 5, pp. 542-557, 2019.

CHAPTER 8

Microwave Drilling of Polymer Based Composite: Challenges and Opportunities

Gaurav Kumar[1,*], Apurbba Kumar Sharma[2] and **Mukund Kumar[3]**

[1] *Department of Mechanical Engineering, NIT Uttarakhand, Srinagar-246174, India*

[2] *Department of Mechanical & Industrial Engineering, IIT Roorkee, Roorkee-246174, India*

[3] *Department of Mechanical Engineering, BIT Mesra, Ranchi - 835215, India*

Abstract: Microwave drilling is an advanced machining process in which electromagnetic energy converted into thermal energy with the help of a metallic concentrator is used to create the desired shape in the work material. High strength electric field developed around the tooltip ionizes the dielectric media around the tooltip and results in plasma formation. High-temperature plasma ablates the material just beneath the tool tip to create the desired hole in the workpiece. In the present research work, micro-hole drilling on thermoset and thermoplastic-based composites using microwave energy in the air and transformer oil has been investigated. The drilling characteristics have been investigated in terms of the heat-affected zone, and overcut; a comparison has been made in air and transformer oil. The study revealed that drilling in the presence of dielectric-like transformer oil reduces the defects like HAZ and overcut significantly. It was also observed that thermal damage was more in thermoset-based composites as compared to thermoplastic-based composites.

Keywords: Composite, Heat affected zone, Microwave drilling, Micro-hole, Overcut.

INTRODUCTION

Difficult-to-machine materials like glass and polymer composite are gaining vast popularity these days due to their wide application in aerospace, naval, automotive, MEMS and other industries. Drilling of micro-hole in polymer composite using conventional drilling methods is a challenging task due to its unique physical and mechanical properties [1 - 4]. Polymer composite often experiences machining damage like fuzzing, matrix cracking, spalling, fiber pull-out, excessive dust, tool wear, thermal degradation, and delamination due to their inherent anisotropy and heterogeneity [4]. Thus, it becomes crucial to minimize

---

Corresponding author Gaurav Kumar Department of Mechanical Engineering, NIT Uttarakhand, Srinagar-246174, India; Email: grv.kmr@nituk.ac.in

Amar Patnaik, Albano Cavaleiro, Malay Kumar Banerjee, Ernst Kozeschnik & Vikas Kukshal (Eds.)

machining damage for better surface integrity of the machined product. Above mentioned machining damage associated with conventional machining resulted in the development of advanced machining methods like AWJM, LBM, *etc*. Though the above-mentioned non-conventional methods have made significant strides in the field of micromachining of glass and plastic fiber composites, the use of water with AWJM and USM for machining of composite reduces the strength of composite significantly. Laser machining has shown significant potential in the field of micromachining, but defects like heat-affected zone and taper are a serious concern [5].

Microwave drilling has drawn the wide attention of researchers in recent times due to its Omni-machining characteristics, high machining rate, and eco-friendly characteristics [6 - 16]. It has been successfully used to drill a hole in materials like metals, glass, ceramics, *etc*., irrespective of their electrical conductivity [7 - 16]. Jerby *et al*. reported the successful drilling of the hole on various materials like ceramics, concretes, *etc*., using a coaxial near-field radiator-based microwave drill experimental setup in the year 2002 [7 - 10]. But, the leakage of high-power microwave radiation was a significant concern as the complete machining process was taking place in an open environment. A significant improvement in the microwave drilling experimental setup has occurred since 2002. Later on, Lautre *et al*. used a gravity-fed microwave drilling set-up in which a concentrator made of metal concentrated the electromagnetic energy at its tooltip in a domestic microwave applicator which ensured no leakage of radiation [11, 12]. But, the defect around the drilled micro-hole in the glass was significant. Later on, Gaurav *et al*. changed the media around the tool by immersing the tool and workpiece inside a dielectric media [13 - 16]. The defect around the hole was reduced significantly, but the process still needed improvement to drill a hole with an almost negligible defect. Later on, Gaurav *et al*. used controlled feed to drill a hole in the glass. Subsequently, the effect of feed rate, tool-workpiece gap, immersion depth, and shape of the tool on heat-affected zone, material removal rate, roundness, and overcut was studied. Tool having a conical tip outperformed tool having a cylindrical tip. It was observed that a particular machining gap, i.e., ~ 300 μm, helped in reducing the roundness error and thermal damage in glass drilling, whereas overcut and material removal rate got increased as compared to zero tool work gap due to the removal of glass residue from the material removal zone. Moreover, it was observed that an increase in feed rate (up to 1.2 mm/s) and immersion depth (up to 45 mm) affects the thermal damage and overcut inversely [14]. However, defects were not completely eliminated from the machining zone. In the Year 2021, it was reported that dielectric in dynamic mode minimized the defect to a greater extent as compared to static dielectric and air. Drilling at lower power using dynamic dielectric helped in minimizing defect further, but a decrease in power in the case of stagnant dielectric and air increased the defect significantly. However, the machining time was less in the stagnant dielectric as compared to the dynamic dielectric due to the formation of a concentrated plasma zone in the stagnant dielectric. Cracks around the hole were less in the case of dynamic dielectric as compared to stagnant dielectric due to the low value of thermal stress. Thermal stress was observed to be low in the case of dynamic dielectric due to the flushing out of excess thermal energy from the machining zone as compared to stagnant dielectric [16].

The micro-hole drilling on composite laminates using non-conventional methods like laser machining, Electric discharge drilling, Abrasive Jet Machining, *etc.* has been reported by various researchers [1 - 4, 17 - 22]. Ravinder *et al.* successfully reported the drilling of micro-holes in carbon-fiber reinforced composite using Electric discharge drilling [3]. But, the inherent limitation of the EDD process is that the fibre should be of conductive nature. Feng *et al.* reported the successful drilling of holes in carbon fiber-reinforced epoxy composite without fiber pull-out and taper using rotary ultrasonic machining (RUM). However, chipping was observed around the hole. Further, the deformation of the composite during machining caused fluctuation in thrust forces, increasing the surface roughness [18]. Besides, excessive tool wear and comparatively higher machining time are observed in drilling hole in composite using ultrasonic machining [18, 19]. Non-conventional machining processes like laser machining and AJM can be used to drill holes all types of composite, but thermal damage and taper in the case of laser machining and stray cutting in the case of AJM is a serious concern [20 - 22].

Microwave drilling has shown a significant potential to drill a hole in all types of materials regardless of their conductivity. The present work was focused on drilling a micro hole in glass fiber and natural fiber reinforced polymer composite at a controlled feed rate in air and transformer oil and compares the defect like HAZ and overcut in different dielectric media like air and transformer oil.

MATERIALS AND METHODOLOGY

Glass fiber and natural fiber-reinforced composites (40×30×4 mm) of rectangular shape were used as a test specimen for the present study. Glass fiber reinforced epoxy composite was developed by hand lay-up process using glass fiber and epoxy, whereas natural fiber reinforced polypropylene composite was prepared using an injection molding process. Tungsten carbide (Diameter: 500 µm) has been used as a tool material due to its good thermo-physical properties. The microwave drilling set-up, as evident from Fig. (1) comprises a domestic microwave oven that acts as a source for the microwave, a metallic concentrator, which is used to concentrate the microwave energy at its tip, a tool holder which is used to hold the metallic concentrator and a dielectric container to store the dielectric as shown in Fig. (**1**). The tool and workpiece are immersed in a dielectric media to a depth known as immersion depth inset of Fig. (**1**). The desirable feed to the tool is given with the help of a stepper motor (torque capacity 10.1 Kg-cm, current 2.8A) driven by the motor driver, which is controlled by MACH-3 CNC software (Demo version) through the breakout board. The motor driver has been powered by an industrial DC power source (24 V, 20A). The mechanism of microwave drilling, as shown in the inset of Fig. (**1**), has been discussed by Gaurav *et al*. [4, 11]. In the microwave drilling process, the electromagnetic energy gets converted into thermal energy with the help of a thin metallic concentrator. In the presence of an electromagnetic field, free electrons in a thin metallic concentrator concentrate at the sharp corner of the tool. The focused electron creates a high-intensity electric field that ionizes the surrounding material, resulting in plasma. The heat energy released by the plasma produced at the tool's tip is enormous. Thermal energy interacts with the workpiece's surface right beneath it, ablates/melts it, and creates a hole of the required shape. Following that, the tool is fed forward using a motor drive, as shown in Fig. (**1**). The machining condition has been tabulated in Table **1**.

Table 1. Parameters of the process and their ranges.

Parameter	Workpiece	
	Thermoplastic based Composite	Thermoset based Composite
Tool	Tungsten Carbide	Tungsten Carbide
Dielectric	Air, Transformer oil	Air, Transformer oil
Feed rate	0.20, 0.70, 1.2 mm/sec	0.20, 0.70, 1.2 mm/sec
Tool-workpiece gap	0 μm	0 μm
Immersion depth	0, 20, 40 mm	0, 20, 40 mm

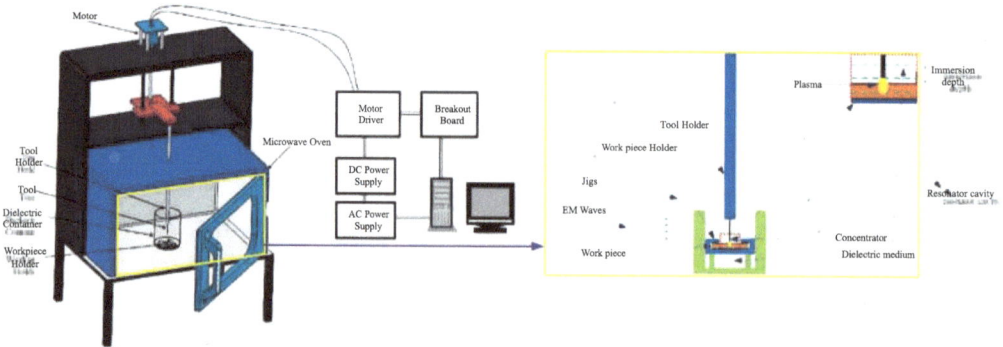

Fig. (1). Image of the experimental set-up of microwave drilling.

MEASUREMENT AND CHARACTERIZATION

The results of the drilled hole have been evaluated in terms of thermal damage and overcut. ImageJ 1.51k software was used to analyse hole and glass optical images. A stereo-zoom microscope was used to take the optical image of the specimen drilled. Thermal damage was obtained by calculating the area of the circle encompassing the burnt polymer and fiber, and resolidified layers. Overcut was calculated as the ratio of the difference between the hole's diameter and the concentrator's diameter to the concentrator's diameter using following equation **1**:

$$\text{Overcut} = \frac{D_h - D_c}{D_c} \times 100 \qquad (1)$$

where, D_h = Diameter of the hole in the workpiece

D_c = Diameter of the concentrator used to drill the hole

RESULTS AND DISCUSSION

Microwave drilling was used to produce holes in a glass fibre-reinforced epoxy composite and a natural fibre-reinforced polypropylene-based composite. The effect of process parameters such as feed rate and immersion depth on process responses such as thermal damage and overcut has been analysed and described in great detail in the next section.

THERMAL DAMAGE

The variation of thermal damage with respect to immersion depth has been shown in Fig. (**2**). It can be observed that a very high thermal damage was observed at lower immersion depth. As polypropylene has a very low melting point (160 °C), even little interaction with plasma induces a very large thermal damage in the composite Fig. (**3a**). But, when drilling was performed by immersing the workpiece and tool in a dielectric medium like transformer oil to an immersion depth of 20 mm and 40 mm, defects like thermal damage and overcut were reduced significantly, as shown in Figs. (**2 - 3**). Higher immersion depth reduced the thermal damage due to better cooling and decreased electromagnetic energy [13, 14]. The quality of the hole improved significantly at higher immersion depth, as evident from Figs. (**3 - 4**). Further, it can be observed that defects like fuzzing, matrix cracking, spalling, fiber pull out, excessive dust, and delamination can't be observed around the hole drilled, which shows the significant potential of microwave drilling. However, thermal damage is present around the hole, which can be eliminated by further improvement in the process.

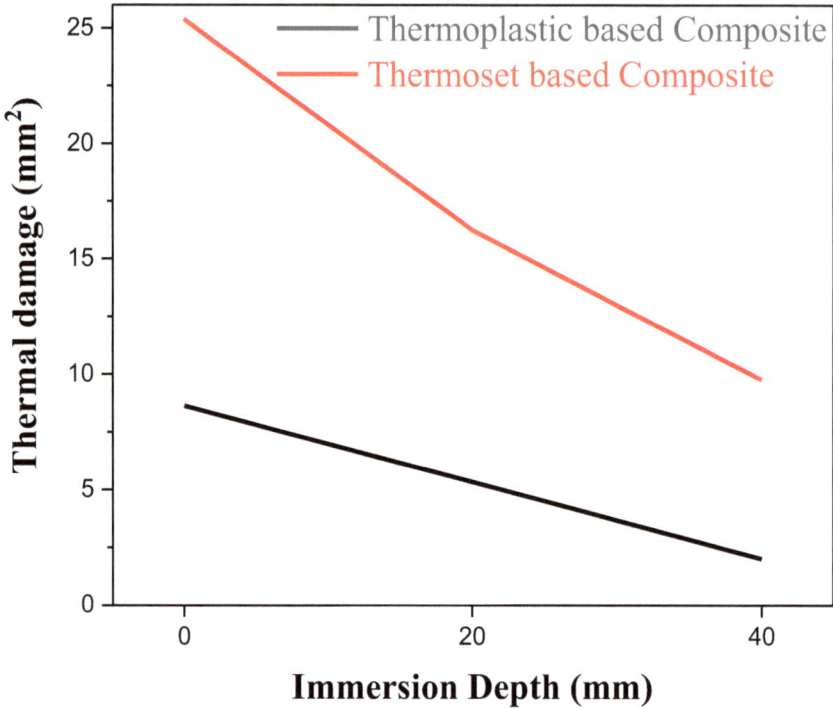

Fig. (2). Variation of thermal damage with respect to the immersion depth.

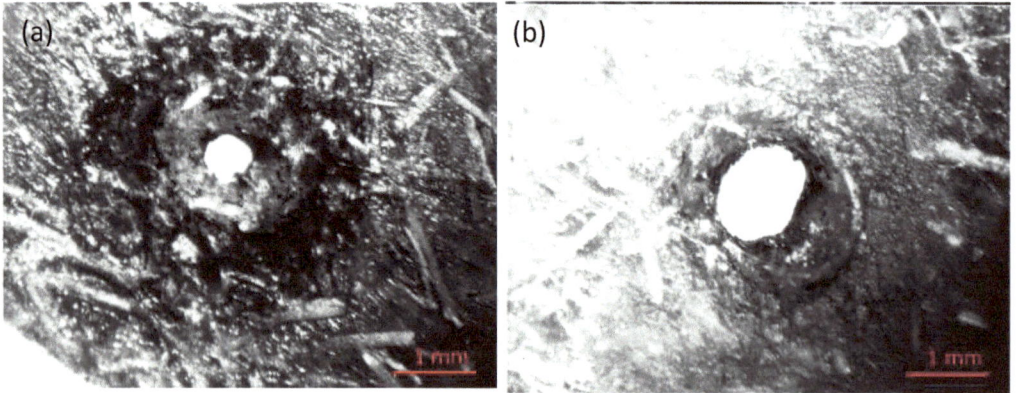

Fig. (3). Hole drilled in thermoplastic-based composite at an immersion depth of: (a) 0 mm (b) 40 mm

Fig. **(5)** depicts the variation of thermal damage as a function of feed rate. Thermal damage diminishes with increasing feed rate in both thermoplastic and thermoset-based composites, owing to a shorter interaction time between plasma produced at the concentrator tip and the composite at higher feed rates. Microwave drilling showed better results in the thermoplastic-based composite than thermoset-based composite, as evident by Figs. (2-5). This happens because thermoplastics like polypropylene melt at lower temperatures, whereas thermoset doesn't melt once it solidifies. Therefore, the thermoset allows the concentrator to move forward only after it gets burnt, as shown in Fig. (4). Due to this, the interaction time between plasma formed at the tip of the concentrator and composite increases, which in turn increases the machining time. Additionally, as demonstrated in Figs. (6 and 7), thermal damage in thermoset-based composites increases further if the reinforcement changes from natural fibre to glass fibre because glass fibre ablates at a greater temperature than natural fibre. Due to the high melting point temperature, glass fiber poses higher resistance to plasma than a natural fibre, increasing the interaction time between plasma formed at the tip and workpiece. Besides it, glass melt in the machining zone also hinders the movement of the concentrator and further increases the machining time. In comparison to natural fibre-based epoxy composites, a longer contact period causes more thermal damage around the hole in glass fibre-based epoxy composites. Consequently, thermal damage is less in the thermoplastic-based composite as compared to the thermoset-based composite.

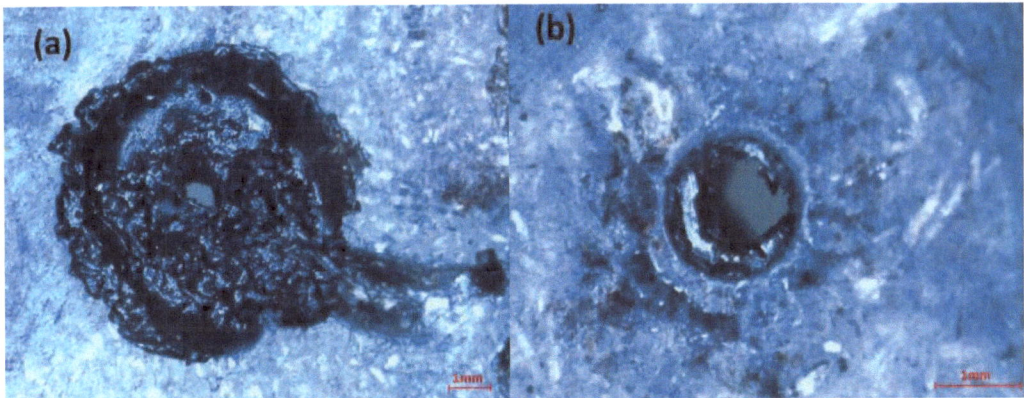

Fig. (4). Hole drilled in natural fiber reinforced epoxy composite at an immersion depth of: (a) 0 mm (b) 40 mm

Overcut

The effect of overcut with respect to immersion depth is seen in Fig. (7). Overcut is observed to increase as immersion depth is increased. The plasma zone becomes narrower as the immersion depth increases. At higher immersion depths, the low-diameter plasma zone forms a hole with a lower diameter, and so the overcut reduces. Further, it can be observed that the overcut is less in the thermoset-based composite as compared to the thermoplastic-based composite. Polypropylene ablates very fast as compared to thermoset due to its very low melting point (160 °C). As a result, more material is removed from a thermoplastic-based composite as compared to a thermoset-based composite over the same interaction period. As a result, thermoplastic-based composites have more overcut than thermoset-based composites.

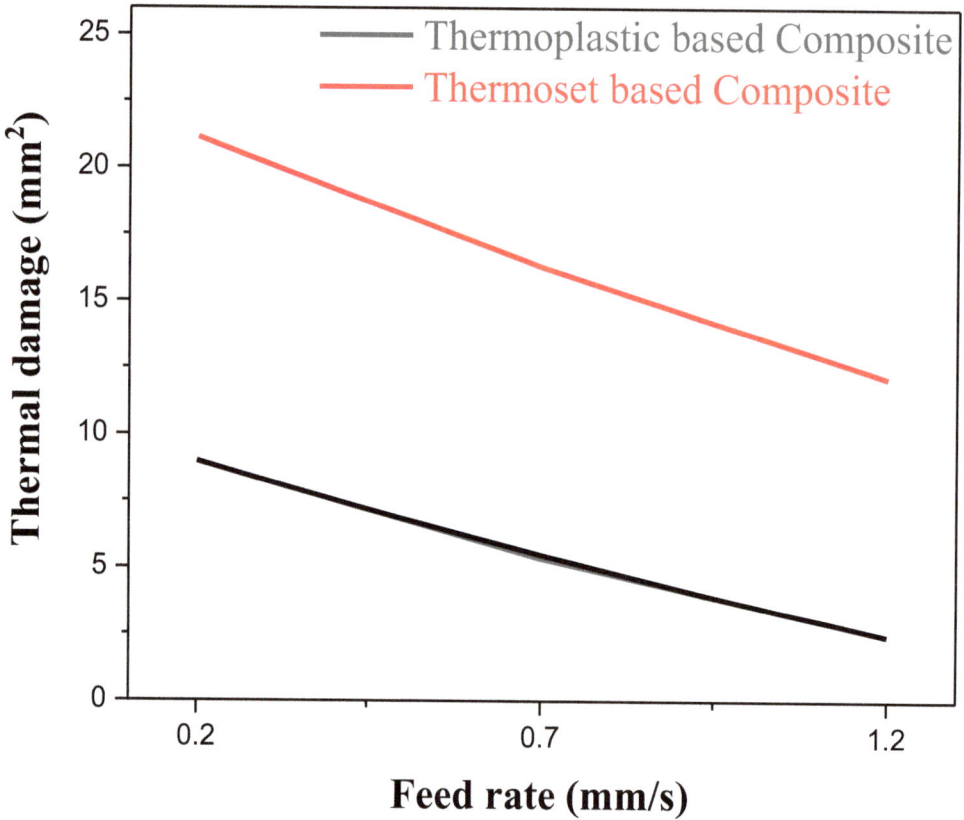

Fig. (5). Variation of thermal damage with respect to feed rate.

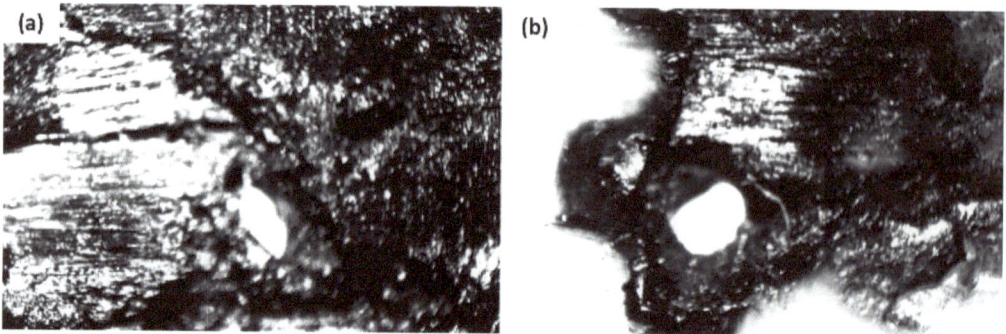

Fig. (6). Hole drilled in glass fiber reinforced epoxy composite in (a) air (b) transformer oil.

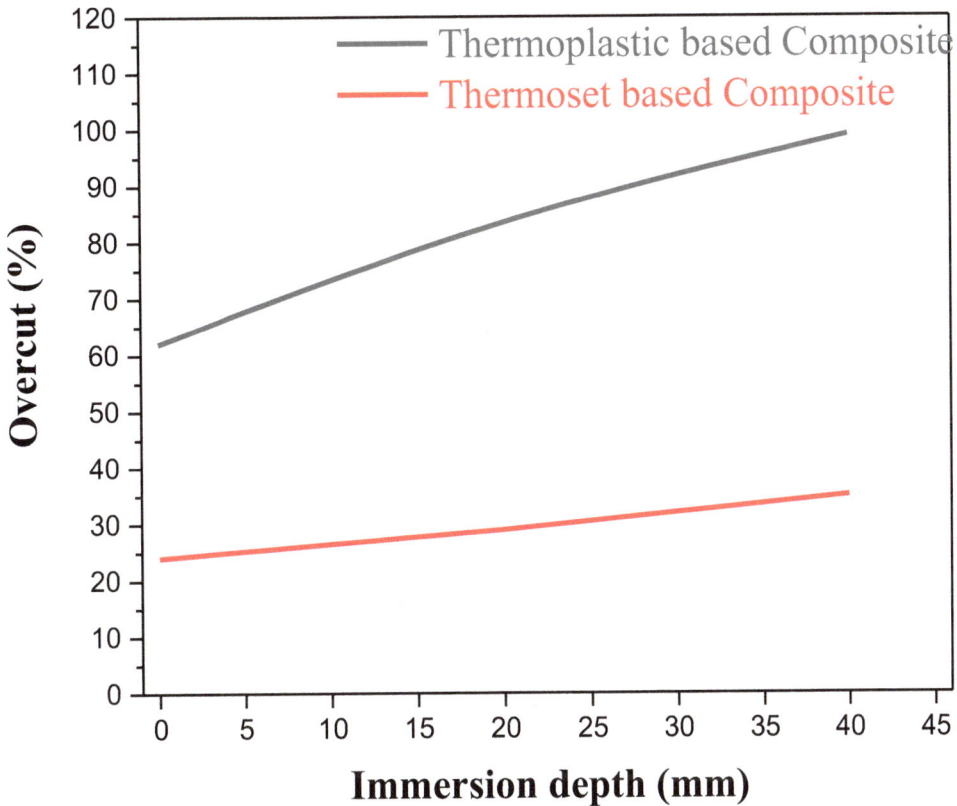

Fig. (7). Variation of overcut with respect to the immersion depth.

Fig. (**8**) depicts the relationship between overcut and feed rate. With an increase in feed rate, the amount of overcut reduces. The interaction period between plasma and workpiece decreases as the feed rate increases. As the interaction time is shortened, less material is removed from the workpiece's surface, resulting in a smaller diameter hole being drilled in the workpiece. However, tool wear increases significantly at a higher feed rate. Besides it, the problem of jamming of the tool and bending of the tool also increases at a higher feed rate.

OPPORTUNITIES AND CHALLENGES

There is hardly any literature available on microwave drilling of polymer composite. The above study has shown significant potential in the drilling of polymer composite using a microwave. The followings are the opportunities and challenges in microwave drilling of composites that can be explored in the future:

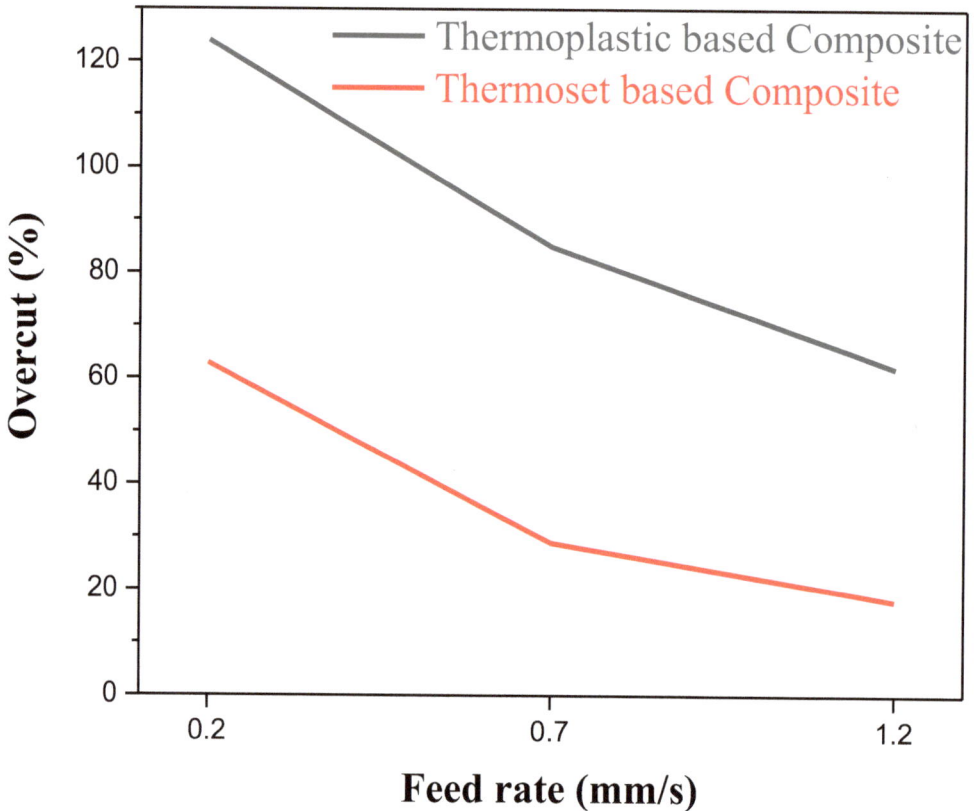

Fig. (8). Variation of overcut with respect to feed rate.

- Effect of input power on the surface integrity of the hole drilled.
- Effect of input power, feed rate, *etc*., on material removal
- Effect of volume fraction of reinforcement on process efficiency and hole quality.
- Drilling of defect-free holes in the thermoset-based composite.

CONCLUSION

Microwave drilling has been used successfully to drill holes in thermoplastic and thermoset-based composites. However, the hole drilled can be improved significantly by improving the process. Based on the above study, the conclusions are as follows:

- Microwave drilling was successfully used to drill a hole in composites.
- Dielectric assisted in the reduction of thermal damage around the hole drilled in

the workpiece.
- Thermal damage was observed to be more in the case of thermoset-based composites as compared to thermoplastic-based composites.
- High immersion depth decreases the heat-affected zone due to the confinement of plasma, but at the same time, it increases the overcut.
- Higher feed rate reduces the thermal damage, and overcut due to less interaction time.

REFERENCES

[1]   A. Hejjaji, D. Singh, S. Kubher, D. Kalyanasundaram, and S. Gururaja, "Machining damage in FRPs: Laser versus conventional drilling", *Compos., Part A Appl. Sci. Manuf.*, vol. 82, pp. 42-52, 2016.

[2]   K. Debnath, I. Singh, and A. Dvivedi, "Rotary mode ultrasonic drilling of glass fiber-reinforced epoxy laminates", *J. Compos. Mater.*, vol. 49, no. 8, pp. 949-963, 2015.

[3]   R. Kumar, A. Kumar, and I. Singh, "Electric discharge drilling of micro holes in CFRP laminates", *J. Mater. Process. Technol.*, vol. 259, pp. 150-158, 2018.

[4]   V. Gaitonde, S.R. Karnik, J.C. Rubio, A.E. Correia, A.M. Abrao, and J.P. Davim, "Analysis of parametric influence on delamination in high-speed drilling of carbon fiber reinforced plastic composites", *J. Mater. Process. Technol.*, vol. 203, no. 1-3, pp. 431-438, 2008.

[5]   A.K. Dubey, and V. Yadava, "Laser beam machining—A review", *Int. J. Mach. Tools Manuf.*, vol. 48, no. 6, pp. 609-628, 2008.

[6]   C. Leonelli, and T.J. Mason, "Microwave and ultrasonic processing: Now a realistic option for industry", *Chem. Eng. Process.*, vol. 49, no. 9, pp. 885-900, 2010.

[7]   E. Jerby, V. Dikhtyar, O. Aktushev, and U. Grosglick, "The microwave drill", *Science*, vol. 298, no. 5593, pp. 587-589, 2002.

[8]   E. Jerby, and A.M. Thompson, "Microwave drilling of ceramic thermal□barrier coatings", *J. Am. Ceram. Soc.*, vol. 87, no. 2, pp. 308-310, 2004.

[9]   Y. Meir, and E. Jerby, "Localized rapid heating by low-power solid-state microwave drill", *IEEE Trans. Microw. Theory Tech.*, vol. 60, no. 8, pp. 2665-2672, 2012.

[10]   E. Jerby, Y. Nerovny, Y. Meir, O. Korin, R. Peleg, and Y. Shamir, "A silent microwave drill for deep holes in concrete", *IEEE Trans. Microw. Theory Tech.*, vol. 66, no. 1, pp. 522-529, 2017.

[11]   N.K. Lautre, A.K. Sharma, P. Kumar, and S. Das, "A photoelasticity approach for characterization of defects in microwave drilling of soda lime glass", *J. Mater. Process. Technol.*, vol. 225, pp. 151-161, 2015.

[12]   N.K. Lautre, A.K. Sharma, S. Das, and P. Kumar, "On crack control strategy in near-field microwave drilling of soda lime glass using precursors", *J. Therm. Sci. Eng. Appl.*, vol. 7, no. 4, p. 041001, 2015.

[13]   G. Kumar, and A.K. Sharma, "Role of dielectric fluid and concentrator material in microwave drilling of borosilicate glass", *J. Manuf. Process.*, vol. 33, pp. 184-193, 2018.

[14]   G. Kumar, and A.K. Sharma, "On processing strategy to minimize defects while drilling borosilicate glass with microwave energy", *Int. J. Adv. Manuf. Technol.*, vol. 108, no. 11, pp. 3517-3536, 2020.

[15]   G. Kumar, R.R. Mishra, and A.K. Sharma, "On finite element analysis of material removal rate in microwave drilling of borosilicate glass", *Mater. Today Proc.*, vol. 41, pp. 759-764, 2021.

[16]   G. Kumar, R.R. Mishra, and A.K. Sharma, "On defect minimization during microwave drilling of borosilicate glass at 2.45 GHz using flowing dielectric and optimized input power", *J. Therm. Sci. Eng. Appl.*, vol. 13, no. 3, p. 031021, 2021.

[17]   R. Suresh, K.S. Reddy, and K. Shapur, "Abrasive jet machining for micro-hole drilling on glass and GFRP composites", *Mater. Today Proc.,* vol. 5, no. 2, pp. 5757-5761, 2018.

[18]   Q. Feng, W.L. Cong, Z.J. Pei, and C.Z. Ren, "Rotary ultrasonic machining of carbon fiber-reinforced polymer: Feasibility study", *Mach. Sci. Technol.,* vol. 16, no. 3, pp. 380-398, 2012.

[19]   F.D. Ning, W.L. Cong, Z.J. Pei, and C. Treadwell, "Rotary ultrasonic machining of CFRP: A comparison with grinding", *Ultrasonics,* vol. 66, pp. 125-132, 2016.

[20]   M.H. El-Hofy, and H. El-Hofy, "Laser beam machining of carbon fiber reinforced composites: A review", *Int. J. Adv. Manuf. Technol.,* vol. 101, no. 9, pp. 2965-2975, 2019.

[21]   P.F. Mayuet Ares, J.M. Vázquez Martínez, M. Marcos Bárcena, and A.J. Gámez, "Experimental study of macro and microgeometric defects in drilled carbon fiber reinforced plastics by laser beam machining", *Materials,* vol. 11, no. 8, p. 1466, 2018.

[22]   D.K. Jesthi, R.K. Nayak, B.K. Nanda, and D. Das, "Assessment of abrasive jet machining of carbon and glass fiber reinforced polymer hybrid composites", *Mater. Today Proc.,* vol. 18, pp. 3116-3121, 2019.

CHAPTER 9

Parametric Evaluation in Context to the Functional Role of Eco-Friendly Water Vapour Cutting Fluid Through Chip Deformation Analysis in HSM Of Inconel 718

Ganesh S. Kadam[1,2,*] and **Raju Shrihari Pawade**[1]

[1] *Babasaheb Ambedkar Technological University, Lonere, Maharashtra, India*
[2] *SIES Graduate School of Technology, Navi Mumbai, Maharashtra, India*

Abstract: Demand for increased production rates, better quality, and incorporation of green manufacturing practices has been continually challenging the manufacturers. This could be feasible by adopting high-speed machining (HSM) using eco-friendly cutting fluids but with careful process control. On these lines, the current paper explores process characteristics of the exotic superalloy Inconel 718 being turned at high speeds with tooling as coated carbide inserts and eco-friendly cutting fluid as water vapour. The experiments were carried out by varying three process parameters, viz. cutting speed, feedrate as well as water vapour pressure, following central composite design based on response surface methodology. A special tool holder with an in-built fluid supply channel was used to facilitate precise delivery of water vapour cutting fluid onto the machining zone. The process mechanics has been analyzed with the aid of the chip deformation coefficient as the same is a crucial indicator revealing the cutting fluid performance in machining as a result impacting the surface integrity, tool wear, machinability, etc. Analysis revealed that the response surface quadratic model for the chip deformation coefficient was statistically significant. The feedrate, vapour pressure, and the interaction between feedrate and vapour pressure were highly dominating factors influencing the chip deformation coefficient, with contributions of around 23.41%, 25.33% and 21.49%, respectively. An increase in vapour pressure was highly beneficial in lowering the chip deformation coefficient on account of water vapour's better penetrability and performance into the machining zone. Overall usage of cutting fluid as water vapour within feasible HSM parametric ranges can be notably beneficial.

Keywords: Chip deformation coefficient, Eco-friendly, High-speed machining, Inconel 718, Water vapour.

* **Corresponding author Ganesh S. Kadam:** Babasaheb Ambedkar Technological University, Lonere, Raigad, Maharashtra, India; Email: gskadam@ymail.com

Amar Patnaik, Albano Cavaleiro, Malay Kumar Banerjee, Ernst Kozeschnik & Vikas Kukshal (Eds.)

INTRODUCTION

Manufacturers worldwide have constantly been seeking methods for increasing productivity. In machining, productivity can be appreciably enhanced by incorporating high-speed machining (HSM). HSM basically involves carrying machining at those cutting speeds as well as feed rates that are much higher than that of their conventional counterparts. Superalloy Inconel 718, a nickel nickel-based alloy, possesses excellent properties like good corrosion resistance, strength at high-temperature and lower thermal conductivity. This makes it suitable for critical applications in aerospace, defense, gas turbines, nuclear reactors, chemical plants, marine equipment, etc [1]. However, Inconel 718 is also known as difficult-to-machine material as a result of its properties leading to poor machinabilities, like preserved strength during machining, exorbitant strain rate sensitivity, and poor thermal conductivity [2]. The majority of the problems during machining primarily originate on account of higher cutting temperatures, and the control over it can be exercised by proper selection as well as delivery of cutting fluids. Also, due to stringent environmental laws imposed by governments worldwide for manufacturing industries, there imposes a necessity to use eco-friendly cutting fluids. A variety of researchers have focused their attention on the machining of Inconel 718, wherein different cutting fluids and application methods ranging from conventional to eco-friendly grades like wet [3 - 10], high-pressure jet cooling [7, 8, 10 - 12], minimum quantity lubrication (MQL) [9, 11, 13 - 17], cryogenic cooling [3, 4, 9, 11, 15], *etc.* have been explored, and the benefits noted in the form of better surface finish reduced cutting forces and improved tool life.

During the 1990s, Podgorkov and Godlevski proposed water vapour as a new possible cutting fluid during machining [18]. Water vapour is cheap as it can be made easily available and further absolutely pollution-free too. Till date, very few studies have been reported on using cutting fluid as water vapour specifically for machining steels [18 - 20], Ti6Al4V [21] and Inconel 718 [22]. Water vapour basically serves as a lubricant and thus coolant in machining. Water vapour forms a lubrication film of low shear strength on the underside of chip, thus alleviating friction at the chip-tool interface and hence reducing cutting temperatures also. It can be concluded from past work that water vapour as cutting fluid in machining aids in lowering friction coefficient, cutting forces, and tool wear which, as a result, improves surface integrity in comparison to other machining environments like dry, wet, compressed air and gases. Even though having known the importance of water vapour in contrast to other machining environments, its functional role towards lubrication performance needs to be assessed especially in the case of HSM of Inconel 718. It becomes investigatory to understand how water vapour's functional performance varies under different parametric ranges as a negligible amount of work has been carried out on the same. The present paper discusses the same within the HSM regime for Inconel 718 with a due focus on exploring the effects of process parameters on the resulting chip deformation coefficient as the helpful monitoring element. Initially, a wise selection of cutting tools and process parameters has been done and further followed by vigorous experimentation adopting a design-of-experiments methodology. This is followed by an analysis of the results with due focus on reasoning and explanation of the same. Finally, keeping a broad view, conclusions from the study and the scope for further work have been made.

MATERIALS AND METHODOLOGY

Keeping in view the need for productive and sustainable practices by the manufacturers, it was decided to incorporate HSM of Inconel 718 using eco-friendly cutting fluid as water vapour. The experimental design established on RSM (response surface methodology) was adapted for conducting and analyzing experiments. The work particularly involved experiments at high-speed turning formulated on CCD (central composite design) of RSM. The three process parameters, cutting speed, feed rate and water vapour pressure, were varied. The levels of these process parameters are given in Table **1** [23]. The selection of these machining parameters was through wisdom from literature and past experience. Chip deformation coefficient was selected as the response as it is a crucial indicator for describing the status of lubrication in machining as well as analyzing machining characteristics [24].

Table 1. Process parameters along with their corresponding levels.

Levels	Coded Levels	Cutting Speed V_c (m/min)	Feed f (mm/rev)	Pressure P (bar)
1	-2	72.96	0.05	0.20
2	-1	90	0.07	0.30
3	0	115	0.10	0.45
4	+1	140	0.13	0.60
5	+2	157.04	0.15	0.70

Round bar specimens of Inconel 718 were employed as the work material with 25 mm diameter and 100 mm length. Inconel 718 had a chemical composition comprising of Ni-Cr-Fe-Nb-Ti-Co-Al-Si-C to be 54.95-17.90-16.54-4.85- 0.92-0.92-0.52-0.08-0.03 taken in order. The cutting tool used was carbide inserts with PVD TiAlN coating having specification CNMG120408MS and grade KCU10 from Kennametal. From recent research [23, 24], it was learned that using tool cutting fluid delivery assists in precise fluid jetting onto the insert tip and further enhances the cutting fluid's performance in the machining zone. Hence special tool holder PCLNL2525M12HP (make Sandvik) having an inbuilt channel and nozzle for cutting fluid delivery was used. The experiments on turning were carried out on a production-grade CNC lathe from Micromatic Ace and model being Jobber XL. In order to supply the cutting fluid as water vapour into the core machining zone, the sophisticated steam generation device was brought into operation (Fig. 1). For eliminating any of the effects of workpiece inhomogeneity on experimental results, a skin cut of 1 mm was taken over the full length of the workpiece prior to every experiment. The water vapour pressure could be precisely controlled and fixed at a specific level due to the aid of sophisticated controls embedded in the steam generation device. A fresh cutting tip was utilized for each experiment. During machining, the chips were carefully collected experiment-wise. Post experimentation, the collected chips were morphologically analyzed, wherein measurements of their thickness (t_c) were accomplished with the help of a tool maker's microscope from Mitutoyo, model TM505. For calculating the chip thickness before the cut (t_u), the following equation was helpful [25],

$$t_u = f \cdot \sin K_r \tag{1}$$

where, f = feedrate and K_r = principal cutting edge angle

Now having known cut and uncut chip thicknesses, the chip deformation coefficient can be calculated from the following equation [25],

$$\zeta = \frac{t_c}{t_u} \tag{2}$$

where, ζ = chip deformation coefficient, t_c = cut chip thickness and t_u = uncut chip thickness

Replication and re-measurements of all the experiments were carried out for data validation. Design Expert software was utilized for statistical analysis and to generate the 3D surface graphs.

Fig. (1). (a) Steam generation device and supply of water vapour onto the CNC turning lathe's machining zone from hose, and **(b)** Jet of water vapour directed precisely onto insert tip through special tool holder.

RESULTS AND DISCUSSION

One of the crucial indicators to understand the performance status of any cutting fluid during machining is basically the chip deformation coefficient. Whenever cutting fluid access the machining zone, particularly into the chip and tool interface, it provides a lubricating film between the chip and tool. Additionally, the pressure of cutting fluid may also assist in the lifting of the chip away from the tool. Both these aspects enable towards reduction of friction occurring in-between chip and tool; as a consequence, the machining becomes easy as chips glide smoothly over a tool and are thinner in size. Considering the dimensions of thinner chips and correlating it to uncut chip thickness, an inference may be drawn stating corresponding chip deformation coefficient shall be smaller. Hence lower chip deformation coefficient as a consequence of easy gliding of thinner chips also implies that tool wear will be minimal, minimum cutting forces, better surface finish and lower cutting temperatures, all of which imply better machinability; needless to say, opposite of the same shall imply the inverse. As a result study of chip deformation coefficient can enable systematic monitoring of machining characteristics and cutting fluid performance. In the current research work, as water vapour being supplied in the form of a jet acted as cutting fluid, its functional performance in context to process parameter variations can be systematically analyzed by examining the chip deformation coefficient. Table **2** represents analysis of variance (ANOVA) for the chip deformation coefficient. The quadratic model of the response surface for the chip deformation coefficient was found to be statistically significant.

Table 2. ANOVA for chip deformation coefficient.

Source	Sum of Squares	DoF	Mean Square	F value	P-value
Model	2.31	9	0.26	8.09	0.0023
A	0.16	1	0.16	4.92	0.0537
B	0.61	1	0.61	19.36	0.0017
C	0.66	1	0.66	20.89	0.0013
AB	0.002	1	0.002	0.080	0.7841
AC	0.002	1	0.002	0.087	0.7753
BC	0.56	1	0.56	17.82	0.0022
A^2	0.20	1	0.20	6.44	0.0318
B^2	0.13	1	0.13	4.13	0.0727
C^2	0.013	1	0.013	0.42	0.5355

A – cutting speed, B – feedrate, C - pressure.

The empirical model for chip deformation coefficient (ζ) constituting its actual factors is as given below,

$$
\begin{aligned}
\zeta = {} & 1.38051 + 0.047896(V_c) - 15.16457(f) \\
& - 5.59120(P) + 0.023667(V_c \times f) \\
& - (4.93333 \times 10^{-3})(V_c \times P) + 59.00(f \times P) \\
& - (1.90308 \times 10^{-4})(V_c^2) - 105.83793(f^2) \\
& - 1.34224(P^2)
\end{aligned}
\tag{3}
$$

From ANOVA, it is revealed that the feedrate, pressure, as well as the interaction between feedrate and pressure, are the most important factors significantly influencing the chip deformation coefficient. The percentage contribution of feedrate, pressure, and interaction between feedrate and pressure on the resulting chip deformation coefficient was 23.41%, 25.33% and 21.49%, respectively. However, all the factors, viz. cutting speed, feedrate and pressure, as well as their interactions, impact the resulting chip deformation coefficient. The same is clearly noticed by observing 3D surface graphs shown in Fig. (2-4) for the chip deformation coefficient.

The effect of cutting speed and that of feedrate on the resulting chip deformation coefficient is presented in Fig. (2). As may be observed, with an increase in cutting speed irrespective of the feedrate, the chip deformation coefficient goes on increasing. The increase of chip deformation coefficient with an increase in cutting speed is mainly on account of the thickening of chips, particularly at cutting speeds that are higher. This phenomenon occurs due to a decrease in the effectiveness of water vapour jet, especially at higher cutting speeds, due to reduced jet penetrability and, thus, poor lubrication. Further in the case of feedrate irrespective of the cutting speed, the reverse trend is observed. The chip deformation coefficient decreases for increased feed rates at all the values of cutting speeds. This trend is more noticeable for lower cutting speeds and gradually diminishes for higher cutting speeds. Thinning and widening of chips are found to occur at higher feedrates. Thus the decrease of chip deformation coefficient with an increase in feedrate is on account of the reduction in chip thickness.

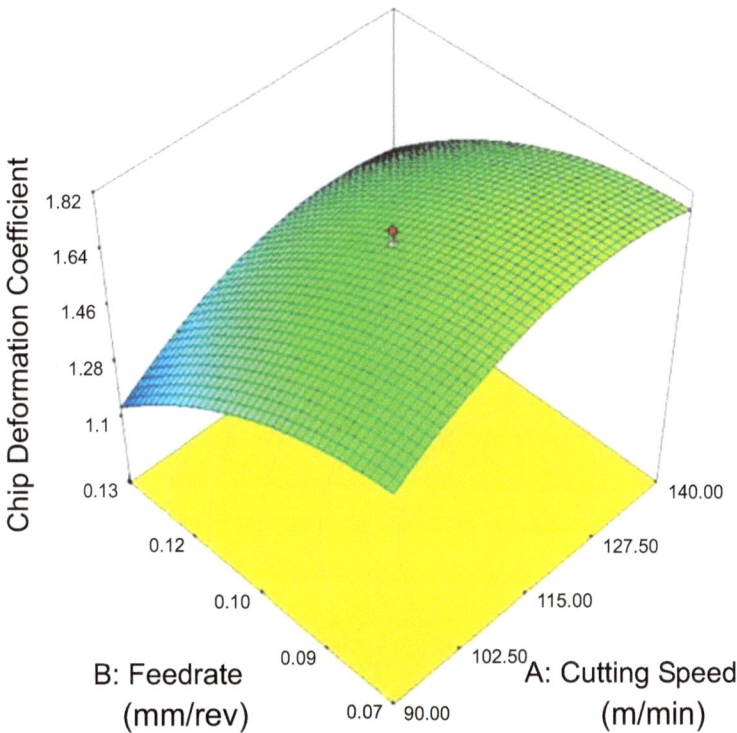

Fig. (2). Cutting speed and feedrate's effect on chip deformation coefficient.

Fig. (3) reveals the chip deformation coefficient as a result of cutting speed as well as pressure's effect on the same. In context to cutting speed, the chip deformation coefficient is observed to be increasing with a corresponding increase in cutting speed for all pressure ranges. The justification for this effect is the same as discussed above. However, this trend at a higher pressure for all cutting speeds shows lower values of chip deformation coefficient, which is mainly on account of more effectiveness of the water vapor jet's lubrication, particularly at higher pressure. Now consider the effect of pressure; the chip deformation coefficient got decreased with a corresponding increase in pressure at all ranges of cutting speed. It ultimately may be attributed to thinning of chips for higher levels of pressure. The water vapour jet accesses more precisely into the machining zone, particularly at high pressures, and provides effective lubrication, which alleviates the friction at the chip-tool interface. Hence chip thickness reduces on account of easy chip gliding over the tool face, thus leading to a lower chip deformation coefficient. The effectiveness of water vapour in providing the necessary lubrication is more evident at lower cutting speeds as inferred from the low values of chip deformation coefficient in contrast to that at higher cutting speeds, and the same may be ascribed to the diminishing of effectiveness of water vapour towards lubrication at higher cutting speeds.

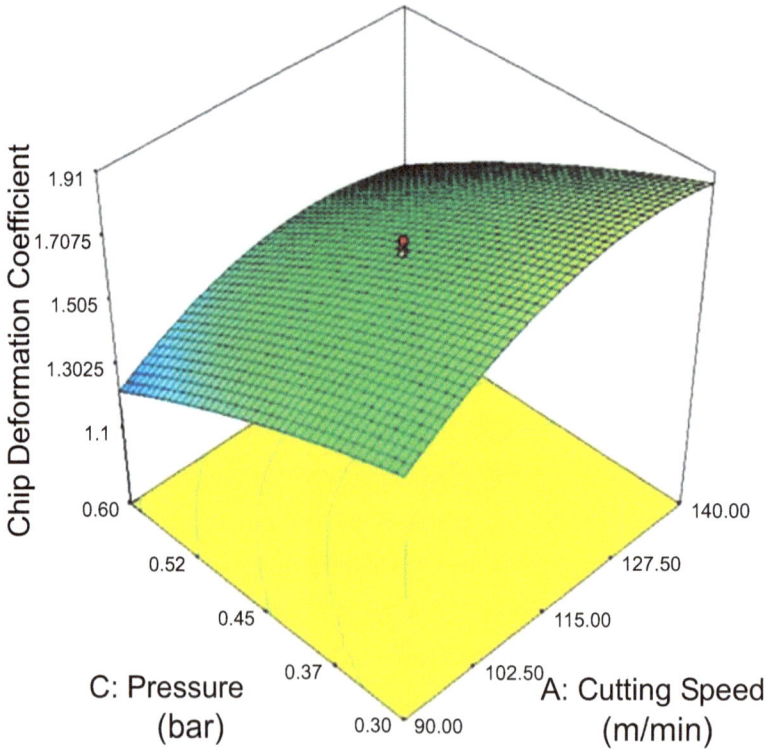

Fig. (3). Cutting speed and pressure's effect on chip deformation coefficient.

Fig. (**4**) represents the effect of feedrate as well as pressure on the resulting chip deformation coefficient. As can be observed, at low regimes of pressure, the chip deformation coefficient decreases with an increase in feedrate. This effect of decreased chip deformation coefficient at higher feed rates is due to thinning and widening of chips, as explained earlier. But for high-pressure regimes, the above trend almost ceases, and the chip deformation coefficient values are almost constant at all feedrates. Considering the effect of pressure, for lower feedrate values, the chip deformation coefficient notably decreases with an increase in pressure. However, at higher values of feedrate, the chip deformation coefficient is unaffected and thus remains more or less constant with an increase in pressure. This can be attributed to thinner and wider chips at higher feedrates. Lifting of chips from the tool face and thus alleviating the tool-chip interface friction is more pronounced due to the effective lifting force exerted by the water vapour jet on such chips. Further increment in pressure does not seem to alter the chip deformation coefficient at high feedrates as the lifting of chips, and corresponding lubrication almost remains constant, and the same as that found at lower pressure.

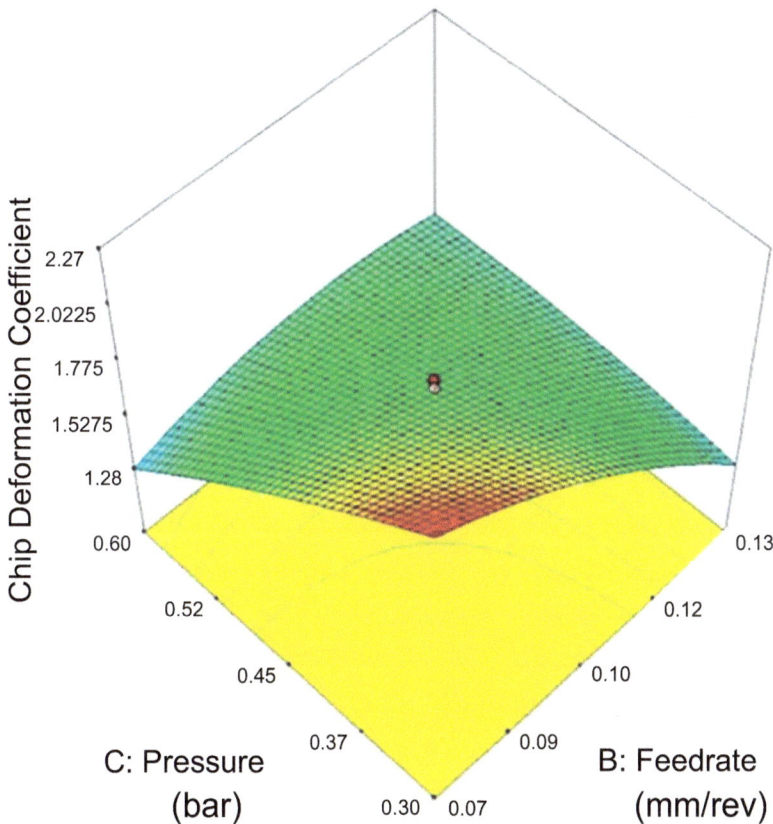

Fig. (4). Feedrate and pressure's effect on chip deformation coefficient.

Finally, a model adequacy check was carried out through a few sample tests. Table **3** shows the comparison between the predicted values of the chip deformation coefficient from the empirical model and the experimentally observed values. It is clearly seen that the predicted values of the chip deformation coefficient are very much close to experimentally obtained values. Overall the error is less than 7% for all the confirmatory tests. Thus the model can be accurately used for the prediction of chip deformation coefficient within the range of process parameters. This has indirect implications on the resulting surface integrity and the associated aspects. Enabling a lower chip deformation coefficient shall provide a feasible regime of process parameters that may correspondingly lead to lower surface roughness, lower cutting forces, lower tool wear, better machinability, and, thus, better surface integrity.

Table 3. Results of confirmatory test.

Test	V_c (m/min)	F (mm/rev)	P (bar)	$\zeta^\prime$	ζ	Error (%)
1.	90	0.07	0.3	2.026	2.151	6.16
2.	115	0.15	0.45	1.263	1.338	5.94
3.	140	0.07	0.3	2.241	2.294	2.35
4.	90	0.13	0.3	1.036	1.004	3.10
5.	115	0.1	0.7	1.330	1.405	5.64

$\zeta^\prime$ - predicted values, ζ - observed values.

CONCLUSION

The experimental investigations lead to the arrival of the below conclusions.

- The chip deformation coefficient enables us to interpret the lubrication status in machining, and thus the lower values of the same may enable us to achieve lower cutting temperatures, reduced cutting forces, better tool life, lower surface roughness and hence overall better surface integrity.
- Very high cutting speeds may be detrimental in machining, as evident from the higher chip deformation coefficient since the water vapour cutting fluid has limited access to the machining zone on account of poor penetrability.
- Higher water vapour pressures are helpful in machining, as evident from correspondingly yielded lower chip deformation coefficient values since the same facilitates better water vapour cutting fluid penetration at the chip-tool interface and hence better lubrication.
- The feedrate, pressure, as well as the interaction between feedrate and pressure, were the dominant factors influencing the chip deformation coefficient, with a contribution of 23.41%, 25.33% and 21.49%, respectively.
- The quadratic model of a response surface for the chip deformation coefficient was found to be statistically significant, and this model can be precisely used for

predicting the chip deformation coefficient within the range of explored process parameters.

- Overall, water vapour supplied at a higher pressure in the HSM range proves advantageous in Inconel 718 machining, thus fulfilling requirements of higher productivity, better quality and green manufacturing.
- The current paper is restricted to understanding machining mechanics with the aid of chip deformation coefficient only; however, the same can be extended further to surface integrity and tool wear aspects, particularly for water vapor as cutting fluid, to get greater insight.
- In context to future scope, a comparative analysis of cutting fluids and their implications on machining aspects like surface integrity, cutting forces, and tool wear/tool life can be given due consideration.

ACKNOWLEDGEMENT

Authors humbly acknowledge the financial assistance rendered by the MHRD Government of India towards the acquisition of a production-grade CNC lathe under TEQIP Phase-I and a sophisticated steam generation device under TEQIP Phase-II for developing the setup of experimentation in DBATU, Lonere.

REFERENCES

[1] R.S. Pawade, S.S. Joshi, and P.K. Brahmankar, "Effect of machining parameters and cutting edge geometry on surface integrity of high-speed turned Inconel 718", *Int. J. Mach. Tools Manuf.*, vol. 48, pp. 15-28, 2008.

[2] A.U.H. Mohsan, Z. Liu, and G.K. Padhy, "A review on the progress towards improvement in surface integrity of Inconel 718 under high pressure and flood cooling conditions", *Int. J. Adv. Manuf. Technol.*, vol. 91, pp. 107-125, 2017.

[3] N. Khanna, C. Agrawal, M. Dogra, and C.I. Pruncu, "Evaluation of tool wear, energy consumption, and surface roughness during turning of Inconel 718 using sustainable machining technique", *J. Mater. Res. Technol.*, vol. 9, pp. 5794-5804, 2020.

[4] M. Hribersek, F. Pusavec, J. Rech, and J. Kopac, "Modeling of machined surface characteristics in cryogenic orthogonal turning of Inconel 718", *Mach. Sci. Technol.*, vol. 22, pp. 829-850, 2018.

[5] X.M. Anthony, M. Manohar, M.M. Patil, and P. Jeyapandiarajan, "Experimental investigation of work hardening, residual stress and microstructure during machining Inconel 718", *J. Mech. Sci. Technol.*, vol. 31, pp. 4789-4794, 2017.

[6] R.P. Zeilmann, F. Fontanive, and R.M. Soares, "Wear mechanisms during dry and wet turning of Inconel 718 with ceramic tools", *Int. J. Adv. Manuf. Technol.*, vol. 92, pp. 2705-2714, 2017.

[7] R. Polvorosa, A. Suarez, L.N. Lopez de Lacalle, I. Cerrillo, A. Wretland, and F. Veiga, "Tool wear on nickel alloys with different coolant pressures: Comparison of Alloy 718 and Waspaloy", *J. Manuf. Process.*, vol. 26, pp. 44-56, 2017.

[8] Z. Fang, and T. Obikawa, "Turning of Inconel 718 using inserts with cooling channels under high pressure jet coolant assistance", *J. Mater. Process. Technol.*, vol. 247, pp. 19-28, 2017.

[9] V. Tebaldo, G. Gautier di Confiengo, and M.G. Faga, "Sustainability in machining: Eco-friendly turning of Inconel 718 Surface characterisation and economic analysis", *J. Clean. Prod.*, vol. 140, pp. 1567-1577, 2017.

[10]   A. Suarez, L.N. Lopez de Lacalle, R. Polvorosa, F. Veiga, and A. Wretland, "Effects of high-pressure cooling on the wear patterns on turning inserts used on alloy IN718", *Mater. Manuf. Process.,* vol. 32, pp. 678-686, 2017.

[11]   B.C. Behera, H. Alemayehu, S. Ghosh, and P.V. Rao, "A comparative study of recent lubri-coolant strategies for turning of Ni-based superalloy", *J. Manuf. Process.,* vol. 30, pp. 541-552, 2017.

[12]   P. Hoiera, U. Klement, N.T. Alagan, T. Beno, and A. Wretland, "Flank wear characteristics of WC-Co tools when turning Alloy 718 with high-pressure coolant supply", *J. Manuf. Process.,* vol. 30, pp. 116-123, 2017.

[13]   Y.V. Deshpande, A.B. Andhare, and P.M. Padole, "Experimental results on the performance of cryogenic treatment of tool and minimum quantity lubrication for machinability improvement in the turning of Inconel 718", *J. Braz. Soc. Mech. Sci. Eng.,* vol. 40, pp. 1-21, 2018.

[14]   H. Hegab, U. Umer, M. Soliman, and H.A. Kishawy, "Effects of nano-cutting fluids on tool performance and chip morphology during machining Inconel 718", *Int. J. Adv. Manuf. Technol.,* vol. 96, pp. 3449-3458, 2018.

[15]   A. Bagherzadeh, and E. Budak, "Investigation of machinability in turning of difficult-to-cut materials using a new cryogenic cooling approach", *Tribol. Int.,* vol. 119, pp. 510-520, 2018.

[16]   B.C. Behera, D.S Chetan, S. Ghosh, and P.V. Rao, "Spreadability studies of metal working fluids on tool surface and its impact on minimum amount cooling and lubrication turning", *J. Mater. Process. Technol.,* vol. 244, pp. 1-16, 2017.

[17]   A. Marques, M.P. Suarez, W.F. Sales, and A.R. Machado, "Turning of Inconel 718 with whisker-reinforced ceramic tools applying vegetable-based cutting fluid mixed with solid lubricants by MQL", *J. Mater. Process. Technol.,* vol. 266, pp. 530-543, 2019.

[18]   J. Liu, R. Han, and Y. Sun, "Research on experiments and action mechanism with water vapor as coolant and lubricant in Green cutting", *Int. J. Mach. Tools Manuf.,* vol. 45, pp. 687-694, 2005.

[19]   J. Liu, R. Han, L. Zhang, and H. Guo, "Study on lubricating characteristic and tool wear with water vapor as coolant and lubricant in green cutting", *Wear,* vol. 262, pp. 442-452, 2007.

[20]   L. Junyan, L. Huanpeng, H. Rongdi, and W. Yang, "The study on lubrication action with water vapor as coolant and lubricant in ANSI 304 stainless steel", *Int. J. Mach. Tools Manuf.,* vol. 50, pp. 260-269, 2010.

[21]   R.S. Pawade, D.S.N. Reddy, and G.S. Kadam, "Chip segmentation behaviour and surface topography in high-speed turning of titanium alloy (Ti-6Al-4V) with eco-friendly water vapour", *Inter. J Mach. and Mach. Mate.,* vol. 13, pp. 113-137, 2013.

[22]   G.S. Kadam, and R.S. Pawade, "Surface integrity and sustainability assessment in high-speed machining of Inconel 718 – An eco-friendly green approach", *J. Clean. Prod.,* vol. 147, pp. 273-283, 2017.

[23]   G.S. Kadam, Investigations on Surface Integrity in High-Speed Machining of Inconel 718 under different Machining Environments, 2019.

[24]   S.K. Ganesh, and S.P. Raju, "Chip deformation aspects in relative Eco-friendly HSM of Inconel 718", *Procedia Manuf.,* vol. 20, pp. 35-40, 2018.

[25]   G. Boothroyd, and W.A. Knight, "Fundamentals of Machining and Machine Tools". CRC Press-Taylor & Francis Group: New York, 2006.

CHAPTER 10

Parametric Analysis and Modeling of die-sinking Electric Discharge Machining of Al6061/SiC Metal Matrix Composite Using Copper Electrode

Bipul Kumar Singh[1,*], **Ankit Kumar Maurya**[1], **Sanjay Mishra**[1] and **Anjani Kumar Singh**[1]

[1] *Mechanical Engineering Department, Madan Mohan Malaviya University of Technology Gorakhpur, Uttar Pradesh, India*

Abstract: Aluminum-based metal matrix composites (MMC) are widely used in modern industries due to their lightweight, high strength, and superior hardness. In this study, silicon carbide (SiC) reinforced MMC has been fabricated using the stir casting method. Die-sinking EDM of fabricated MMC was performed using a copper (Cu) electrode. Experiments were carried out using the response surface methodology of box-behnken design (BBD) (RSM). The response surface plot was used to do parametric analysis on the effect of peak current (I_p), gap voltage (V_g), pulse-on-time (T_{on}), and duty factor(τ) on material removal rate (MRR) and surface roughness (Ra) using a second order regression model. The interaction effect of current with a pulse on time and duty factor has a substantial effect on MRR, while the interaction of current and voltage has a major impact on Ra, according to ANOVA. The increase of current increases both MRR and Ra. In the case of pulse-on-time, the value of Ra begins to decrease after 150 μs when the machining is performed at low voltage (40 V).

Keywords: ANOVA, EDM, MMC, MRR, RSM, Stir Casting, Surface Roughness.

INTRODUCTION

Aluminum-based MMCs are widely used in automobiles, aerospace, military, marine, *etc.* It has improved physical properties like high strength with low weight, enhanced stiffness, and hardness with improved tribological properties. Silicon carbide (SiC) is used for reinforcement due to its high hardness. Stirred casting technique is applied for the fabrication of composite material. Matrix and reinforcement were mixed in the ratio of 90wt% and 10 wt%, respectively. The molten mixture aluminum and reinforcement were mixed effectively using four

* **Corresponding author Bipul Kumar Singh:** Mechanical Engineering Department, Madan Mohan Malaviya University of Technology Gorakhpur, Uttar Pradesh, India; Email: bipulkumarsingh742@gmail.com

-blade stirrers. The fabricated composite was machined using a die-sinking electric discharge machine (EDM). EDM precisely removes the material from the electrically conductive workpiece through the spark generated between the interelectrode gaps. The sparking in EDM is regulated by DC pulses, and the direct current ionizes dielectric fluid within the interelectrode gap. Due to high heat, the workpiece material melts or even vaporizes. When the electric current is switched off the plasma channel collapses and ejects the re-solidified material in the form of debris through flushing.

Phate *et al.* [1] explored the wire-EDM machining of SiC-reinforced Aluminium MMC with different wt % of reinforcement, *i.e.*, 0%, 15% and 20%. MRR and Ra were studied in relation to feed, speed, and electric-related input parameters. Parametric analysis of Al7075 reinforced with 10% wt Boron Carbide (B_4C) was investigated by Gopalakannan *et al.* [2]. The analysis reveals that pulse on, and current affects the MRR, EWR and Ra. Shandilya *et al.* [3] proposed that the feed rate of wire and pulse-off duration were the highly significant and least significant factors, respectively, during wire EDM of SiC-reinforced Al6061. S Debnath *et al.* [4] developed a hybrid composite by the ex-situ method. Multiple regressions were used for the analysis of parameters for machining like pulse-on-time, current, pulse-off-time and voltage. Rizwee *et al.* [5] explained EDM parameters should be optimized for EDM machining of MMC and compared the prediction accuracy of artificial neural network (ANN) and RSM model. They proposed that ANN has better accuracy than RSM. Arunkumar *et al.* [6] developed Mg/SiC composite by powder metallurgy and discovered that it currently influences the output responses TWR, MRR, and Ra. Hourmond *et al.* [7] investigated the EDM of hybrid composite $AlMg_2Si$ using a copper tool with positive polarity. Using RSM, an experimental model for MRR and EWR was created, and it was discovered that the interaction effect of voltage and current had the greatest impact on these two parameters.

Khajuria *et al.* [8] developed Al2024/Al_2O_3 MMC by stir casting method using 5wt% reinforcement. EDM was performed for input parameter voltage, pulse-o--time, reinforcement wt% and current. Authors found that increasing pulse-o--time increases MRR, and increasing wt% of reinforcement, decreases MRR. Singh *et al.* [9] investigate the EDM of Inconel 601 and characterize the effect of machining settings on MRR and Ra by RSM. They discovered that current has a direct impact on Ra and MRR. Shihab *et al.* [10] performed W-EDM on friction stirred welded 5754 Al alloy. ANOVA indicates that pulse-on-time is the most effective factor for kerf width, Ra and MRR.

This study developed a second order regression model for MRR and Ra for die-sinking EDM of Al6061/SiC MMC using a copper electrode. Experiments were performed using the BBD approach of RSM, and the effect of major interaction factors on output responses was investigated using parametric analysis. The effect of peak current, gap voltage, pulse-on-time, and duty factor, as well as variations in output response about these input parameters, have been explored and analysed using response surface plots.

EXPERIMENTAL PROCEDURE

Material Preparation

In this research work, Aluminum 6061 rods were used as a matrix, and silicon carbide (SiC) powder was used as reinforcement. The weight fraction of silicon carbide reinforcement is 10 wt%. The composite has been formulated by the stir-casting method. Al6061 rods are cut into small pieces and then melted in a graphite crucible up to 800°C. The SiC particles are preheated in a crucible up to 750°C before mixing so that there is no moisture content remains in reinforcement which leads to porosity after mixing. SiC is preheated for 1 hour. After melting aluminum 6061 alloy, SiCp is mixed with the help of a stirrer for 15 minutes. After proper mixing of matrix and reinforcement, the molten mixed material is poured into a mold cavity to get the desired shape.

Tool Selection

A cylindrical shape copper tool of diameter 14 mm was used as an electrode. The negative polarity electrode was mounted vertically with a workpiece. Kerosene oil acts as dielectric fluid due to its greater flash point, better transparency of fluid, good dielectric strength, low specific gravity, and low viscosity.

Equipment Used

Machining of the developed MMC has been performed on ELEKTRA Pulse S-50 ZNC, die sinking EDM produced by Electronica Machine Tools Limited, Pune, as shown in Fig. (**1**). It has a maximum peak current of 50A and pulse duration of 4000µs. In the present research work, four process parameters, *i.e.*, I_p, V_g, T_{on} and τ, are used for the investigation of MRR and Ra. MRR was determined using Eq. (1). Digital weighing equipment with a precision of 0.01g is used to determine the weight of the workpiece. The workpiece is cleaned with a clean cloth and dried so that dielectric fluid and dirt are removed from the surface before weighing. The change in weight of the sample composite before EDM machining and after machining is the total material removed during the machining of workpiece. The MRR in g/min was calculated by dividing the change in material weight by the machining time [11].

Fig. (1). Die-sinking sparks EDM.

$$MRR(gram/min) = \frac{(\text{Initial wt of workpiece} - \text{Final wt of workpiece})(\text{grams})}{\text{Machining time(min)}} \qquad (1)$$

Roughness of the surface (Ra) is an important component of surface integrity which affects the primary function of the component. Here, Ra has been determined by a surface roughness tester (Taylor Hobson) Fig. **(2)**.

Fig. (2). Surface Roughness Tester.

DESIGN OF EXPERIMENT AND PARAMETERS

The experiment design is an extremely useful tool to conduct experiments scientifically. Here, experiments were performed based on the BBD approach of RSM. RSM helps to develop a relation between output response and several input factors considering the non-linearity of the process. Even though many RSM designs are available, the BBD approach has been chosen because it does not require the condition that all input parameters are at their maximum or minimum levels simultaneously [12]. So, it helps to avoid those experiments which are to be performed at extreme conditions, which may produce unsatisfactory results. BBD helps to develop a mathematic model using experimental trials at three levels of input parameters. The number of experimental trials (N) in BBD is given by $N=2m\,(m-1)+C_o$, (where m is the parameters considered for the experiment and C_o is centre points). A total number of experimental trials in the central composite design (CCD) approach of RSM is $N=2^m+2m+C$. Therefore, the BBD approach is comparatively more efficient than CCD. Based on the pilot experiment, the levels of input parameters were determined. Table **1** shows the values of the input parameters. The ratio of T_{on} to T_{off} can be used to calculate the duty factor.

Table 1. Process parameters for EDM and its levels.

Input Parameters	Symbols	Unit	Levels		
			-1	0	+1
Peak current	I_p	Ampere	4	8	12
Gap voltage	V_g	Volts	40	50	60
Pulse-on-time	T_{on}	µs	100	150	200
Duty factor	τ	-	4	5	6

Table 2. BBD Matrix and observation data.

Exp Trial	I_p	V_g	T_{on} (µs)	τ	MRR (g / min)	R_a (µm)
1	4	40	150	5	0.028368794	6.6
2	12	40	150	5	0.110091743	9.09
3	4	60	150	5	0.043333333	5.94
4	12	60	150	5	0.161849711	10.5
5	8	50	100	4	0.0625	7.5
6	8	50	200	4	0.080597015	8.06
7	8	50	100	6	0.111650485	7.3
8	8	50	200	6	0.101503759	7.66
9	4	50	150	4	0.033823529	5.98
10	12	50	150	4	0.111111111	10
11	4	50	150	6	0.033613445	5.72
12	12	50	150	6	0.161490683	9.2
13	8	40	100	5	0.079113924	7.5

(Table 2) cont.....

Exp Trial	I_p	V_g	T_{on} (µs)	τ	MRR (g / min)	R_a (µm)
14	8	60	100	5	0.103448276	6.9
15	8	40	200	5	0.064516129	8
16	8	60	200	5	0.104651163	9.1
17	4	50	100	5	0.051044084	5.22
18	12	50	100	5	0.115384615	9.12
19	4	50	200	5	0.035426731	6.65
20	12	50	200	5	0.147727273	9.58
21	8	40	150	4	0.054404145	7.3
22	8	60	150	4	0.072072072	7.73
23	8	40	150	6	0.091176471	8.53
24	8	60	150	6	0.106481481	7.66
25	8	50	150	5	0.061016949	8.3
26	8	50	150	5	0.078291815	8.4
27	8	50	150	5	0.094594595	8.2

RESULTS AND DISCUSSION

Experimental trials based on the BBD approach and corresponding values of MRR, and R_a are shown in Table **2**. In order to minimize the experimental errors, each experiment was repeated thrice, and measurements related to the determination of output responses were also repeated thrice so that the error due to assignable causes can be minimized. The average value of these repetitions was used for the development of a regression model using the BBD approach of RSM. The accuracy of the developed model has been evaluated using the statistical approach, and significant factors for each output response have been identified using the ANOVA approach. Parametric analysis using the RSM plots has been discussed in the subsequent section.

ANOVA for MRR

Table **3** shows the ANOVA for MRR. The low value of S and high value of R^2 (Table **4**) for the developed model shown in Eq. (2) implies that the model is adequate and that experimental data well fitted the model. ANOVA analysis shows that the interaction of I_p with T_{on} and τ has a significant effect on the MRR.

Table 3. ANOVA for MRR.

Source	DF	Adjt SS	Adjt MS	F-Value	P-Value
Model	14	0.035650	0.002546	27.96	0.000
Linear	4	0.033541	0.008385	92.07	0.000
I_p	1	0.028231	0.028231	309.99	0.000
V_g	1	0.002246	0.002246	24.66	0.000
T_{on}	1	0.000011	0.000011	0.12	0.739
τ	1	0.003053	0.003053	33.52	0.000
Square	4	0.000292	0.000073	0.80	0.546
I_p*I_p	1	0.000090	0.000090	0.99	0.339
V_g*V_g	1	0.000031	0.000031	0.34	0.573
$T_{on}*T_{on}$	1	0.000273	0.000273	3.00	0.109
$\tau*\tau$	1	0.000034	0.000034	0.37	0.555
2-Way Interaction	6	0.001817	0.000303	3.32	0.036
I_p*V_g	1	0.000338	0.000338	3.72	0.078
I_p*T_{on}	1	0.000575	0.000575	6.31	**0.027**
$I_p*\tau$	1	0.000640	0.000640	7.03	**0.021**
V_g*T_{on}	1	0.000062	0.000062	0.69	0.424
$V_g*\tau$	1	0.000001	0.000001	0.02	0.904
$T_{on}*\tau$	1	0.000199	0.000199	2.19	0.165
Error	12	0.001093	0.000091		
Lack-Of-Fit	10	0.000529	0.000053	0.19	0.973
Pure Error	2	0.000564	0.000282		
Total	26	0.036743			

Table 4. MRR Model Overview.

S	R-sqr	R-sqr(adjt)	R-sqr(predt)
0.0095431	97.03%	93.56%	88.25%

The predicted value of MRR in this model is given below, in which only significant factors are considered.

$$
\begin{aligned}
\text{MRR (gm/min)} = \; & 0.263 - 0.0283\ I_p - 0.001008\ T_{on} - 0.00376\ V_g \\
& - 0.0103\ \tau + 0.000060\ I_p*T_{on} + 0.000230\ I_p*V_g \qquad (2) \\
& + 0.00316\ I_p*\tau
\end{aligned}
$$

ANOVA for R$_a$

Different terms of the ANOVA (Table **5**) for R$_a$ show that the developed model can predict the R$_a$ with fair accuracy. A high value of R^2 and R^2 (adjt) with a low value of S (Table **6**) corroborate the model's suitability. A high p-value of lack-o--fit further validates the high prediction capability of the developed model.

Table 5. ANOVA for Ra

Source	DF	Adjt SS	Adjt MS	F-Value	P-Value
Model	14	44.3610	3.1686	27.01	0.000
Linear	4	40.6976	10.1744	86.73	0.000
I$_p$	1	38.0920	38.0920	324.72	0.000
V$_g$	1	2.5300	2.5300	21.57	0.001
T$_{on}$	1	0.0547	0.0547	0.47	0.508
τ	1	0.0208	0.0208	0.18	0.681
Square	4	1.1291	0.2823	2.41	0.107
I$_p$*I$_p$	1	0.2945	0.2945	2.51	0.139
V$_g$*V$_g$	1	0.6960	0.6960	5.93	0.031
T$_{on}$*T$_{on}$	1	0.0331	0.0331	0.28	0.605
τ*τ	1	0.6721	0.6721	5.73	0.034
2-Way Interaction	6	2.5343	0.4224	3.60	0.028
I$_p$*V$_g$	1	0.2352	0.2352	2.01	**0.011**
I$_p$*T$_{on}$	1	1.0712	1.0712	9.13	0.182
I$_p$*τ	1	0.0729	0.0729	0.62	0.446
V$_g$*T$_{on}$	1	0.7225	0.7225	6.16	**0.029**
V$_g$*τ	1	0.0100	0.0100	0.09	0.775
T$_{on}$*τ	1	0.4225	0.4225	3.60	0.082
Error	12	1.4077	0.1173		
Lack-Of-Fit	10	1.3877	0.1388	13.88	0.069
Error	2	0.0200	0.0100		
Total	26	45.7687			

Table 6. R$_a$ Model Overview.

S	R-sqr	R-sqr(adjt)	R-sqr(predt)
0.3425	96.92%	93.34%	82.44%

The predicted value of surface roughness by RSM with significant factors is given below in the equation.

$$Ra(\mu m) = -11.93 + 0.384\ Ip + 0.0247\ Ton + 0.017\ Vg + 5.55\ \tau$$
$$- 0.000144\ Ton*Ton - 0.355\ \tau*\tau - 0.001212\ Ip*Ton \tag{3}$$
$$+ 0.01294\ Ip*Vg + 0.000850\ Ton*Vg$$

Parametric Analysis for MRR and R $_a$

Fig. (3) demonstrates the effect of I_p and T_{on} on MRR. It is observed that MRR increases very rapidly with the rise of peak current when machining is performed at V_g= 50 V and τ=5. The figure also reveals that during die sinking EDM of a given MMC, an increase of pulse on time at a fixed value of peak current enhances the MRR. The increase of current increases the diameter of the channel produced by discharge energy and the energy content of each pulse, resulting in higher melting of the workpiece [13]. Thus, the MRR increases with peak current. The figure also depicts that when pulse on time increases at a lower peak current, MRR shows a declining trend between 100 µs-150 µs, but after this duration, the MRR does not show any appreciable change with respect to pulse on time.

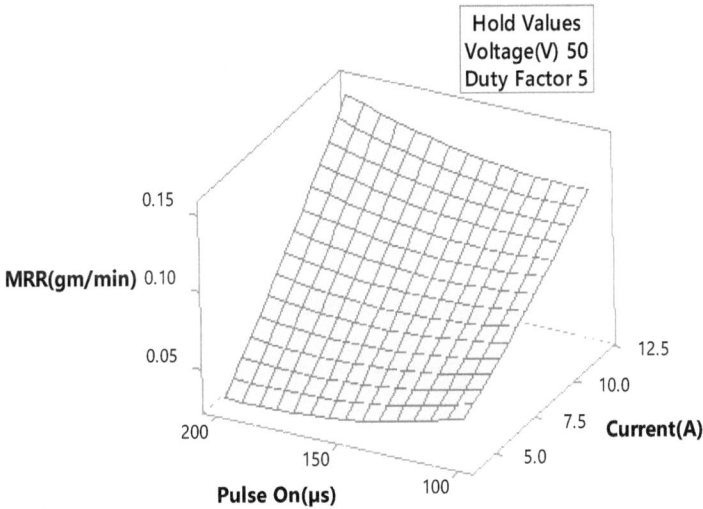

Fig. (3). Response of I_p and T_{on} on MRR.

Fig. (4) demonstrates the result of the current and duty factor on MRR. An increase in duty factor at a certain value of current led to higher MRR. The magnitude of this variation is very small at a lower current (5 A), but at a high value of current, MRR shows a significant increase when the duty factor increases from 4 to 6. At a higher duty factor, the thermal energy generation capacity increases results in a higher MRR [14]. The RSM plots in Fig. (**4**) indicate that the highest MRR occurs when the machining is performed at the highest duty factor using the highest current during experimentation (12.5 A).

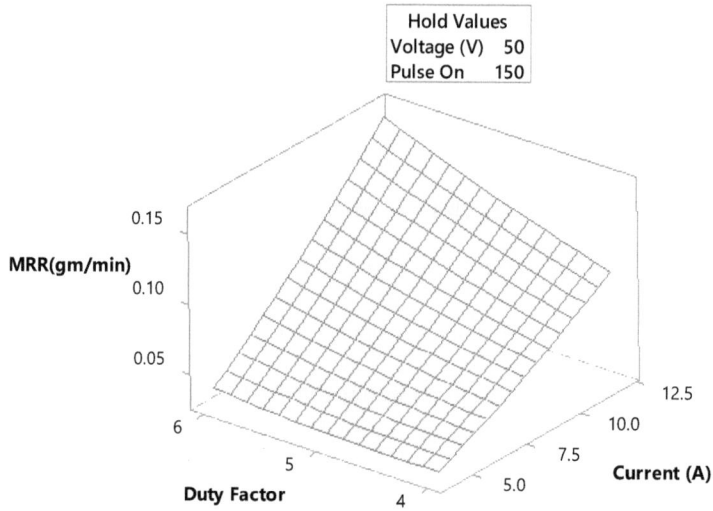

Fig. (4). Response of I_p and τ on MRR.

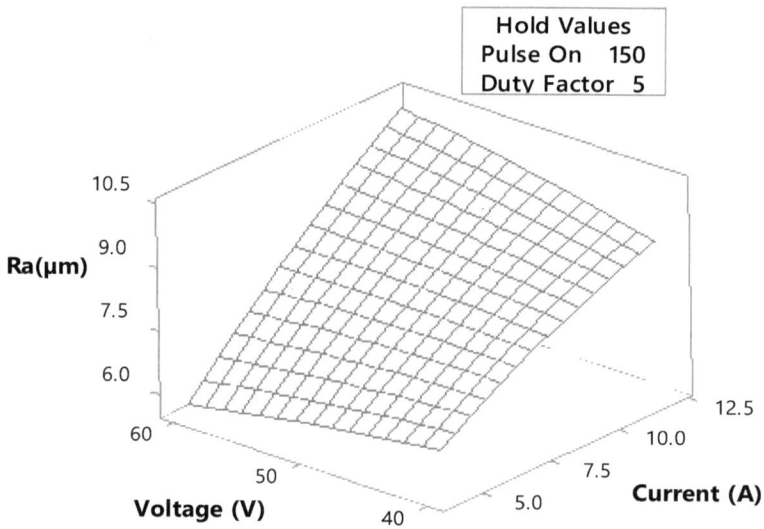

Fig. (5). Response of current and voltage on Ra.

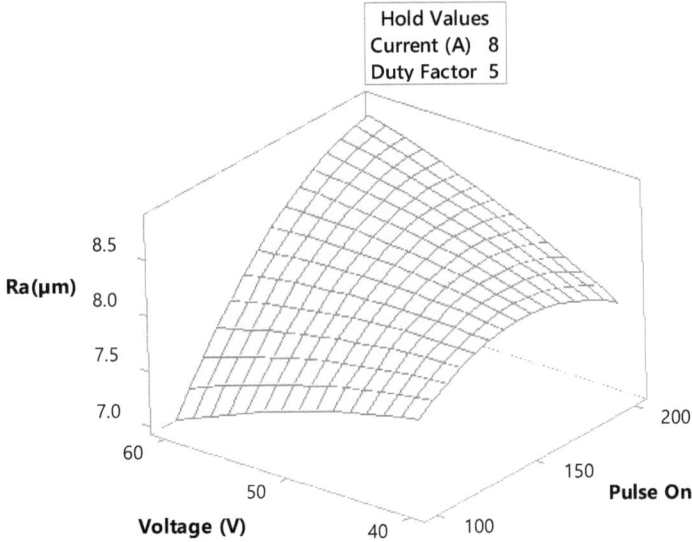

Fig. (6). Response of T_{on} and voltage on R_a.

The influence of current and voltage on the Ra is depicted in Fig. (**5**). It is observed that Ra increases with the rise of current and voltage. At higher currents, the speed of discharge and generation of heat energy between the tool and workpiece is more, therefore, the high ionic discharge erodes the workpiece surface quickly, thereby giving higher Ra. The increase in voltage applied during EDM also increases the discharge gap between the electrode and the workpiece [15]. Therefore, an increase in voltage at a lower current decreases the Ra, but as the current increases during machining, the surface roughness deteriorates, thus increasing the Ra.

Fig. (**6**) deliberates the response of T_{on} and V_g on the surface roughness during EDM of Al6061/SiC Metal Matrix Composite. It is observed that when the voltage rises at a low pulse on time (100 μs), surface roughness decreases. But as the value of pulse on time increases, Ra also increases with voltage. Similarly, an increase of T_{on} at low voltage (40 V) decreases Ra after 150 μs. The figure also reveals that at higher voltage, the increase of pulse on time increases the Ra, but at low voltage, the surface roughness begins to show decreasing trend at higher pulse on time.

CONCLUSION

In this study, Al6061/SiC MMC has been developed using the stir casting method. Experiments were conducted based on the BBD approach of RSM. Parametric analysis using response surface plots delivers the following significant conclusions for die-sinking EDM of MMC.

a. For MRR and Ra, the proposed second-order regression model using the RSM technique is adequate, and the experimental data well fitted the model.
b. The ANOVA results for MRR and Ra show that the interaction of I_p with T_{on} and τ has a substantial impact on MRR, whereas the interaction of V_g with I_p and T_{on} has a large impact on Ra.
c. The increase of current increases both MRR and Ra.
d. The increase of T_{on} and τ increases MRR at a certain value of current, and the rate of variation in MRR increases with current.
e. I_p and T_{on} are more influencing parameters for MRR, and V_g and I_p are more influencing factors for Ra.
f. In the case of T_{on}, the value of Ra begins to decrease after 150 µs when the machining is performed at low voltage (40 V).

ACKNOWLEDGEMENT

We wish to acknowledge the financial support received under the Research Promotion scheme of the All India Council for Technical Education (AICTE), Government of India.

REFERENCES

[1]   M. R. Phate, and S. B. Toney, "Modeling and prediction of WEDM performance parameters for Al/SiCp MMC using dimensional analysis and artificial neural network", *Engineering Science and Technology, an International Journal,* 2018.

[2]   S. Gopalakannan, T. Senthilvelan, and S. Ranganathan, "Modeling and optimization of EDM process parameters on machining of Al 7075-B₄C MMC using RSM", *International Conference on Modeling, Optimization and Computing* 2012, pp. 685-690.

[3]   P. Shandilya, P.K. Jain, and N.K. Jain, "Prediction of surface roughness during wire electrical discharge machining of SiCp/6061 Al metal matrix composite", *Int. J. Industrial and Systems Engineering,* vol. 12, no. 3, pp. 301-315, 2012.

[4]   S. Debnatha, R.N. Raib, and G.R.K. Sastryc, "A study of Multiple regression analysis on die sinking Edm machining of ex-situ developed Al-4.5cu-SiC composite", *7th International Conference of Materials Processing and Characterization* 2018, pp. 5195-5201.

[5]   M. Rizwee, and P. Sudhakar Rao, "Recent advancement in electric discharge machining of metal matrix composite materials," In: Materials Today: Proceedings", *International Conference on Newer Trends and Innovation in Mechanical Engineering: Materials Science,* 2020

[6]   L. Arunkumar, and B.K. Raghunath, "Electro discharge machining characteristics of Mg/SiCP Metal matrix composites by Powder Metallurgy (P/M) Techniques", *International Journal of Engineering and Technology,* vol. 5, no. 5, pp. 0975-4024, 2013.

[7]   M. Hourmand, A.A.D. Sarhan, S. Farahany, and M. Sayuti, "Microstructure characterization and maximization of the material removal rate in nano-powder mixed EDM of Al-Mg2Si metal matrix composite ANFIS and RSM approaches", *Int. J. Adv. Manuf. Technol.,* no. Dec, pp. 2723-2737, 2018.

[8]   A. Khajuria, R. Bedi, B. Singh, and M. Akhtar, "EDM machinability and parametric optimization of 2014Al/Al2O3 composite by RSM", *Int. J. Machining and Machinability of Materials,* vol. 20, no. 6, pp. 536-555, 2018.

[9]   N. Singh, B.C. Routara, and D. Das, "Study of machining characteristics of Inconel 601in EDM using RSM", *7th International Conference of Materials Processing and Characterization* 2018, pp. 3438-3449.

[10]   S.K. Shihab, "Optimization of WEDM Process Parameters for Machining of Friction-Stir-Welded 5754 Aluminum Alloy Using Box-Behnken Design of RSM", *Arab. J. Sci. Eng.,* no. March, pp. 5017-5027, 2018.

[11]   P. Malhotra, R.K. Tyagi, and N.K. Singh, "Experimental investigation and effects of process parameters on EDM of Al7075/SiC composite reinforced with magnesium particles", *International Conference on Mechanical and Energy Technologies,* 2019

[12]   C. Douglas Montgomery, *Design and Analysis of Experiments.* Wiley: Delhi, India, 2006.

[13]   A. Dey, V.R. Reddy Bandi, and K.M. Pandey, "Wire electrical discharge machining characteristics of AA6061/cenosphere aluminium matrix composites using RSM", *International Conference on Processing of Materials, Minerals and Energy* 2016, pp. 1278-1285.

[14]   R. Baghel, H.S. Mali, and S.K. Biswas, "Parametric optimization and surface analysis of diamond grinding-assisted EDM of TiN-Al2O3 ceramic composite", *Int. J. Adv. Manuf. Technol.,* no. March, pp. 1183-1192, 2018.

[15]   M.K. Pradhan, and C.K. Biswa, "Investigating the effect of machining parameters on EDMed components a RSM approach", *Jixie Gongcheng Xuebao,* vol. 7, no. 1, pp. 47-64, 2010.

CHAPTER 11

A Comprehensive Review on Application of Spark Discharge Method (SDM) for Production of Nanoparticles

Mudit K. Bhatnagar[1], **Siddharth Srivastava**[1], **Vansh Malik**[1], **Mamatha Theetha Gangadhar**[1] and **Mohit Vishnoi**[1,*]

[1] *Department of Mechanical Engineering, JSS Academy of Technical Education Noida, Uttar Pradesh, India*

Abstract: Nanoparticles encompass great potential in the current era due to their small size. They are employed in a myriad of applications, from biotechnology to manufacturing and energy applications. The production of nanoparticles, therefore, has been a focus of interest for researchers since its inception. Amongst non-conventional methods for nanoparticle production, Spark Discharge has emerged as an effective and viable method. This review encapsulates various experiments and works done over the years on the application of the spark discharge method for the production of nanoparticles and postulates the prospects of future work in the field. Different ways to control nanoparticle size by altering different parameters such as dielectric medium spark frequency, the gap between electrodes, and energy per spark and flow rate have been explored. Contrast has been drawn between conventional and non-conventional processes of nanoparticle production. In conclusion, new non-conventional techniques and hybrid techniques for nanoparticle production with spark discharge methods have been discussed, along with the applications of nanoparticles in emission control, cooling and lubrication.

Keywords: Energy Application, Ferro-fluids, Lubrication, Nanoparticles, Spark Discharge method.

INTRODUCTION

Nanoparticles are tiny particles of any metal oxide, alloy, semiconductor, composite, or Multi-Walled Carbon Nanotube of the order of 10-1000nm. They can also be defined as particulate dispersions. In addition, with their small size, they are employed in a wide array of applications due to their unique thermal and mechanical properties [1].

---
[*] **Corresponding author Mohit Vishnoi:** Department of Mechanical Engineering, JSS Academy of Technical Education Noida, Uttar Pradesh, India; E-mail: vishnoi.mohit06@gmail.com

Amar Patnaik, Albano Cavaleiro, Malay Kumar Banerjee, Ernst Kozeschnik & Vikas Kukshal (Eds.)

In biomedical, they are used in the form of nano gel for cancer therapy [2]. In addition, new green methods to synthesize the nanoparticles were explored using living cells which irrefutably proved to be less toxic and hazardous [3, 4]. Metallic nanoparticles' ever-increasing popularity in the field of thermal and energy applications is attributed to their high conductivity and mechanical endurance, which could be explained by their high heat-carrying capacity used in automobiles and supercapacitors [5, 6]. Quantum dots, which are zero-dimensional nanoparticles, are extensively used in optoelectronics, hydrogen evolution, and energy applications [7]. Nanoparticles are also used in nanocomposites to form electrodes and membranes in supercapacitors and hydrogen fuel cells [8]. Pertaining to their vast use, the efficient production of nanoparticles is crucial. This review articulates the production of nanoparticles by Spark Discharge Method (SDM). In addition to the aforementioned applications, the review's focus would be on their efficient use in lubrication, cooling, and emission control.

Phenomena like lubrication, heat transfer, sealing, combustion, catalysis, and mass diffusion are key to a proper functioning mechanical or physical system [9, 10]. These are the properties that govern most of the machining process. The addition of NP is one of the easiest ways to enhance these properties, which would further contribute to a positive and more effective outcome [11, 12]. SDM is a top-down thermal method that breaks down larger clusters of solid particles in the form of smaller nanoparticles by the action of successive spark discharges. As no chemical process is involved, it was observed that SDM produced nanoparticles with enhanced efficiency and rate. It is similar to Laser Ablation, which is entirely environmentally friendly with no chemical precursors required and leaving zero waste products [13, 14].

NANOPARTICLE (NP) PRODUCTION THROUGH SDM

Experimental Apparatus

The Experimental setup varies from one material to another depending upon the properties required in NP and their method of quenching and flow rate of dielectric or inert gas.

The typical setup consists of four major parts (i) A magnetic stirrer along with a stirring bar; (ii) A dielectric medium of water or ethanol to form the gold nanoparticle suspension and to facilitate the dispersion of gold nanoparticles; (iii) A power supply to regulate a stable pulse voltage between electrodes to ionize the medium; (iv) The servo motor system to control the gap between electrodes.

As the name suggests, the energy liberated during Spark discharge is used for the production of NP. This is the same procedure the first time used by Svedberg [15] to produce colloidal suspensions; the electrodes of the required material are placed under a dielectric or inert gas medium with a particular spark gap to maintain a repetitive spark between the electrodes [16]. Plasma consisting of a spark reaches a temperature near 20,000°K for a few seconds. This temperature is sufficient for the boiling and nucleating of droplets of metal. These tiny droplets further solidify into NP. The properties of metal depend upon the grain sizes that are directly influenced by the Quenching ability of the dielectric used. The dielectric medium/liquid used may be organic, aqueous, or cryogenic, depending upon the property of NP required.

In the working of SDM, various researches were conducted. Hallberg *et al.* used hydrogen assisted spark discharge method to produce metallic nanoparticles. The formed nanoparticles proved to be different from other nanoparticles as the nanoparticles formed through this method had a reduced tendency towards oxidation [17]. Controlling the composition of nanoparticles is still a concern for many researchers. Kohut *et al.*, in their paper, concluded that the tenability of nanoparticles was achievable through the use of SDM [18]. In the experimental papers by Sabzehparvar *et al.* [19], Sahu *et al.* [20], and Tseng *et al.* [21], it was observed that SDM could be used to achieve varied forms of nanoparticles such as super magnetic nickel oxide, aluminum NPs, and Nano-Ag colloid by varying the process conditions. These researches aided in forming a comprehensive path to understanding the effects of various process parameters (input parameters) on the formation of nanoparticles. These are explained in the following section.

Material Selection for NP Production

There is no definite restriction of Material in conducting or semiconducting metals and their alloys [22, 23]. Multiple kinds of research have already been done to produce NP using Spark Discharge Method (SDM); some of their results are given in Table **1**.

Table 1. Synthesis of NP of different Materials.

S. No.	Material Class	System	Notes	Reference
1.	**Metal**	W, Cu, Ag, Nb, Pd, Mg, Sb	Special precautions with regard to O_2 needed in the case of Mg	[24, 25]
		Au	For the 5nm range, go with ethanol	[1]

(Table 1) cont.....

S. No.	Material Class	System	Notes	Reference
2.	Alloy	Fe/Cu, Fe/Zn	Two different electrodes	[26]
		Cr/Co	Alloy electrodes	[27]
		SnSb, Cu_2Sb	Sintered electrodes	[28]
		Cu–W, Ag–Cu, Pt– Au	Non-alloying metals	[29]
3.	Semicon-ductor	Si, Ge	Doped electrodes	[30, 31]
4.	Oxide	CdO, Fe_2O_3	3% O_2 in Ar	[32]
		MgO	99.999% Ar	[25]
5.	Carbon composite	Ag/C	Metal & graphite rods. Various morphologies	[33]
		Cu/C	2 ppm C_2H_2 in Ar. Amorphous carbon shell	[34]
6.	Particulate composite	Mg with Pd or Nb catalyst	2 Sparks in series	[25, 35]

Effects of Process Parameters on Nanoparticle Production

Zeta Potential

It is defined as the potential difference between the surface of a solid particle and the liquid. The zeta potential distribution of gold colloid having a negative charge in ethanol which was higher than −50 mV and −30 mV in water, was enough to restrain particles interacting with one other hand, ergo, remain a stable sample for this particle size. The measurement of Zeta potential (Malvern NanoZS90) is called laser Doppler Electrophoresis (LDE). Although the intensity of the dispersion light is directly proportional to the sixth power of the size of tiny particles of Colloidal gold in ethanol and has the small collective intensity of the scattered light. In addition, even particle size distribution of colloidal gold in water leads to increased intensities with attenuated bandwidth [1]. The size of the NP can be determined or set as per the requirements by altering specific parameters, which govern the rate of agglomeration and initial rate concentration of NP in the experimental apparatus.

Energy per Spark

Through their study, Pfeiffer *et al.* [14] concluded that primarily the size of NP depends upon the initial vapor quantity or initial concentration, which was determined by the energy provided per spark.

Flow Rate

It has been stated by Efimov *et al.* that the rate of agglomeration can be reduced by increasing the flow velocity of the inert gas up to 5.4m/s, which helped in obtaining weakly agglomerated NP of reduced size. The relation between NP size, NP concentration, and flow rate/airflow velocity, which was determined through the experimentation, showed that at low airflow velocity (1.4m/s), the particle concentration was highest for low particle diameter, decreasing as the particle size increased [36].

Di-electric

It can be easily concluded from the experimental results of Tseng *et al.* [1] that dielectric properties played a vital role not only in decreasing the size of NP but also in altering the grain size-based properties of NP. The gold NP was produced under the dielectric effect of ethanol and deionized water with dielectric constants of 24.3 and 80, respectively. But the size obtained in ethanol was 1/6th of the size obtained in water. Moreover, the shape obtained in water was elliptical, while a nearly spherical shape was obtained in ethanol. This was due to the high-temperature spark discharge process, resulting in ethanol medium breakdown into various tiny and active derivatives leading to surface modification. As a result, the Zeta potential of gold particles increased, and smaller NPs were obtained. Farrukh *et al.* also obtained similar results in the case of La/SnO_2-TiO_2 NP [37].

The other possible explanation for this is that after the nucleation of droplets, oxygen from both the di-electrics is absorbed on the gold surface. These oxygen atoms make H-bonds with the hydrogen atoms present in water and ethanol molecules. The steric hindrance of the molecules around NP prevents the agglomeration of particles and maintains the nano-size of the particles.

Gap Between Electrodes

There is a possibility of governing the size of NP smaller than or equal to 10 nm under self-breakdown of spark discharge gap. However, this method is not used because of its drawbacks like unstable voltage, instability of dispersing composition, and low-performance particle production in a single electrode gap [36].

Spark Frequency

It is the number of striking spark impulses on the electrode per second. It was observed that the increase in spark frequency particle concentration also increased with an increase in the mean diameter of the produced NP [36].

AMALGAMATION OF SDM WITH OTHER PROCESSES: HYBRID PROCESSES FOR NANOPARTICLE PRODUCTION

The nanoparticle can be prepared by application of other processes in conjugation with SDM. These processes were termed hybrid processes of nanoparticle production as they added two or more processes for nanoparticle production. In this review, we focus on processes used with SDM for enhanced efficiency of synthesizing method [38]. Hybrid processes can be broadly understood as processes, which use two or more processes to perform one of the two functions:

- Ameliorate thermal ablation for the production of nanoparticles
- Improve particle separation or agglomeration depending upon the Top-down or Bottom-up method, respectively.

Processes such as the Electrochemical Discharge method and laser-assisted SDM are processes that aid in ameliorating thermal ablation. Khosrozadeh and Shabgard, in their experimental research, used Ultrasonic assisted EDM (UEDM) and Power Mixed EDM (PM-UEDM) to find their effects on the machining of titanium alloy. It was observed that by incorporating UEDM and PMEDM, the efficiency of the machining process was enhanced [39]. Multiple kinds of research were conducted to elucidate the effects of hybrid processes on the production of nanoparticles [40 - 42], though not much research has been conducted on hybrid processes like Gas-Phase Synthesis, Laser ablation, and chemical EDM for the production of Nanoparticles. This is a future scope and potential field of research.

COMPARISON BETWEEN SPARK DISCHARGE METHOD (SDM) AND CONVENTIONAL METHODS OF NANOPARTICLE (NP) PRODUCTION

Conventional Methods of Nanoparticle Production

Conventional methods of nanoparticle production include both mechanical and chemical processes [43, 44]. Widely used conventional methods for nanoparticle production are elucidated below:

Mechanical Ball Milling

In the mechanical ball milling process, larger clusters of material are broken down into smaller nanoparticles. A spinning ball mill with metallic clusters rotates at varying speeds to provide varied-sized nanoparticles at the end of the process [45].

Vacuum Deposition and Vaporization

The vacuum deposition method is the process, where the atomization of elements, compounds, or alloys takes place by a thermal process at 500°C and is stored in a vacuum at 10 to 0.1 MPa [38]. This process produces nanoparticles of the order 1 to 100nm.

Sol-gel Method

A popular bottom-up (sol-gel) method incorporates the use of a colloidal solution formed by solid particulates suspended in a solvent (liquid phase). As it is a wet chemical process, it takes time to form nanoparticles ultimately. Pertaining to its simplicity, it is widely used for producing nanoparticles [44].

Sputtering

Sputtering is a top-down method, which means breaking down larger particulates or clusters into nanoparticles. It incorporates the disintegration of a surface by the action of gas ions. This is also called ion sputtering. This is an emerging method for the production of nanoparticles [46].

Laser ablation

Laser ablation, like the Spark discharge method, is a widely used thermal top-down process. The formation of nanoparticles through this process primarily takes place by the irradiating laser on the surface of the metal, which further vaporizes the substance into nanoparticles [47].

Pyrolysis

Pyrolysis, which is also called Flame Pyrolysis, is a bottom-up approach to the production of nanoparticles. Its simplicity makes it the most used method for the production of nanoparticles on an industrial level. This process involves burning a substance, primarily a liquid (precursor), in a furnace, which is later collected [48].

These were some of the most used Top-down and Bottom-up methods for the production of nanoparticles. More such processes are present, but their low efficiency, complicity, or low rate of nanoparticle production, are not as widely used as the aforementioned methods [49 - 52].

Advantages of SDM over Conventional Methods

The Spark Discharge method holds merit and pros over conventional processes of nanoparticle production. Research showed that SDM produced nanoparticles at accelerated rates compared to other methods, primarily the ball milling method and chemical precipitation, which required longer durations of time. It can further be concluded that as SDM is a thermal process, the size of nanoparticles could be easily controlled by varying the intensity of spark discharge. Further, the juxtaposition of SDM with other processes consecutively enhances the nanoparticle production method.

APPLICATIONS OF NANOPARTICLES IN EMISSION CONTROL, COOLING AND LUBRICATION

In the automobile section, NP possesses the excellent potential and has proved to be very useful for emission control measures of an IC Engine. Attributing to different chemical and physical properties, various NP is being used to control emission and enhance engine efficiency. A brief study on this topic is done by Soudagara *et al.* [53]. Some NP and their application in IC Engine are tabulated below in Table **2**.

Table 2. Application of NP in IC engine.

Nanoparticles	Property Enhanced	References
Aluminum	More conductive and shorter ID up to 35%	[11, 54, 55]
Cobalt	Reduces NO_x and Fuel consumption	[56]
Magnalium (Al-Mg)	It acts as a heat sink, reduces temperature, avoids hot spots and reduces NO_x and increases the thermal efficiency	[57]
Manganese	Reduces polycyclic aromatic HC emission	[58]
TiO_2	Due to its photoelectric catalytic property to generate H from H_2O	[59]
Al_2O_3	CO reduction	[60]
Mg	To overcome the cold start problem Reduces flash point	[61, 62]
CeO_2	Highly catalytic properties led to proper combustion and reduced emission and fuel consumption.	[12]

Ferrofluids are the suspension of magnetic NP (Fe_3O_4) 5% coated with surfactant 10% (generally Tetramethyl Ammonium Hydroxide) immersed in carrier fluid 85% by volume with no signs of agglomeration. The fluid gets magnetized and shows valleys and peaks because of the externally applied magnetic field. It further shows Brownian motion in the absence of an external magnetic field. Ferrofluids alone are used in many applications, such as lubrication [58], sealing, and Heat Exchangers, to name a few.

According to Allen *et al.* [54], FF can also be used to enhance most properties of IC Engine mentioned in the table when used along with Diesel and Biodiesel [55]. Gan *et al.* [56] have a patent for removing oil film from open water bodies using FF. The researchers thoroughly observed heat exchangers [11, 54, 55]. They concluded that by the NP addition of Al and Fe_3O_4, the thermal conductivity of the fluids had increased, which turned to reduce the weight and increase the cooling system capacity.

LIMITATIONS OF SDM FOR NP PRODUCTION

The thorough study done by Soudagara *et al.* [53] showed that the NP emitted into the environment caused new problems, such as silver NP that changed into silver sulfide in due course of time. Silver sulfide with water in ponds, rivers, and canals results in hazardous effects as well as affected aquatic life. Al forms the passivation layer of Al_2O_3 after reacting with oxygen in the air, which is insoluble in water. These compounds either deposit on the ground or floats on the water's surface. 40% of Ce particles contribute to soot formation and accumulate in soil or on the roadside.

Moreover, the most crucial matter of concern constitutes the degradation or ingestion of such hazardous compounds into the biological system. These compounds stay in the introduced system for a very long time. In human beings or other animals, once inhaled, they get deposited in the lungs. Due to their nano size, they can easily penetrate nearly any cell membrane and get transmitted to other vital organs such as the heart, liver, intestines, and brain [53]. This not only disturbs cells' biochemical environment but also threatens the life of the individual. Immune systems continuously work on recovery, but due to the insoluble property of some NP like CeO_2, it takes nearly 700 days for its complete removal from human lungs. In the meantime, it may cause respiration problems. The primary bloodstream may reach the liver and intestines, where it may further cause inflammatory intestine, illnesses, and intestinal cancer. According to Kocheril *et al.* [63] and Ramanan *et al.* [64], more than 1% of the use of Fe_3O_4 magnetic particles in IC Engines may corrode Engine Parts and reduce the life of the Engine.

CONCLUSION

Nanoparticles pave the path for a better future. This review articulates the spark discharge method for nanoparticle production. The following points were observed after an extensive review of various experimentations:

- The spark discharge method is a Top-down thermal method for nanoparticle production. Literature is present on the production of nanoparticles through SDM, although experimentation with a combined usage of SDM with other processes is yet to be explored more intensely.
- Effects of various process parameters on the production of NP with SDM were observed. It was concluded that Spark frequency and Spark energy were the most influential parameters which affected the size and shape of nanoparticles.
- Various methods, in addition to SDM, were briefly discussed. SDM had certain advantages, such as an accelerated rate of nanoparticle production and easier tunability compared with other nanoparticle production methods.
- The application of nanoparticles in various fields was discussed, focusing on their use as ferrofluids in IC engines. It was observed that due to their thermal and mechanical properties, they were able to augment the efficiency of IC engines in terms of Brake Power and Energy consumption. Some limitations, primarily inhalation of NPs along with other environmental hazards, were observed.

ACKNOWLEDGEMENTS

We thank the Department of Mechanical Engineering, JSS Academy of Technical Education, Noida, for constant support and motivation.

REFERENCES

[1]   K.H. Tseng, C.Y. Liao, J.C. Huang, D.C. Tien, and T.T. Tsung, "Characterization of gold nanoparticles in organic or inorganic medium (ethanol/water) fabricated by spark discharge method", *Mater. Lett.,* vol. 62, no. 19, pp. 3341-3344, 2008.

[2]   S. Iyer, and A. Das, "Responsive nanogels for anti-cancer therapy", *Mater. Today Proc.,* vol. 44, pp. 2330-2333, 2021.

[3]   S.S. Salem, and A. Fouda, "Green synthesis of metallic nanoparticles and their prospective biotechnological applications: An overview", *Biol. Trace Elem. Res.,* vol. 199, no. 1, pp. 344-370, 2021.

[4]   E. Hoseinzadeh, P. Makhdoumi, P. Taha, H. Hossini, J. Stelling, and M. Amjad Kamal, "A review on nano-antimicrobials: Metal nanoparticles, methods and mechanisms", *Curr. Drug Metab.,* vol. 18, no. 2, pp. 120-128, 2017.

[5]   H. Bishwakarma, and A.K. Das, "Synthesis of zinc oxide nanoparticles through hybrid machining process and their application in supercapacitors", *J. Electron. Mater.,* vol. 49, no. 2, pp. 1541-1549, 2020.

[6]   J.M. Mahdi, S. Lohrasbi, and E.C. Nsofor, "Hybrid heat transfer enhancement for latent-heat thermal energy storage systems: A review", *Int. J. Heat Mass Transf.,* vol. 137, pp. 630-649, 2019.

[7]   S. Srivastava, V. Malik, and M. Vishnoi, "Study on effects of quantum dots in optoelectronics and hydrogen evolution: A review", *Mater. Today Proc.,* vol. 45, pp. 5672-5677, 2021.

[8]   V. Malik, S. Srivastava, M.K. Bhatnagar, and M. Vishnoi, "Comparative study and analysis between solid oxide fuel cells (sofc) and proton exchange membrane (PEM) fuel cell–A review", *Mater. Today Proc.,* vol. 47, pp. 2270-2275, 2021.

[9]   K. Yamamoto, T. Imaoka, M. Tanabe, and T. Kambe, "New horizon of nanoparticle and cluster catalysis with dendrimers", *Chem. Rev.,* vol. 120, no. 2, pp. 1397-1437, 2019.

[10]   M.K. Bhatnagar, M. Rai, M. Ashraf, O. Kapoor, T.G. Mamatha, and M. Vishnoi, "Efficiency enhancement of heat exchanger using inserts and nano-fluid-A review", *Mater. Today Proc.,* vol. 44, pp. 4399-4403, 2021.

[11]   H. Tyagi, P.E. Phelan, R. Prasher, R. Peck, T. Lee, J.R. Pacheco, and P. Arentzen, "Increased hot-plate ignition probability for nanoparticle-laden diesel fuel", *Nano Lett.,* vol. 8, no. 5, pp. 1410-1416, 2008.

[12]   V. Sajith, C.B. Sobhan, and G.P. Peterson, "Experimental investigations on the effects of cerium oxide nanoparticle fuel additives on biodiesel", *Adv. Mech. Eng.,* vol. 2, p. 581407, 2010.

[13]   D.D. Kumar, and A.V. Arasu, "A comprehensive review of preparation, characterization, properties and stability of hybrid nanofluids", *Renew. Sustain. Energy Rev.,* vol. 81, pp. 1669-1689, 2018.

[14]   T.V. Pfeiffer, J. Feng, and A. Schmidt-Ott, "New developments in spark production of nanoparticles", *Adv. Powder Technol.,* vol. 25, no. 1, pp. 56-70, 2014.

[15]   A.E. Berkowitz, and J.L. Walter, "Ferrofluids prepared by spark erosion", *J. Magn. Magn. Mater.,* vol. 39, no. 1-2, pp. 75-78, 1983.

[16]   S. Shahidi, A. Jamali, S. Dalal Sharifi, and H. Ghomi, "In-situ synthesis of CuO nanoparticles on cotton fabrics using spark discharge method to fabricate antibacterial textile", *J. Nat. Fibers,* vol. 15, no. 6, pp. 870-881, 2018.

[17]   R.T. Hallberg, L. Ludvigsson, C. Preger, B.O. Meuller, K.A. Dick, and M.E. Messing, "Hydrogen-assisted spark discharge generated metal nanoparticles to prevent oxide formation", *Aerosol Sci. Technol.,* vol. 52, no. 3, pp. 347-358, 2018.

[18]   A. Kohut, L.P. Villy, A. Kéri, Á. Bélteki, D. Megyeri, B. Hopp, G. Galbács, and Z. Geretovszky, "Full range tuning of the composition of Au/Ag binary nanoparticles by spark discharge generation", *Sci. Rep.,* vol. 11, no. 1, pp. 1-0, 2021.

[19]   M. Sabzehparvar, F. Kiani, and N.S. Tabrizi, "Spark discharge generation of superparamagnetic nickel oxide nanoparticles", *Mater. Today Proc.,* vol. 5, no. 7, pp. 15821-15827, 2018.

[20]   RK Sahu, and SS Hiremath, "Synthesis of aluminium nanoparticles in a water/polyethylene glycol mixed solvent using μ-EDM", In: *InIOP conference series: Materials science and engineering 2017 Aug 1* vol. 225. , 2018, no. 1, p. 012257.

[21]   K.H. Tseng, Y.S. Lin, Y.C. Lin, D.C. Tien, and L. Stobinski, "Deriving optimized PID parameters of nano-Ag colloid prepared by electrical spark discharge method", *Nanomaterials,* vol. 10, no. 6, p. 1091, 2020.

[22]   Y. Tang, Y. Lai, D. Gong, K.H. Goh, T.T. Lim, Z. Dong, and Z. Chen, "Ultrafast synthesis of layered titanate microspherulite particles by electrochemical spark discharge spallation", *Chemistry,* vol. 16, no. 26, pp. 7704-7708, 2010.

[23]   K.H. Tseng, and J.C. Huang, "Pulsed spark-discharge assisted synthesis of colloidal gold nanoparticles in ethanol", *J. Nanopart. Res.,* vol. 13, no. 7, pp. 2963-2972, 2011.

[24]   V.A. Vons, A. Anastasopol, W.J. Legerstee, F.M. Mulder, S.W. Eijt, and A. Schmidt-Ott, "Low-temperature hydrogen desorption and the structural properties of spark discharge generated Mg

nanoparticles", *Acta Mater.,* vol. 59, no. 8, pp. 3070-3080, 2011.

[25]   N.S. Tabrizi, M. Ullmann, V.A. Vons, U. Lafont, and A. Schmidt-Ott, "Generation of nanoparticles by spark discharge", *J. Nanopart. Res.,* vol. 11, no. 2, pp. 315-332, 2009.

[26]   D.E. Evans, R.M. Harrison, and J.G. Ayres, "The generation and characterisation of elemental carbon aerosols for human challenge studies", *J. Aerosol Sci.,* vol. 34, no. 8, pp. 1023-1041, 2003.

[27]   N.S. Tabrizi, Q. Xu, N.M. Van Der Pers, U. Lafont, and A. Schmidt-Ott, "Synthesis of mixed metallic nanoparticles by spark discharge", *J. Nanopart. Res.,* vol. 11, no. 5, pp. 1209-1218, 2009.

[28]   U. Lafont, L. Simonin, N.S. Tabrizi, A. Schmidt-Ott, and E.M. Kelder, "Synthesis of nanoparticles of Cu, Sb, Sn, SnSb and Cu2Sb by densification and atomization process", *J. Nanosci. Nanotechnol.,* vol. 9, no. 4, pp. 2546-2552, 2009.

[29]   N.S. Tabrizi, Q. Xu, N.M. Van Der Pers, and A. Schmidt-Ott, "Generation of mixed metallic nanoparticles from immiscible metals by spark discharge", *J. Nanopart. Res.,* vol. 12, no. 1, pp. 247-259, 2010.

[30]   S. Kala, M. Rouenhoff, R. Theissmann, and F.E. Kruis, *Synthesis and film formation of monodisperse nanoparticles and nanoparticle pairs. InNanoparticles from the Gasphase.* Springer: Berlin, Heidelberg, 2012, pp. 99-119.

[31]   V.A. Vons, L.C. de Smet, D. Munao, A. Evirgen, E.M. Kelder, and A. Schmidt-Ott, "Silicon nanoparticles produced by spark discharge", *J. Nanopart. Res.,* vol. 13, no. 10, pp. 4867-4879, 2011.

[32]   C. Roth, G.A. Ferron, E. Karg, B. Lentner, G. Schumann, S. Takenaka, and J. Heyder, "Generation of ultrafine particles by spark discharging", *Aerosol Sci. Technol.,* vol. 38, no. 3, pp. 228-235, 2004.

[33]   J.H. Byeon, and J.W. Kim, "Production of carbonaceous nanostructures from a silver-carbon ambient spark", *Appl. Phys. Lett.,* vol. 96, no. 15, p. 153102, 2010.

[34]   H. Förster, and W. Peukert, *In-situ synthesis and functionalisation of metal nanoparticles in a spark discharge process.* InInternational Aerosol Conference, 2010.

[35]   A. Anastasopol, T.V. Pfeiffer, A. Schmidt-Ott, F.M. Mulder, and S.W. Eijt, "Fractal disperse hydrogen sorption kinetics in spark discharge generated Mg/NbOx and Mg/Pd nanocomposites", *Appl. Phys. Lett.,* vol. 99, no. 19, p. 194103, 2011.

[36]   A.A. Efimov, V.V. Ivanov, A.V. Bagazeev, I.V. Beketov, I.A. Volkov, and S.V. Shcherbinin, "Generation of aerosol nanoparticles by the multi-spark discharge generator", *Tech. Phys. Lett.,* vol. 39, no. 12, pp. 1053-1056, 2013.

[37]   M.A. Farrukh, I. Muneer, K.M. Butt, S. Batool, and N. Fakhar, "Effect of Dielectric Constant of Solvents on the Particle Size and Bandgap of La/SnO₂□TiO₂ Nanoparticles and Their Catalytic Properties", *J. Chin. Chem. Soc. (Taipei),* vol. 63, no. 12, pp. 952-959, 2016.

[38]   N. Rajput, "Methods of preparation of nanoparticles-a review", *Int. J. Adv. Eng. Technol.,* vol. 7, no. 6, p. 1806, 2015.

[39]   B. Khosrozadeh, and M. Shabgard, "Effects of hybrid electrical discharge machining processes on surface integrity and residual stresses of Ti-6Al-4V titanium alloy", *Int. J. Adv. Manuf. Technol.,* vol. 93, no. 5, pp. 1999-2011, 2017.

[40]   J.H. Byeon, and Y.W. Kim, "An aerosol-seed-assisted hybrid chemical route to synthesize anisotropic bimetallic nanoparticles", *Nanoscale,* vol. 4, no. 21, pp. 6726-6729, 2012.

[41]   J.H. Byeon, and Y.W. Kim, "Hybrid gas-phase synthesis of nanoscale Fe–SiO 2 core–shell agglomerates for efficient transfection into cells and use in magnetic cell patterning", *RSC Advances,* vol. 3, no. 33, pp. 13685-13689, 2013.

[42]   WS Chu, CS Kim, HT Lee, JO Choi, JI Park, JH Song, KH Jang, and SH Ahn, "Hybrid manufacturing in micro/nano scale: a review", *International journal of precision engineering and manufacturing-green technology.,* vol. 1, no. 1, pp. 75-92, 2014.

[43]   H.W. Xian, N.A. Sidik, S.R. Aid, T.L. Ken, and Y. Asako, "Review on preparation techniques, properties and performance of hybrid nanofluid in recent engineering applications", *Journal of Advanced Research in Fluid Mechanics and Thermal Sciences.*, vol. 45, no. 1, pp. 1-3, 2018.

[44]   G. Amoabediny, F. Haghiralsadat, S. Naderinezhad, M.N. Helder, E. Akhoundi Kharanaghi, J. Mohammadnejad Arough, and B. Zandieh-Doulabi, "Overview of preparation methods of polymeric and lipid-based (niosome, solid lipid, liposome) nanoparticles: A comprehensive review", *International Journal of Polymeric Materials and Polymeric Biomaterials.*, vol. 67, no. 6, pp. 383-400, 2018.

[45]   SA Ealia, and MP Saravanakumar, "A review on the classification, characterisation, synthesis of nanoparticles and their application", In: *In IOP conference series: Materials science and engineering.* vol. 263. , 2017, no. 3, p. 032019.

[46]   T. Charinpanitkul, K. Faungnawakij, and W. Tanthapanichakoon, "Review of recent research on nanoparticle production in Thailand", *Adv. Powder Technol.*, vol. 19, no. 5, pp. 443-457, 2008.

[47]   P.G. Jamkhande, N.W. Ghule, A.H. Bamer, and M.G. Kalaskar, "Metal nanoparticles synthesis: An overview on methods of preparation, advantages and disadvantages, and applications", *J. Drug Deliv. Sci. Technol.*, vol. 53, p. 101174, 2019.

[48]   SL Pal, U Jana, PK Manna, GP Mohanta, and R Manavalan, "Nanoparticle: An overview of preparation and characterization", In: *Journal of applied pharmaceutical science.*, 2011, no. 30, pp. 228-34.

[49]   A. Jaworek, "Micro-and nanoparticle production by electrospraying", *Powder Technol.*, vol. 176, no. 1, pp. 18-35, 2007.

[50]   V.J. Mohanraj, and Y. Chen, "Nanoparticles-a review", *Trop. J. Pharm. Res.*, vol. 5, no. 1, pp. 561-573, 2006.

[51]   AV Rane, K Kanny, VK Abitha, and S Thomas, "Methods for synthesis of nanoparticles and fabrication of nanocomposites", In: *In Synthesis of inorganic nanomaterials*, 2018, pp. 121-139.

[52]   J. Feng, X. Guo, N. Ramlawi, T.V. Pfeiffer, R. Geutjens, S. Basak, H. Nirschl, G. Biskos, H.W. Zandbergen, and A. Schmidt-Ott, "Green manufacturing of metallic nanoparticles: A facile and universal approach to scaling up", *J. Mater. Chem. A Mater. Energy Sustain.*, vol. 4, no. 29, pp. 11222-11227, 2016.

[53]   M.E. Soudagar, N.N. Nik-Ghazali, M.A. Kalam, I.A. Badruddin, N.R. Banapurmath, and N. Akram, "The effect of nano-additives in diesel-biodiesel fuel blends: A comprehensive review on stability, engine performance and emission characteristics", *Energy Convers. Manage.*, vol. 178, pp. 146-177, 2018.

[54]   C. Allen, G. Mittal, C.J. Sung, E. Toulson, and T. Lee, "An aerosol rapid compression machine for studying energetic-nanoparticle-enhanced combustion of liquid fuels", *Proc. Combust. Inst.*, vol. 33, no. 2, pp. 3367-3374, 2011.

[55]   D. Jackson, D. Davidson, and R. Hanson, "Application of an aerosol shock tube for the kinetic studies of n-dodecane/nano-aluminum slurries", *In 44th AIAA/ASME/SAE/ASEE joint propulsion conference exhibit* 2008, pp.4767.

[56]   Y. Gan, and L. Qiao, "Combustion characteristics of fuel droplets with addition of nano and micron-sized aluminum particles", *Combust. Flame,* vol. 158, no. 2, pp. 354-368, 2011.

[57]   D. Ganesh, and G. Gowrishankar, "Effect of nano-fuel additive on emission reduction in a biodiesel fuelled CI engine", *In2011 International conference on electrical and control engineering 2011* 2011, pp. 3453-3459.

[58]   R. Fazliakmetov, and G.S. Shpiro, "Selection and manufacture technology of antismoke additives for diesel fuel and boiler fuels oils", *Izdetal Stvo Neft I Gaz.*, vol. 4, pp. 43-55, 1997.

[59]   S. Ichikawa, "Photoelectrocatalytic production of hydrogen from natural seawater under sunlight", *Int.*

J. Hydrogen Energy, vol. 22, no. 7, pp. 675-678, 1997.

[60] A.C. Sajeevan, and V. Sajith, "Synthesis of stable cerium zirconium oxide nanoparticle–diesel suspension and investigation of its effects on diesel properties and smoke", *Fuel,* vol. 183, pp. 155-163, 2016.

[61] A. Keskin, M. Gürü, and D. Altıparmak, "Biodiesel production from tall oil with synthesized Mn and Ni based additives: effects of the additives on fuel consumption and emissions", *Fuel,* vol. 86, no. 7-8, pp. 1139-1143, 2007.

[62] M. Gürü, A. Koca, Ö. Can, C. Çınar, and F. Şahin, "Biodiesel production from waste chicken fat based sources and evaluation with Mg based additive in a diesel engine", *Renew. Energy,* vol. 35, no. 3, pp. 637-643, 2010.

[63] R. Kocheril, and J. Elias, "Fuel efficiency enhancement by addition of nano sized magnetised ferro particles in cooling system of internal combustion engines", *Mater. Today Proc.,* vol. 21, pp. 722-726, 2020.

[64] M.V. Ramanan, and D. Yuvarajan, "Emission analysis on the influence of magnetite nanofluid on methyl ester in diesel engine", *Atmos. Pollut. Res.,* vol. 7, no. 3, pp. 477-481, 2016.

CHAPTER 12

Natural Fiber-Reinforced Polymer Composite: A Review

Satendra Singh[1,*] and **Pankaj Kumar Gupta**[1]

[1] *Department of Mechanical Engineering, Malaviya National Institute of Technology Jaipur, Rajasthan, India*

Abstract: The manufacturing industry uses a variety of materials, including pure metals, alloys and composites. Due to the inability of pure metals to meet the demands of modern products, a transition in materials from pure metals to composites is taking place. Composite materials are invented to attain the desired properties, including lightweight, high strength, creep resistance, high corrosion resistance, fatigue resistance, high-temperature resistance and high wear resistance. Natural plant fibers, such as flax, hemp, kenaf, jute, sisal, coir and cotton, are a reliable source for producing composites because they have various advantages over synthetic fibers, including cheaper cost, low specific gravity, biodegradability, lightweight, fewer health hazards, availability, low-grade greenhouse emissions and high flexibility. Natural fiber-reinforced polymer composites (NF-RPC) are commonly utilized in automotive applications because they are lighter in weight, resulting in lower fuel consumption and greenhouse gas emissions. The mechanical properties of NF-RPC, such as tensile strength, Young's modulus, flexural strength, hardness and many others, are affected by several factors, for example, fiber aspect ratio, the weight percentage of fiber, different orientations of fiber, usage of the fabrication process, chemical compositions of fiber and different pre-treatments of fiber. Therefore, in this article, some specific applications, mechanical properties, fabrication techniques of NF-RPC, and methods to enhance the properties of natural fibers, have been discussed.

Keywords: Polymer matrix composites, Natural plant fibers classifications, Natural fiber reinforced composites.

INTRODUCTION

The term "composite material" refers to a substance that is made up of two or more chemically different constituents that result in unique properties of both individual constituents. The matrix and reinforcement are the two major constituents of composite materials. The matrix categories composite materials

---

* **Corresponding author Satendra Singh:** Department of Mechanical Engineering, Malaviya National Institute of Technology Jaipur, Rajasthan, India; Email: singhid750@gmail.com

Amar Patnaik, Albano Cavaleiro, Malay Kumar Banerjee, Ernst Kozeschnik & Vikas Kukshal (Eds.)

into three groupings, for example metal-matrix composites, ceramic-matrix composites and polymer-matrix composites.

Based on reinforcement, composite materials are also classified into three groupings, such as particulate-reinforced composite, structural-reinforced composite and fiber-reinforced composite [1 - 4]. A pure polymer frequently lacks the mechanical strength needed for use in a variety of applications. The load-carrying component of these materials is fibers, which offer rigidity and strength, whereas polymer matrices keep the fibers aligned (position and orientation). They also shield them from the environment and other potential harm. There are numerous types of natural fibers and synthetic fibers as per their source of origin, as shown in Fig (1) [5 - 7]. Natural fibers were first used 3000 years ago in antique Egypt in composite approaches, wherein clay and straw were mixed to make walls. Polymer matrix materials reinforced through natural plant fibers have acquired a lot of curiosity in the past decade, from many companies and universities [8]. Moreover, natural plant fibers are used because of several desirable factors, such as low cost, biodegradability, lightweight, fewer health risks, availability, low-grade carbon emissions and great adaptability. As a result, when compared to synthetic fiber-reinforced polymer composites and metal alloys, NF-RPC is in high demand [9, 10]. NF-RPC is thus used in a variety of engineering fields, for example, aerospace, marine construction, vehicles, sports, transportation, biomedical, packaging and structural applications [11 - 13]. Natural plant fibers, on the other hand, have several disadvantages, such as low thermal stability, non-water resistance, seasonal quality changes, aggregate formation tendency during processing, and many more. Therefore, an appropriate extraction technique, physical pre-treatment, alkaline pre-treatment and chemical pre-treatment are necessary to improve fiber quality and compatibility with the matrix, which further enhances the mechanical, physical and thermal properties of NF-RPC [14 - 17]. NF-RPC is typically made *via* hand lay-up techniques, resin transfer molding (RTM), injection molding, extrusion and compression molding [18].

There are some natural plant fibers, for example, flax, pineapple, jute, hemp, coir (coconut) and kenaf, that can be obtained from their source of origin, as shown in Fig (2).

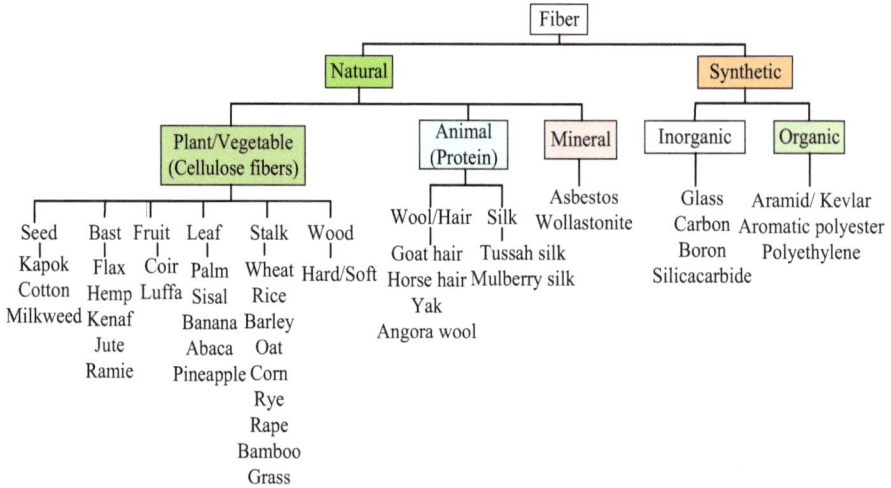

Fig. (1). Classification of fiber [5, 7].

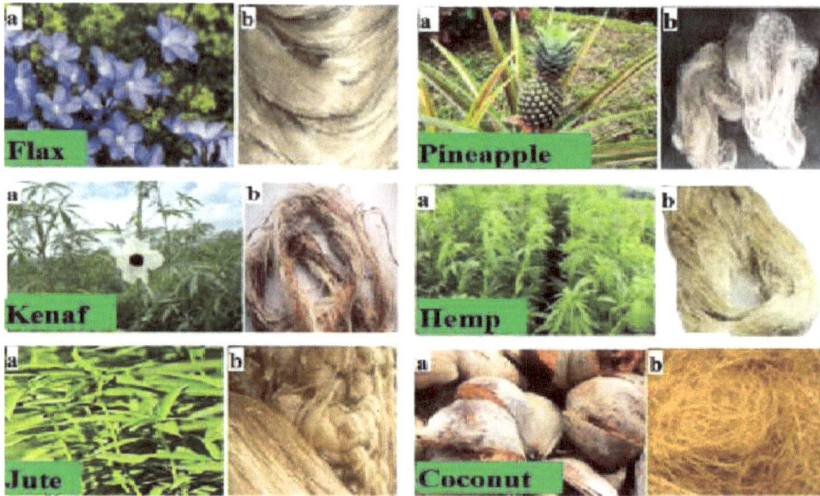

Fig. (2). Some natural plant fibers with their source of origin **(a)** Plant **(b)** Fiber [9 - 11, 16].

NATURAL FIBER-REINFORCED POLYMER COMPOSITE

In modern times, all research focuses on sustainability. As to environmental concerns and the limited amount of conventional resources of energy, there is a need to use non-conventional resources of energy for sustainable development. So, there is a replacement of synthetic fibers, for example, carbon, glass and kevlar, with natural fibers, for example, hemp, flax, jute, sisal, kenaf, wheat and rice, as reinforcement in composites [19 - 21]. The different types of polymer matrix materials are used with natural plant fibers to fabricate the NF-RPC, as shown in Fig (**3**).

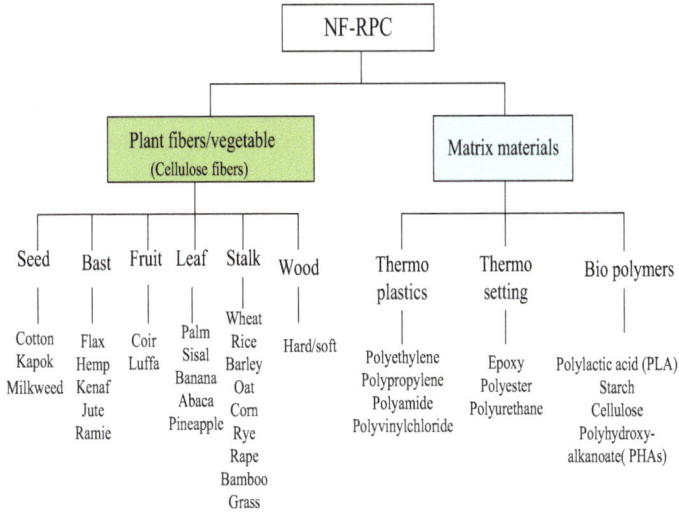

Fig. (3). Classification of natural plant fibers with polymer matrix materials [5, 21].

APPLICATIONS OF NATURAL FIBER-REINFORCED POLYMER COMPOSITE

Natural fiber composites have been used in automobiles since the 1940s, when Henry Ford created the initial composite car consuming hemp fiber. NF-RPC are mostly used in automobile applications because they are lighter in weight which further, results in dropping fuel consumption and lesser releases of greenhouse gases. There are some specific commercial NF-RPC that are used in automobile applications, as shown in Table **1** [22].

Table 1. Some specific NF-RPC automobile applications [22].

S.NO	Product trade name	Material used	Applications
1	FlexForm (Non-woven mat)	Blast fibers, for example, flax, kenaf, sisal, jute, hemp and polyester with polypropylene	Door panels, side and back walls, seatbacks, headliners, pillars, door bolsters and load floor (presently in manufacture for Ford, Nissan, Chrysler (DCX), Mercedes-Benz GM and Honda).
2	NF-EP	Fiber mat (hemp or flax) and epoxy resin	Carrier for enclosed door boards (presently in manufacture for BMW).
3	Fibrit	Wood fiber and acrylic resin	Carrier for enclosed door boards, enclosed or foamed instrument boards, enclosed inserts (presently in manufacture for Opel, Mercedes-Benz).

(Table 1) cont.....

S.NO	Product trade name	Material used	Applications
4	EcoCor	Bast fibers, for example, hemp, kenaf, flax and polypropylene	Carrier for soft and hard armrests, covered parts for instrument panels, enclosed inserts and seat back boards (presently in manufacture for Ford, Mercedes-Benz, Dodge, Opel, Porsche and Volkswagen).

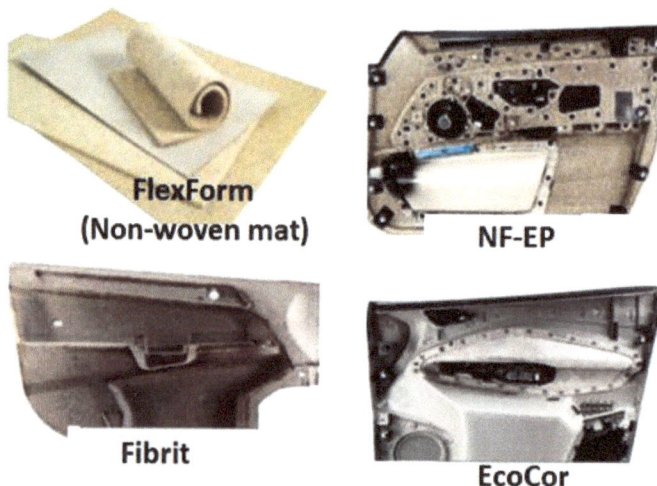

Fig. (4). Commercial NF-RPC used in automobile applications [22].

L. Kerni *et al.* [20] conducted a review of various kinds of natural fibers which can be consumed in polymer matrix composites as reinforcement. Natural fibers have the capability to replace synthetic fibers as reinforcement in polymer matrix composites as an outcome of their magnificent properties, for example, lightweight, low cost, user-friendly, high wear resistance, high flexibility, lower environmental impact and re-usability. Therefore, NF-RPC is used in automobile, aerospace, infrastructure and structural applications. Natural fibers' low moisture resistance and wettability limit their use in composites toward some level, however which can be resolved with chemical pre-treatment.

Y. Singh *et al.* [23] performed experimental work on the fabrication of coir/carbon fiber-hybrid composites that can be used in helmet shell, flooring and roofing applications. Moreover, natural fibers composites, for example, jute, banana and sisal hybrid reinforced composites, have been found to have long-term strength in industrialized security helmet usages. The hybrid reinforced composite fabricated of jute, banana and sisal fibers is stronger and reduces the mass by half. In industrialized security helmets, it is consequently safer to use hybrid composite material instead of AS4 polyphenylene sulphide plastic.

P. Dey and S. Ray [24] reviewed green composites and their applications. To fabricate the green composites, the natural resources are used with a process or methods which create lower environmental impact with negligible creation of waste materials. Bast, seed, leaf, grass, fruit and wood are the source of natural plant fibers. Different types of physical properties that can be considered while choosing natural fibers for composites are fiber aspect ratio, strength, thickness, surface chemistry, density and chemical composition. Green composites are mostly used in short-lifetime products like disposable tableware, packaging and toys for kids.

STUDIES ON METHODS/TECHNIQUES TO ENHANCE THE PROPERTIES OF NATURAL FIBERS

D. Wang *et al.* [14] performed experimental work to prepare wheat straw reinforced polypropylene (pp) composites built on improved nano-TiO_2 particles. The objective of incorporating TiO_2 nanoparticles into wheat straw fibers was to improve their performance to provide UV-protective features and the mechanical strength of the composites. Silane coupling agents, KH550 or A171, were used for the surface modification of nano-TiO_2. In comparison to untreated or A171-improved nano-TiO_2, samples cured through KH550-improved nano-TiO_2 demonstrated a more excellent UV-shielding property. The mechanical properties of samples cured with KH550-improved nano-TiO_2 particles were enhanced due to the homogenous pore structures and strong interfacial bonding among the fillers and matrix. Hence, the composite samples cured with KH550-improved nano-TiO_2 particles had the best mechanical properties compared to the other composite samples.

B. Ismail *et al.* [25] performed experimental work to optimize the performance of insulating materials built with wheat straw, gypsum plaster and lime composites consuming natural supplements. The goal of this experimental work was to decline the thermal conductivity of a wheat straw bio-based composite while preserving adequate mechanical strength to sustain its own mass. The optimization method was established in three dissimilar ways by consuming two supplements, such as hemoglobin from pork blood and casein, which were acquired from a renewable and decomposable source. The thermal conductivity was reduced by the accumulation of 5% of hemoglobin in the lime. But, this addition has induced a severe decline in the mechanical strength of the composite. The incorporation of 5% casein and 5% hemoglobin caused a denser composite that had three times the mechanical strength of the reference specimen but had extremely high conductivity. Hence, using 5% hemoglobin and 2.5% casein, a balance between thermal conductivity and mechanical strength was produced.

S. Sathish *et al.* [26] studied chemical pre-treatments and extraction techniques of natural fibers in composites. To increase the quality of fibers and their compatibility with the matrix, as well as the mechanical, physical and thermal properties of natural fiber composites, an appropriate extraction procedure and chemical treatment are required. Natural fibers were treated with a variety of chemical treatments, including alkaline, silane, benzoyl peroxide, acetic acid, potassium permanganate, polymer coating and stearic acid. There are three types of fiber extraction processes, for example, water retting, dew retting and mechanical decortications.

M. Li *et al.* [27] reviewed strategies to enhance natural fiber-reinforced composite performance. Mechanical performance is reduced because of various factors associated with natural fiber composites, for example, diverse attributes of natural fiber, which leads to the larger variety of fiber quality, high water absorption, lower thermal stability, incompatibility due to hydrophilicity and accumulation tendency in the hydrophobic polymer matrix. So, numerous other approaches, for example, fiber chemical treatment, hybridization and systemic arrangement, and advancements in manufacturing processes like 3D printing, have been used to overcome these drawbacks. It was concluded that several properties of NF-RPC, for example, chemical stability, compatibility of natural fiber with the polymer matrix materials, thermal and mechanical properties, were improved by hybridization, coupling agents and chemical modification.

M. Chougana *et al.* [28] examined the effect of wheat straw pre-treatment on the tensile properties of the wheat straw reinforced-polylactic acid composite. Some pre-treatment techniques, such as H+S and H+M were used as a hybrid of steam (S), hot water (H) and microwave (M). There were desired variations in the micro-structure of the wheat straw-internode as the expansion of parenchyma and decreasing the thickness of epidermis, which enhanced the deeper and more effective infiltration of matrix in the liquid form. These preferred variations lead to superior bonding and, successively, the higher tensile strength of composites. The mechanical properties of wheat straw reinforced-polylactic acid matrix composites, for example, modulus of elasticity, toughness and tensile strength, were drastically improved with H+S pre-treatment.

FABRICATION TECHNIQUES OF NATURAL FIBER-REINFORCED POLYMER COMPOSITE

S. Witayakran *et al.* [22] conducted a review based on various types of the fabrication process that can be used to make NF-RPC according to recent automotive applications. Injection molding, spray lay-up, RTM, compression molding, extrusion and hand lay-up techniques are mostly used to produce natural fiber composites for automotive usages. The selection of fabrication techniques is decided according to various factors, such as economic, lower environmental impact, and physical, mechanical, thermal properties of fibers and matrix materials.

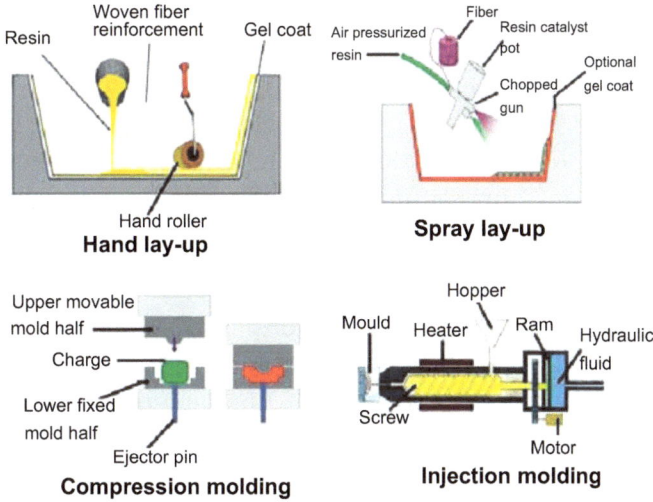

Fig. (5). Some fabrication techniques of NF-RPC [29].

K. Srinivas *et al.* [29] reviewed injection molding and other techniques to fabricate the NF-RPC. Injection molding is a technique in which a liquefied polymer is enforced into a mold cavity under high pressure. Polymer matrix material in the arrangement of granules is fed into an injection moulding machine through a hopper. The material is then forced into the split mold *via* a feeding system and forward with the help of a feeding screw to fill the mold cavity. The mildew tool is mounted on a movable platen, which opens when the part has solidified and ejects the component using ejector pins. This method produces a nice surface polish. This is appropriate for larger volumes. However, the tensile strength of this technique is lower than that of other thermoset techniques.

M.S. Huda *et al.* [30] discussed the difference between hand lay-up and RTM with their merits and demerits. In hand lay-up, the resin is impregnated by hand rollers/brushes with fibers. So, this process is widely used because of several advantages, for example, lower tooling cost, easy to use and use of long fibers as compared to spray lay-up. Whereas, in RTM, the resin is injected with one or numerous ports depending upon the size of the composite. The complex shapes, closer tolerances, and high production rate can be achieved with RTM as compared to hand lay-up techniques.

INVESTIGATIONS ON THE MECHANICAL PROPERTIES OF NATURAL FIBER-REINFORCED POLYMER COMPOSITE

T. Sadik *et al.* [31] evaluated the tensile properties of date palm frond (DPF) fiber-reinforced polymer matrix composites. The tensile strength of the composites increases as the weight fraction of DPF fibers grows from 30 to 70%, as well as the influence of the attentiveness of NaOH cured fibers. As a result, Young's modulus increases, resulting in the most favorable tensile strength at 0.75 mm fiber size in the same fraction. A brittle-like fracture can be seen in the SEM picture of a tensile loading fracture sample of 40:60% composite; a similar fracture was also detected in a sample of 50:50% fiber size of 0.75 mm. A smaller amount of fiber pull-out was detected in 40:60% and 50:50% composites, which could cause tensile strength reduction.

Chandramohan *et al.* [32] investigated the tensile strength and hardness of natural fiber-reinforced composites. In this experimental work, banana, sisal and roselle natural fibers were used either individually or in hybrids as reinforcement with bio epoxy resin matrix material. The hybrid reinforced composites of sisal with banana and roselle with banana fibers showed less percentage elongation (brittle fracture) as compared to sisal with roselle fibers hybrid reinforced composites. So, the tensile strength of sisal with roselle fibers hybrid reinforced composite was maximum because of high percentage elongation or high fiber pull-out was observed. The hardness of sisal and banana hybrid reinforced composite was maximum as compared to other hybrid reinforced composites. "Finally, this research concluded that sisal with roselle fibers as the hybrid is one of the best that can be used for making the various parts of the automobile sector, such as dashboard, door handle, car roof and car door panels".

S. Panthapulakkal *et al.* [33] evaluated the flexural and tensile properties of wheat straw reinforced thermoplastic composites. Poly-propylene was used as a matrix material. The samples for the flexural and tensile tests were prepared as per ASTM D790 and ASTM D638. The mechanical properties of composites were affected by the variation of chemical composition, for example, cellulose, hemicellulose and lignin of prepared wheat straw, with mechanically and chemically processed. The tensile strength, flexural modulus, flexural strength and tensile modulus were maximum for mechanically prepared wheat straw-reinforced polymer composites. Fibers prepared through a chemical process showed improved mechanical properties because surface irregularities were reduced from fiber's surface.

Table 2. Young's modulus and Tensile strength of NF-RPC at the same fiber content and matrix material.

S.NO	NF-RPC	Fiber Content (%)	Young's Modulus (GPa)	Tensile Strength (MPa)	References
1	Polypropylene with jute fiber	30.0	4.2	52.0	10
2	Polypropylene with hemp fiber	30.0	3.9	44.0	34-35
3	Polypropylene with cotton fiber	30.0	4.1	58.5	34-35
4	Polypropylene with sisal fiber	30.0	3.9	45.5	34-35
5	Polypropylene with coir fiber	30.0	2.0	47.8	36
6	Polypropylene with flax fiber	30.0	5.0	52.0	37
7	Polypropylene with kenaf fiber	30.0	5.0	46.0	38

To make the NF-RPC, natural plant fibers, for example, hemp, kenaf, cotton, jute, sisal, flax and coir, were used as reinforcement for the polypropylene matrix material. The fiber content was constant for all composites to compare Young's modulus as well as the tensile strength of composites. It was noticed that the tensile strength was maximum for cotton fiber reinforcement. Whereas, Young's modulus was the maximum for kenaf and flax fiber reinforcement, as shown in above Table **2**.

CONCLUSION

From the prior discussion, it can be decided that natural fibers can replace synthetic fibers as reinforcement in polymer matrix composites due to their outstanding properties, including lightweight, high flexibility, cheap cost, user-friendly, high wear resistance, lower environmental impact, high corrosion resistance and re-usability.

NF-RPCs are mostly used in automobile applications because they are lighter in weight which further results in reducing fuel consumption and lesser emissions of greenhouse gases. Furthermore, natural resources are used in the fabrication of green composites, along with a process or methods that have a lower environmental impact and produce negligible waste materials. Green composites are mostly used in short-lifetime products like disposable tableware, packaging and toys for kids.

Hand lay-up, compression molding, spray lay-up, RTM, extrusion and injection molding are some fabrication processes of polymer matrix composites.

Several factors influence the mechanical properties of NF-RPC, including the aspect ratio of the fiber, weight fraction, different fiber orientations, manufacturing process, pre-treatment techniques, the addition of additives and chemical composition of the fiber. It was noticed that the tensile strength of cotton fiber reinforced polypropylene composite at the same fiber content was maximum, i.e, 58.5 MPa, among other natural plant fibers, for example, kenaf, hemp, jute, flax, coir and sisal. But, Young's modulus of flax and kenaf fiber as reinforcement with polypropylene was maximum, *i.e.,* 5.0 GPa, among other natural plant fibers, for example, jute, cotton, sisal and coir.

Mostly, synthetic polymers, for example, epoxy, polyester, polypropylene, polyethylene and polystyrene, are used to fabricate the NF-RPC. As a result, natural polymers or biopolymers, such as polylactic acid, polyhydroxy-alkanoate, cellulose, and others, have the potential to be combined with natural plant fibers to produce biopolymer composites or NF-RPC in the future.

REFERENCES

[1]   G. Zhengchao, A.P. André, and W.G. Dirk, "Advanced polymer-based composites and structures for biomedical applications", *Eur. Polym. J.,* vol. 149, 2021.110388

[2]   R. Chikesh, A. Zaid, R.K. Sonu, K. Abhishek, K. Abhay, and K. Kaushik, "Experimental investigation and comparative analysis of mechanical properties of cross layer rice straw fibers filled reinforced epoxy biodegradable composite", *Mater. Today Proc.,* vol. 46, no. 1, pp. 340-344, 2021.

[3]   H. Derek, *An introduction of composite materials*, 8th ed.; Cambridge University Press: Cambridge, 1996.

[4]   J.P. Agrawal, *Composite materials, Defence Scientific Information & Documentation Centre (DESIDOC) Ministry of Defence,*DRDO: Delhi, 1990.

[5]   I.M. Fatin, N.K. Kamrun, S. Bijoyee, M.N. Khandakar, and A.K. Ruhul, "A brief review on natural fiber used as a replacement of synthetic fiber in polymer composites", *Materials Engineering Research,* vol. 1, no. 2, pp. 86-97, 2019.

[6]   S. Taj, M.A. Munawar, and S. Khan, "Natural fiber-reinforced polymer composites", *Proceeding of Pakistan Academy of Sciences,* vol. 44, no. 2, pp. 129-144, 2007.

[7]   M. Fathi, S.M. Sapuan, K.A.M.A. Mohd, Y. Nukman, and B. Emin, "Cutting processes of natural fiber-reinforced polymer composites", *Polymers,* vol. 12, no. 6, p. 1332, 2020.

[8]   R. Girimurugan, R. Pugazhenthi, P. Maheskumar, T. Suresh, and M. Vairavel, "Impact and hardness behaviour of epoxy resin matrix composites reinforced with banana fiber/camellia sinensis particles", *Mater. Today Proc.,* vol. 39, no. 6, pp. 373-377, 2021.

[9]   N.M. Nurazzi, M.R.M. Asyraf, A. Khalina, N. Abdullah, H.A. Aisyah, S.A. Rafiqah, F.A. Sabaruddin, S.H. Kamarudin, M.N.F. Norrrahim, R.A. Ilyas, and S.M. Sapuan, "A review on natural fiber reinforced polymer composite for bullet proof and ballistic applications", *Polymers,* vol. 13, no. 4, p. 646, 2021.

[10]   T.D. Ngo, " Natural fibers for sustainable bio-composites", In: *Biomass Conversion & Processing Technologies, InnoTech Alberta (Formerly Alberta Research Council, 1921-2010 and Alberta Innovates Technology Futures*Edmonton, Alberta, Canada,, 2017.

[11]   M.R. Sanjay, G.R. Arpitha, L.N. Laxmana, K. Gopalakrishna, and B. Yogesha, "Applications of natural fibers and its composites: An overview", *Nat. Resour.,* vol. 7, no. 3, 2016.

[12]   S. Dixit, and V.L. Yadav, "Optimization of polyethylene/polypropylene/alkali modified wheat straw composites for packaging application using RSM", *J. Clean. Prod.,* vol. 240, 2019.118228

[13]   J. Hitesh, and J. Piyush, "A review on mechanical behavior of natural fiber reinforced polymer composites and its applications", *J. Reinf. Plast. Compos.,* pp. 1-13, 2019.

[14]   W. Dong, X. Lihui, H. Guangping, H.H.W. Andrew, W. Qingxiang, and C. Wanli, "Preparation and characterization of foamed wheat straw fiber/polypropylene composites based on modified nano-TiO_2 particles", *Compos., Part A Appl. Sci. Manuf.,* vol. 128, p. 105674, 2020.

[15]   F. Omar, K.B. Andrzej, P.F. Hans, and S. Mohini, "Biocomposites reinforced with natural fibers: 2000–2010", *Prog. Polym. Sci.,* vol. 37, no. 11, pp. 1552-1596, 2012.

[16]   U.S. Bongarde, and V.D. Shinde, "Review on natural fiber reinforcement polymer composites", *International Journal of Engineering Science and Innovative Technology,* vol. 3, no. 2, 2014.

[17]   S.Q. Tian, R.Y. Zhao, and Z.C. Chen, "Review of the pre-treatment and bioconversion of lignocellulosic biomass from wheat straw materials", *Renew. Sustain. Energy Rev.,* vol. 91, pp. 483-489, 2018.

[18]   T.P. Sathishkumar, J. Naveen, and S. Satheeshkumar, "Hybrid fiber reinforced polymer composites: A review", *J. Reinf. Plast. Compos.,* vol. 33, no. 5, pp. 454-471, 2014.

[19]   M. Saxena, A. Pappu, R. Haque, and A. Sharma, "Sisal fiber based polymer composites and their applications cellulose fibers", *Bio. Nano. Polym. Compos.,* pp. 589-659, 2011.

[20]   K. Love, S. Sarbjeet, P. Amar, and K. Narinder, "A review on natural fiber reinforced composites", *Mater. Today Proc.,* vol. 28, no. 3, pp. 1616-1621, 2020.

[21]   H.M. Akil, M.F. Omar, A.A.M. Mazuki, S. Safiee, Z.A.M. Ishak, and A.A. Bakar, "Kenaf fiber reinforced composites: A review", *Mater. Des.,* vol. 32, pp. 4107-4121, 2011.

[22]   S. Witayakran, W. Smitthipong, R. Wangpradit, R. Chollakup, and P.L. Clouston, *Natural fiber composites: Review of recent automotive trends.* Materials Science and Materials Engineering, 2017.

[23]   S. Yadvinder, S. Jujhar, S. Shubham, D.L. Thanh, and N.N. Duc, "Fabrication and characterization of coir/carbon fiber reinforced epoxy based hybrid composite for helmet shells and sports-good applications: Influence of fiber surface modifications on the mechanical, thermal and morphological properties", *Materials Research and Technology,* vol. 9, no. 6, pp. 15593-15603, 2020.

[24]   D. Pritam, and R. Srimanta, "An overview of the recent trends in manufacturing of green composites-considerations and challenges", *Mater. Today Proc.,* vol. 5, no. 9, pp. 19783-19789, 2018.

[25]   S. Brahim, B. Naima, and H. Dashnor, "Optimizing performance of insulation materials based on wheat straw, lime and gypsum plaster composites using natural additives", *Constr. Build. Mater.,* vol. 254, 2020.118959

[26]   S. Sathish, N. Karthi, L. Prabhu, S. Gokulkumar, D. Balaji, N. Vigneshkumar, and F.T.S. Ajeem, "A review of natural fiber composites: Extraction methods, chemical treatments and applications", *Mater. Today Proc.,* 2020.

[27]   L. Mi, P. Yunqiao, M.T. Valerie, G.Y. Chang, O. Soydan, D. Yulin, N. Kim, and J.R. Arthur, "Recent advancements of plant-based natural fiber–reinforced composites and their applications", *Compos., Part B Eng.,* vol. 200, 2020.108254

[28]   C. Mehdi, H.G. Seyed, J.A.K. Mazen, and G. Mantas, "Wheat straw pre- treatments using eco-friendly strategies for enhancing the tensile properties of bio-based polylactic acid composites", *Ind. Crops Prod.,* vol. 155, 2020.112836

[29]   K. Srinivas, A.K. Naidu, and M.V.A.R. Bahubalendruni, "A Review on chemical and mechanical properties of natural fiber reinforced polymer composites", *Int. J. Perform. Eng.,* vol. 13, no. 2, pp. 189-200, 2017.

[30]   M.S. Huda, L.T. Drzal, D. Ray, A.K. Mohanty, and M. Mishra, *Natural-fiber composites in the automotive sector.* Properties and Performance of Natural-Fiber Composites, 2008, pp. 221-266.

[31]   S. Tabassum, S. Muthuraman, M. Sivaraj, and S. Rajkumar, "Experimental evaluation of mechanical

properties of polymer matrix composites reinforced with date palm frond fibers from Oman", *Mater. Today Proc.,* vol. 37, pp. 3372-3380, 2021.

[32] S. Chandramohan, and D.J. Bharanichandar, "Natural fiber reinforced polymer composites for automobile accessories", *Am. J. Environ. Sci.,* vol. 9, no. 6, pp. 494-504, 2013.

[33] S. Panthapulakkal, A. Zereshkian, and M. Sain, "Preparation and characterization of wheat straw fibers for reinforcing application in injection molded thermoplastic composites", *Bioresour. Technol.,* vol. 97, no. 2, pp. 265-272, 2006.

[34] C. Burgstaller, and W. Stadlbauer, "Cellulose fibers as reinforcements in polypropylene", *8th Global WPC and Natural Fiber Composites Congress and Exhibition* Stuttgart, Germany, 2010.

[35] B. Christoph, "Investigation on the properties of polypropylene with bio-fillers and natural reinforcements", *International Polyolefins Conference* Houston, Texas, 2013, pp. 24-27.

[36] U.Z. Haydar, and M.D.H. Beg, "Preparation, structure and properties of the coir fiber/polypropylene Composites", *J. Compos. Mater.,* vol. 48, no. 26, pp. 3293-3301, 2014.

[37] K.B. Andrzej, A.M. Abdullah, M.M.L. Gabor, and G. Voytek, "The effects of acetylation on properties of flax fiber and its polypropylene composites", *Express Polym. Lett.,* vol. 2, no. 6, pp. 413-422, 2008.

[38] M. Zampaloni, F. Pourboghrat, S.A. Yankovich, B.N. Rodgers, J. Moore, L.T. Drzal, A.K. Mohanty, and M. Mishra, "Kenaf natural fiber reinforced polypropylene composites: A discussion on manufacturing problems and solutions", *Compos. Part A Appl. Sci. Manuf.,* vol. 38, no. 6, pp. 1569-1580, 2007.

CHAPTER 13

Analysis of Pineapple Leaf Fiber Reinforced Composite Vehicle Bumper with Varying Fiber Volume Fraction

Abhishek Pothina[1] and **Saroj Kumar Sarangi**[1,*]

[1] *Department of Mechanical Engineering, National Institute of Technology Patna, India*

Abstract: This paper presents the analysis of the Pineapple Leaf Fiber (PALF) reinforced composite used as a material for car bumpers. Impact analysis is performed on the modeled front car bumper at different fiber content, *i.e.,* at the difference in the value of the fiber volume fraction, and the results are discussed. The objective is to model a car's rear bumper with considered dimensions, and analyze it by simulating in the circumstances of a crash, *i.e.,* the impact is simulated against a rigid body at speed as per the standards of the vehicle. The natural fiber reinforced composite, which has good specific weight compared to synthetic fiber, results in a reduction in the weight of the whole body, resulting in less weight-to-volume ratio, when compared to the use of synthetic fibers and, therefore, can be considered as a material for car front bumper. There may certainly be a difference in performance, but depending on the required applications, the fiber-matrix bonding, and the aspect ratio can be varied. For PALF, the compensation for the low value of modulus can be done by having a very high aspect ratio, as the composite modulus is influenced by both young's modulus and aspect ratio. In PALF reinforced Composites, the variation of fiber content affects the performance of the composites with less increase in overall weight compared to that of synthetic fibers.

Keywords: Fiber reinforced composite, Natural fiber, Pineapple leaf fiber.

INTRODUCTION

In recent times, fiber-reinforced composites have been used for car bumpers, as they have lightweight, and the conventional aluminum and steel prices are increasing. As the bumper was meant to be lightweight, and should be able to reduce the damage caused to car parts, natural fiber reinforced composites are also being considered as having a few advantages over synthetic fiber like high aspect

---

* **Corresponding author Saroj Kumar Sarangi:** Department of Mechanical Engineering, National Institute of Technology Patna, India; E-mail: sarojksarangi@yahoo.com

Amar Patnaik, Albano Cavaleiro, Malay Kumar Banerjee, Ernst Kozeschnik & Vikas Kukshal (Eds.)

ratio, environmental friendly, and easier attainability. In this simulation, we consider pineapple leaf natural fiber, as it possesses a better aspect ratio than other natural fibers. The matrix considered are polymer matrices, as we will perform the analysis using the same fiber, but with different matrix materials and fiber content to get an optimal result for the specific function of the car bumper.

Car bumper

In an Automotive Vehicle, the bumper is not employed for crashworthiness or occupant protection in a period or cumulative of collisions, but are primarily concentrated on the energy absorption, providing comfort to both passengers and also for the vehicle. For this provision, certain standards are to be followed by the bumper and its arrangements, such as range of clearance and load withstanding competency with regard to speed.

The clearance between the surfaces of the road to that of the bumper normally ranges between 16 inches to 20 inches, and also, the bumper has to, under duress, endure the vehicle speeds of 2mph, 1 mph and 5mph from corner to corner, full width and parked environment crashes correspondingly. In accidents at low speeds, the bumper is one of the methods of protection, which is a place in the rear, *i.e.*, backside and also the front of the automobile body. As these bumpers are designed so as to, in the case of impact of a car, absorb the energy of the said impact.It has been designed with considered materials and shape, such that the above can be achieved with minimum cost. It is to be noted that the bumpers generally are not designed to avoid fracture at high speeds, but at minimum or standard speed, which ranges from 30 kmps to 50 kmps. So, at low speeds, it can prevent injury to passengers and protect other components of the automobile, such as the trunk, exhaust, fuel tank, hood, cooling systems, grille, *etc.* When the vehicle is struck from the front end, most of the energy is absorbed by the front bumper, at low speeds, which leads to the deformation of the bumper, but no energy will be shifted to the passengers of the vehicle, as the impact speed of the vehicle is very low. The same cannot be said in the case of high speeds, as the impact due to the collision of the vehicle may or will cause injury to the occupants and also impose damage to the vehicle. The widest part of the vehicle, which is 100% of the width, is considered to overlap the front impact of the vehicle. The standards of the bumper are given by the National highway traffic safety administration (NHTSA) for the light passenger vehicle. These standards are applicable to the performance requirements for passenger vehicles at low speeds for front and rear accidents.

In recent days, there has been numerous attempt to alter the design and material of the bumper for the improvement in safety, performance of car and expand the aerodynamics of the vehicle. Hence the research for this enhancement is still on going by groups of scientists and researchers.

Manufacturing Method

The process of integration of fibers into the matrix for the fiber-reinforced polymer matrix composite or polymer matrix composite (PMC) is done mainly of two methods; first one being that the fibers and matrix are treated directly into the finished product and the second method makes use of prepeg, which are made in form of sheets by incorporating fibers into the matrix and then stored to be used later to form a laminate structure by methods such as autoclave moulding, compression moulding *etc.*

The thermosetting matrix composite, as being used in this analysis, which is epoxy matrix composite, can be made by different methods such as filament winding, pultrusion, hand layup, spray layup, resin transfer moulding and autoclave moulding. The method that is generally recommended and practiced widely for the fabrication of the required PALF-reinforced PMC for this analysis is the hand layup (HLU) technique.

In the HLU (Fig. **1**) a gel coat is applied on the open mold, followed by application of fiber reinforcement, upon which base resin, which is mixed with catalyst, is applied by pouring and brushing. The layup is built by applying layer upon layer until the desired thickness is obtained.

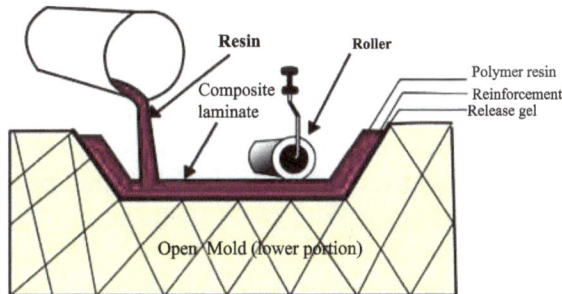

Fig. (1). Hand layup process.

Literature Review

In this section, various research papers related to the characteristics of natural fiber and their composites, along with the design and analysis of car bumpers, are discussed. These are considered the basis for developing the present work.

Davoodi *et al.* [1] studied the composite energy absorber in car bumpers and conducted the experiment in low-speed pedestrian accidents and concluded that the composite's energy absorption is optimal and damage dealt to a pedestrian is minimized. Marzbanrad *et al.* [2] simulated the front car bumper with constraints such as maximum deformation, maximum stress and concluded that using the materials with low young's module and high-strength materials will result in low rigidity and better impact behavior, respectively, as the maximum stress of bumper will be less than that of yield stress. Bumper rigidity and impact force will increase with an increase in the thickness of the bumper and will result in a drop in deflection and stress of the bumper. Kleisner *et al.* [3] simulated the car bumper with respect to the prescribed safety procedures and to design a new composite bumper reinforcement with the objective to preserve or to improve the mechanical properties and to reduce the mass simultaneously and finalized that with a change in the profile of bumper, the performance is within the safety mark and the mass of the bumper is reduced drastically when compared to the steel bumper. Kumar *et al.* [4] performed crash analysis on car bumper with different materials using ANSYS workbench and found that composite reinforced with fiber made of glass bumper is much more suitable in low crash applications when compared to steel bumper, and the factor of safety is also increased by 64% glass compared to that of steel. Ragupathi *et al.* [5] used jute as fiber material-reinforced composite for the improvement of the impact strength of the car bumper and found that natural fiber-reinforced polymer composites perform better and can be used in low-speed vehicles. Doodi *et al.* [6] developed a bumper model and performed the experimental and numerical analysis on it, continued by impact analysis. Godara *et al.* [7] performed the simulation and analysis on the frontal car bumper of automobile, where the bumper is made of carbon fiber reinforced composite. Arbintarso *et al.* [8] studied the use of a PALF-reinforced composite of a traditional bumper of the mini bus at different speeds and concluded that these bumpers can withstand collisions at speeds up to 70KMPH. Díaz-Álvarez *et al.* [9] conducted experiments on a bumper beam made of bio-composite materials and reviewed the energy absorption and residual bending behavior. Muthalagu *et al.* [10] considered the Kevlar and date palm fibers reinforced epoxy composites for the materials of automobile bumper and tensile test; material analysis was performed. Pradeep *et al.* [11] studied and analyzed car bumper with different composite and conventional materials and concluded that the deformation in composite materials is less, when compared to that of conventional materials. Encarnação [12] studied for a road-safe vehicle, the crash energy absorption system for a front part by consideration of multi-objective design optimization. Ramasubbu *et al.* [13] conducted experiments and then also simulated sisal and kenaf hybrid composite bumper using SOLIDWORKS and concluded that the impact strength is better for hybrid bumper when compared to that of kenaf reinforced bumper and is very much suitable for lightweight vehicle applications. Liu *et al.* [14] initially used the theoretical method and then the finite element method to evaluate the Manila hemp fiber's transverse thermal conductivity. Belingardi *et al.* [15] studied the issues regarding manufacturing and crash design with lightweight solutions for vehicle frontal bumper. Chandramohan *et al.* [16] studied the natural fiber-reinforced epoxy composite for its material properties by the use of six different tool materials and used tensile and hardness tests. He showed that the performance is better for hybrid composites and a brittle-like failure is observed in sisal fiber composites under tensile loading conditions. Fast propagating elliptical cracks are observed. Sangilimuthukumar *et al.* [17] studied the usages of pineapple leaf natural fiber reinforced composites in automotive parts and concluded that, as the automotive industry is trying to replace the parts with environmentally friendly materials, PALF is an obvious candidate for such parts. Selwyn and Sunder [18] considered the composite material with polymer matrix materials for automotive bumper applications and studied the formation and characterization along with suitability analysis. Vishwanatha *et al.* [19] simulated the front car bumper using CFRP (Carbon Fiber Reinforced Polymer) composite and subjected to vibration analysis and therefore concluded that the CFRP can be safely used for the front bumper, which resulted in reduced weight, when compared to standard steel bumpers. Ramachandra [20] simulated the carbon fiber reinforced car bumper and concluded that, it is very much suitable for the vehicle in low and medium crash applications. Ilczyszyn *et al.* [21] conducted the analysis on different geometric profiles of hemp fibers for their mechanical properties and after the comparison of the obtained results of experimental and simulation tests, the polygonal cross section was found to be more precise than the circular one. Nilsson and Gustafsson [22] concluded that the cell secondary wall layer of natural fibers. Thuault *et al.* [23] studied the fiber ultra-structure, and said that the fiber made of flax's longitudinal Young's modulus is strongly influenced by S2 layers thicknesses and MFA. Liu *et al.* [24] used theoretical as well as FEM to study the transverse thermal conductivity of the considered manila-hemp fiber for the use in natural fiber reinforced composites, in the solid region. Wang *et al.* [25] concluded that with influence of thermal conductivities of both matrix and fiber volume fraction, there is a change in effective thermal conductivity. Sliseris *et al.* [26] studied for different fiber aspect ratios, defects, and bundles to develop a 3D FEA models in order to simulate flax FRCs for their mechanical properties. Surajarusarn *et al.* [27] conclude that at increasing fiber content, PALMF exhibit better reinforcing efficiency when compared to Kevlar fiber. Todkar *et al.* [28] discuss all the required work in direction to advance the mechanical properties for PALF reinforced polymer composites. Narayana *et al.* [29] analyzed the impact of fiber volume fraction of reinforcing material on the stress-strain relation of bio-composites. DIGIMAT and ANSYS tools are used for emphasis of Multi-scale analysis. Lei Yang *et al.* [30] found that even though the damage mechanisms are the same generally, the initiation, propagation, and position of damage are influenced by fiber distribution. Munde *et al.* [31] studied the mechanical properties of PALF composites, by using Representative Volume Element (RVE) model by micromechanical analysis, which was developed in ANSYS 15 software. Noryani *et al.* [32] estimated the response variables and performed error analysis in order to handpick the finest natural fiber for automotive applications. Tensile strength is found to be more significant for the selection process by this statistical measurement of all selected natural fibers. Mark *et al.* [33] took the natural fibers with potential in composite application development and performed successions of tensile tests. The tensile test is conducted for samples that are being open to moisture environments of room temperature and humidity, which are usually 65% moisture content, 90% moisture content, and also soaked completely. S.S Godara [34] studied the influence of different chemical modifications of fiber surface on different properties of natural fibers composite are reviewed. Different properties of the natural fiber composites, such as the adhesion force between the matrix material and reinforcement surface and fiber strength, are observed to be improved. Ashok *et al.* [35] studied and summarized the dynamic properties of composite materials reinforced with natural fiber. Loss and storage modulus and damping factor are affected within dissimilar matrix materials by interlaced bonds between natural fibers and are temperature dependent. The difference in the length of fiber has affected the dynamic variables, excluding the changes in geometry. Xiong *et al.* [36] used finite element models to review the thermal and micromechanical properties of natural fiber and its composites, thermal properties, and macro shape deformation. Uddandapu [37] studied the impact analysis of a car bumper with consideration of the FMVSS208 standards of the impact analysis simulation and did the simulation at varying speeds by using materials ABS Plastic and Poly Ether Imide in the SOLIDWORKS software, which is Finite Element Analysis software.

Setup Information

As per the Federal Motor Vehicle Safety Standards (FMVSS208) [37], the following setup (Table **1**) is considered for the crash analysis simulation of an automobile bumper;

Table 1. Setup Information.

Type	Value
Velocity	13.33m/s

(Table 1) cont.....

Type	Value
Gravity	9.81m/s
Coefficient of Friction	0
Stiffens of Target	Rigid Body
Critical Damping Ratio	0

Mechanical Properties of the Materials

The Properties of the Pineapple leaf natural fiber and Epoxy Matrix are taken from [31], and are as per below Table **2**.

Table 2. Mechanical Properties of PALF fiber and epoxy matrix.

Mechanical property	PALF fiber	Epoxy matrix
Density (g/cc)	1.526	1.15
Tensile Modulus (GPa)	34.5	3.76
Poisson's Ratio	0.2	0.39
Tensile strength (MPa)	413	-
Elongation at break (%)	1.6	3.5

The objectives of this paper are to model a car rear bumper with the considered dimensions and analyze it by simulating in the circumstances of a crash, *i.e.*, the impact is simulated against a rigid body at speed, as per standards of the vehicle, which is 13.33 meter per second. The bumper will be made of composite material with Pineapple leaf natural fiber with epoxy matrix. The results are observed at different fiber content and are taken as a basis, for whether the PALF-reinforced composite is to be considered for the material of the car bumper.

METHODOLOGY

Analytical Method

The properties of the PALF-reinforced epoxy composite are calculated using the formulas derived from the micromechanical analysis of the composite [31]. The solution is done for different fiber content considered for the analysis.

Density of the Composite

For the Density of the composite, equation (1) is used, (all units are in g/cc).

$$\rho c = \rho f V f + \rho m V m \quad (1)$$

Where

ρc = Density of the composite

ρf = Density of the fiber

ρm = Density of the matrix

Vf = Volume of fiber

Vm = Volume of matrix

Longitudinal Modulus of the Composite

For the Longitudinal Modulus of the composite, equation (2) is used (all units are in GPa).

E1c = EfVf+ EmVm(2)

Where

E1c = Longitudinal Modulus of Composite

Ef = Modulus of Fiber

Em = Modulus of matrix

Transverse Modulus of the Composite

For the Transverse Modulus of the composite, equation (3) is used; (all units are in GPa).

1/E2c = Vf/Ef + Vm/Em(3)

In-plane Poisson's Ratio of the Composite

For the In-plane Poisson's Ratio of the Composite, equation (4) is used;

v12 = vfVf + vmVm(4)

Where

v12 = In-plane Poisson's Ratio of the Composite

vf = Poisson's Ratio of the fiber

vm = Poisson's Ratio of matrix

Modelling and Simulation

The bumper is initially modelled in CATIA V5R20 software. The dimensions of the bumper are from the literature survey. As for the current analysis, the material is PALF reinforced composite, which will result in less weight and Young's Modulus. The modeled bumper using the CATIA V5R20 software is shown in Fig. (2).

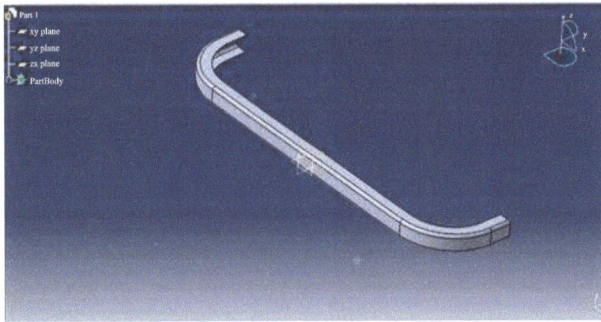

Fig. (2). Car Bumper CATIA Model.

This model (Fig. 2) is then taken into ANSYS R19 software and simulated in the Explicit Dynamics. Initially, the properties that are considered for the analysis as per the Federal Motor Vehicle Safety Standards (FMVSS208) are fed into the engineering materials section against a rigid body. Then, the bumper is given the considered material properties, and meshing is performed, so the mesh is uniform and linear on the whole bumper model, which is shown in Fig. (3).

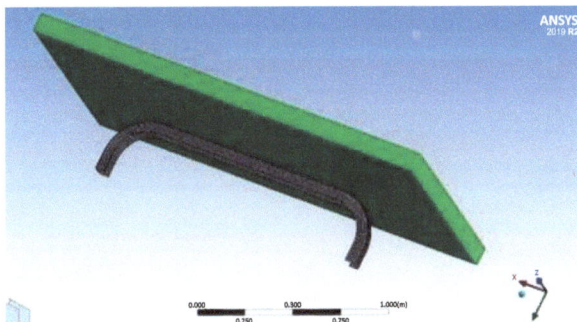

Fig. (3). Meshed Car Bumper Model.

Boundary Conditions

The rectangular section is considered to be a rigid body, which is at a certain distance from the bumper. The bumper is given the velocity as per the standards discussed earlier, which is 13.33 meters per second, and appropriate step time and a number of cycles are also given for the analysis. The standard earth gravity is applied to the body, and the friction coefficient is left at zero itself.

Results and Discussion

When the simulation is completed, the obtained solution of deformation and Equivalent strain are recorded for all the different fiber volume content. The contours obtained are shown as follows (Figs. **4 - 6**):

Fig. (4). Equivalent Stress and Elastic Strain of Car bumper with 30% Vf.

   i. For 30% Fiber Volume Fraction:
  ii. For 40% Fiber Volume Fraction:
iii. For 50% Fiber Volume Fraction:

Fig. (5). Equivalent Stress and Elastic Strain of Car bumper with 40% Vf.

From (Fig. **4–6**), it can be observed that the stress decreases drastically with an increase in fiber content, and the gap is much wider between 40% and 50% of Vf%, than that of 30% and 40%. Therefore, the Vf% value can be maintained at a minimum of 30% for the car bumper for it to give the optimal or preferred performance and can be maintained at 50% for the best performance. The density of the PALF-reinforced epoxy bumper is 2.7g/cc and is far less than that of steel, which is 8.05 g/cc. The variation of Deformation, Equivalent Stress and Equivalent Strain with respect to the considered fiber volume fraction percentage (Vf%) is observed and plotted on the graph below (Figs. **7, 8, 9**, respectively).

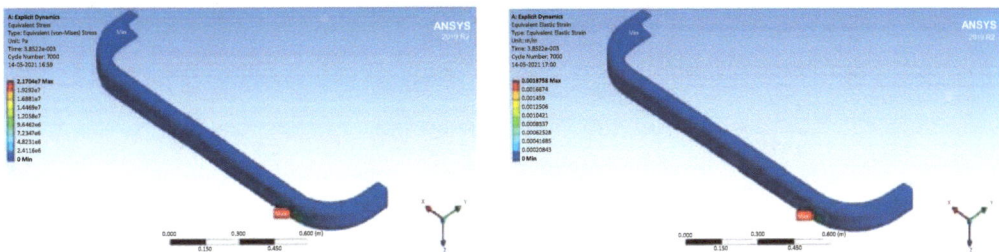

Fig. (6). Equivalent Stress and Elastic Strain of Car bumper with 50% Vf.

From Fig. (**7**), it can be observed that the slope decreases as the V_f % increases, resulting in a decrease of deformation with an increase in V_f %. The difference is deformation, as discussed earlier, is more between 30% and 40%, compared to 40% and 50%. This can mean that the effect of reinforcing material which is PALF, is more noticeable at V_f at a minimum of 40% than that of 30%, and the trend gives better performance with an increase in V_f % even further, but the effect or difference in performance is decreasing.

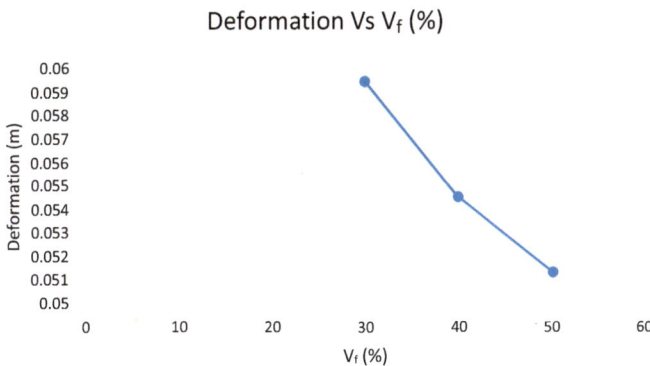

Fig. (7). Effect of changes in deformation (m) with change in Vf%.

From Fig. (**8**), it can be observed that the slope decreases as the V_f % increases, resulting in a decrease of equivalent Stress with an increase in V_f %. The difference is equivalent Stress, as discussed earlier, is between 30% and 40%, as compared to that of 40% and 50%.

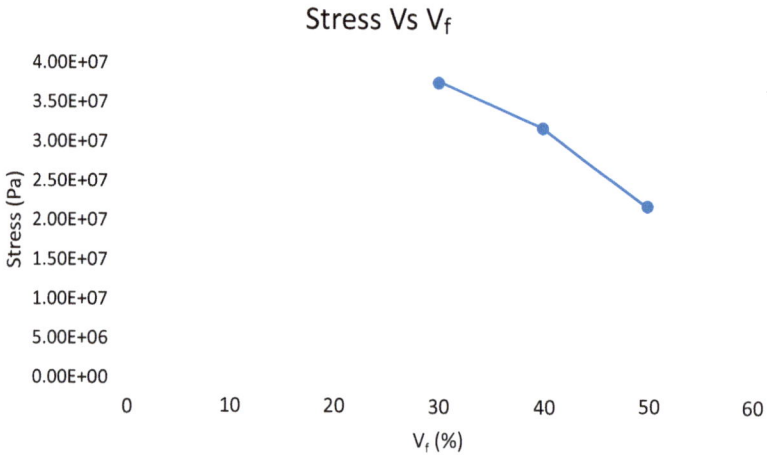

Fig (8). Effect of changes in stress (Pa) with change in Vf%.

This can mean that the effect of reinforcing material which is PALF is more noticeable in terms of equivalent Stress, at V_f at minimum of 40% than that of 30% and trend gives better performance with increase in V_f % even further, but the effect or difference in performance is decreasing.

From (Fig. **9**), it can be observed that the slope decreases as the V_f % increases, resulting in a decrease of equivalent Strain with an increase in V_f %. The difference is equivalent Strain, as discussed earlier, is between 30% and 40%, as compared to that of 40% and 50%.

This can mean that the effect of reinforcing material which is PALF in terms of equivalent strain, is more noticeable at V_f at a minimum of 40% than that of 30%, and the trend gives better performance with an increase in V_f % even further, but the effect or difference in performance is decreasing.

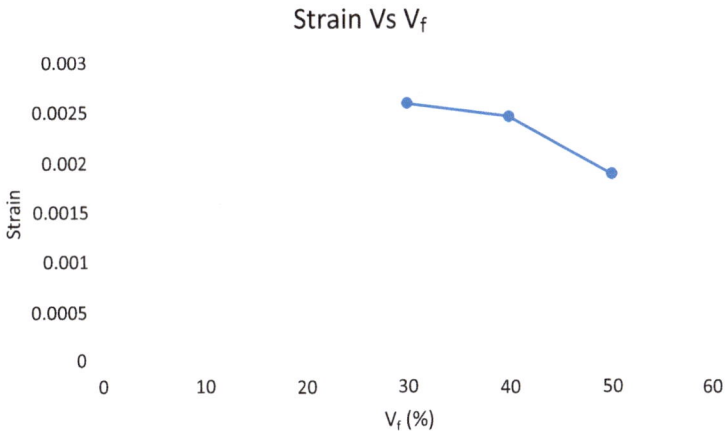

Fig (9). Effect of Equivalent Strain with change in Vf%.

CONCLUSION

It is observed from the results discussed in the previous section that the deformation of the bumper at the standard speed (13.33 mps) is less than 0.01m and is slightly above 0.005m. The stress-induced is also optimal, showing that the natural fiber-reinforced composites are much more capable of being used for the automobile bumper and have significantly less weight than the conventional bumper. The Pineapple leaf fiber, which has a good aspect ratio, compared to other natural fibers, can be very much considered for the reinforcement material for the car bumper, which enables the bumper to have more fiber volume fraction, with a very slight increase in total weight. Therefore, a higher fiber volume fraction can be maintained compared to that of synthetic fibers. The results are considerably better with higher fiber content, and bumper with a fiber volume fraction of 40% performs marginally better than that of 30%, and the performance gets even better with a fiber volume fraction of 50%.

REFERENCES

[1]   M.M. Davoodi, S.M. Sapuan, and R. Yunus, "Conceptual design of a polymer composite automotive bumper energy absorber", *Mater. Des.,* vol. 29, no. 7, pp. 1447-1452, 2008.

[2]   J. Marzbanrad, A.J. Masoud, and S.K. Mahdi, "Design and analysis of an automotive bumper beam in low-speed frontal crashes", *Thin-walled Struct.,* vol. 47, no. 8-9, pp. 902-911, 2009.

[3]   Václav Kleisner, and Robert Zemčík, "Analysis of composite car bumper reinforcement", *Applied and computational mechanics 3, no. 2,* 2009.

[4]   G. Raj Kumar, S. Balasubramaniyam, M. Senthil Kumar, R. Vijayanandh, R. Raj Kumar, and S. Varun, "Crash analysis on the automotive vehicle bumper", *Inter. J. Eng. and Advan. Techno.,* pp. 2249-8958, 2019.

[5]   P. Ragupathi, N.M. Sivaram, G. Vignesh, and D. Selvam. Milon, "Enhancement of impact strength of

a car bumper using natural fiber composite made of jute.i-Manager's", *Jixie Gongcheng Xuebao,* vol. 8, no. 3, p. 39, 2018.

[6]   S. Doddi, K. Channakeshavalu, B. H. Maruthi, and BT. Chandru, "Experimental-numerical modal and impact analysis of car bumper", In: *Mechanical Engineering Dept.* EWIT, Bangalore, Karnataka: India., 2015.

[7]   S. S. Godara, and N. N. Shiv, *Analysis of frontal bumper beam of automobile vehicle by using carbon fiber composite material. Materials Today: Proceedings 26,* vol. 26, pp. 2601-2607, 2020.

[8]   E.S. Arbintarso, M. Muslim, and T. Rusianto, "Simulation and failure analysis of car bumper made of pineapple leaf fiber reinforced composite", *IOP Conference Series: Materials Science and Engineering,* vol. vol. 306, 2018p. 012038

[9]   A. Díaz-Álvarez, L. Jiao-Wang, C. Feng, and C. Santiuste, "Energy absorption and residual bending behavior of biocomposites bumper beams", *Compos. Struct.,* vol. 245, p. 112343, 2020.

[10]   R. Muthalagu, J. Murugesan, S. Sathees Kumar, and B. Sridhar Babu, "Tensile attributes and material analysis of kevlar and date palm fibers reinforced epoxy composites for automotive bumper applications", *Mater. Today Proc.,* 2020.

[11]   G. Pradeep, and P. Chandra Kumar, "Design and experimental analysis on car bumper with composite materials", *Inter. Res. J. Eng. and Techno.,* vol. 5, no. 10, pp. 54-58, 2018.

[12]   T.M.E. Nunes, Multi-objective design optimization of a frontal crash energy absorption system for a road-safe vehicle., 2017.

[13]   T.M.E. Nunes, R. Ramasubbu, and M. Sankaranarayanan, "fabrication of automobile component using hybrid natural fiber reinforced polymer composite", *J. Nat. Fibers,* pp. 1-11, 2020.

[14]   Z. Liu, J. Lu, and P. Zhu, "Lightweight design of automotive composite bumper system using modified particle swarm optimizer", *Compos. Struct.,* vol. 140, pp. 630-643, 2016.

[15]   G. Belingardi, Alem Tekalign Beyene, E. G. Koricho, and Brunetto Martorana, "Lightweight solutions for vehicle frontal bumper: Crash design and manufacturing issues", In: *In Dynamic Response and Failure of Composite Materials and Structures* Woodhead Publishing, 2017, pp. 365-393.

[16]   D. Chandramohan, and J. Bharanichandar, "Natural fiber reinforced polymer composites for automobile accessories", *Am. J. Environ. Sci.,* vol. 9, no. 6, p. 494, 2013.

[17]   J. Sangilimuthukumar, T. Senthil Muthu Kumar, M.C Carlo Santulli, K. Senthilkumar, and S. Suchart, "The use of pineapple fiber composites for automotive applications: A short review", *J. Mat. Sci. Res. Rev.,* pp. 39-45, 2020.

[18]   T.S. Selwyn, "Formation, characterization and suitability analysis of polymer matrix composite materials for automotive bumper", *Mater. Today Proc.,* vol. 43, pp. 1197-1203, 2021.

[19]   R H Vishwanatha, A Ajith, N Anand, K B, PunithRaju, and K. Vasantha, "Design and strength validation of front car bumper using composite material", *International Research Journal of Engineering and Technology,* pp. 5-5, 2018.

[20]   Dr.R. Ramachandra, "Modeling and static analysis of car bumper", *International Journal of Engineering Development and Research,* pp. 5-4, 2017.

[21]   F. Ilczyszyn, A. Cherouat, and G. Montay, "Effect of hemp fibre morphology on the mechanical properties of vegetal fibre composite material", *AMR,* vol. 875–877, pp. 485-489, 2014.

[22]   T. Nilsson, and P.J. Gustafsson, "Influence of dislocations and plasticity on the tensile behaviour of flax and hemp fibres", *Compos Part A Appl Sci,* vol. 38, pp. 1722-1728, 2007.

[23]   A. Thuault, B. Jérôme, E. Sophie, B. Joel, and G. Moussa, "Numerical study of the influence of structural and mechanical parameters on the tensile mechanical behaviour of flax fibres", *J. Ind. Text.,* vol. 44, no. 1, pp. 22-39, 2014.

[24]   K. Liu, H. Takagi, O. Ryosuke, and Y. Zhimao, "Effect of physicochemical structure of natural fiber

on transverse thermal conductivity of unidirectional abaca/bamboo fiber composites", *Compos., Part A Appl. Sci. Manuf.,* vol. 43, no. 8, pp. 1234-1241, 2012.

[25]   M. Wang, Q. Kang, and N. Pan, "Thermal conductivity enhancement of carbon fiber composites", *Appl. Therm. Eng.,* vol. 29, no. 2-3, pp. 418-421, 2009.

[26]   J. Sliseris, Y. Libo, and K. Bohumil, "Numerical modelling of flax short fibre reinforced and flax fibre fabric reinforced polymer composites", *Compos., Part B Eng.,* vol. 89, pp. 143-154, 2016.

[27]   B. Surajarusarn, S. Hajjar-Garreau, G. Schrodj, M. Karine, and A. Taweechai, "Comparative study of pineapple leaf microfiber and aramid fiber reinforced natural rubbers using dynamic mechanical analysis", *Polym. Test.,* vol. 82, p. 106289, 2020.

[28]   S. Todkar, and A. Suresh, "Review on mechanical properties evaluation of pineapple leaf fibre (PALF) reinforced polymer composites", *Compos., Part B Eng.,* vol. 174, p. 106927, 2019.

[29]   K. Narayana, Jagath, and G.B. Ramesh, "Multi-scale modeling and simulation of natural fiber reinforced composites (Bio-composites)", In: *Journal of Physics: Conference Series* vol. 1240. IOP Publishing, 2019, no. 1, p. 012103.

[30]   L. Yang, Y. Yan, Y. Liu, and Z. Ran, "Microscopic failure mechanisms of fiber-reinforced polymer composites under transverse tension and compression", *Compos. Sci. Technol.,* vol. 72, no. 15, pp. 1818-1825, 2012.

[31]   Y.S. Munde, R.B. Ingle, and I. Siva, "Effect of sisal fiber loading on mechanical, morphological and thermal properties of extruded polypropylene composites", *Mater. Res. Express,* vol. 6, no. 8, p. 085307, 2019.

[32]   M. Noryani, M.S Salit, T.M. Mohammad, M.Z. MohdYusoff, and S.Z. Edi, "Material selection of natural fibre using a stepwise regression model with error analysis", *J. Mater. Res. Technol.,* vol. 8, no. 3, pp. 2865-2879, 2019.

[33]   M.C. Symington, W.M. Banks, O.D. West, and R.A. Pethrick, "Tensile testing of cellulose based natural fibers for structural composite applications", *J. Compos. Mater.,* vol. 43, no. 9, pp. 1083-1108, 2009.

[34]   S.S. Godara, "Effect of chemical modification of fiber surface on natural fiber composites: A review", *Mater. Today Proc.,* vol. 18, pp. 3428-3434, 2019.

[35]   R.B. Ashok, C.V. Srinivasa, and B. Basavaraju, "Dynamic mechanical properties of natural fiber composites—a review", *Adv. Compos. Hybrid Mater.,* vol. 2, no. 4, pp. 586-607, 2019.

[36]   X. Xiong, S.Z. Shen, L. Hua, J.Z. Liu, X. Li, X. Wan, and M. Miao, "Finite element models of natural fibers and their composites: A review", *J. Reinf. Plast. Compos.,* vol. 37, no. 9, pp. 617-635, 2018.

[37]   P.K. Uddandapu, "Impact analysis on car bumper by varying speeds using materials ABS plastic and poly ether imide by finite element analysis software solid works", *Int. J. Mod. Eng. Res,* vol. 3, pp. 391-395, 2013.

CHAPTER 14

Experimental studies on Mechanical characteristics of Bamboo Leaf Ash reinforcement with Aluminum 7075 alloy using Rotary Stir Casting Technique

Murahari Kolli[1,*], **Dasari Sai Naresh**[2] and **K. Ravi Prakash Babu**[3]

[1] *Lakireddy Bali Reddy College of Engineering, Mylavaram, Krishna District, Andhra Pradesh, India*

[2] *R&D Mechatronics, Design and Engineering, VEM Technologies PVT Ltd, Hyderabad, Telangana, 502321, India*

[3] *Department of Mechanical Engineering, Prasad V Potluri Siddhartha Institute of Technology, Kanuru, Andhra Pradesh, India*

Abstract: Aluminium metal matrix composites are an exclusive class of materials that have improved performance parameters than their pure metal-based counterparts. These composites are widely used in structural, marine, aviation, defense and mining industries. Numerous synthetically derived hard ceramic reinforcements were widely researched for property enhancement of Aluminum Metal Matrix Composites, but, the exclusivity and the economic concerns of the synthetic reinforcements paved the way for widespread studies of agro-waste-based and industrial waste based Aluminum Metal Matrix Composites. In the current study, Bamboo Leaf Ash, an agro waste derived ceramic reinforcement based Aluminum Metal Matrix Composite; Al 7075/Bamboo Leaf Ash is fabricated using the Liquid Metallurgy Stir Casting technique with varying volume percentages from 2% to 8% of reinforcements in the matrix by weight. Mechanical and Microstructural characterization of the metal matrix composite is performed to ascertain the degree of improvement in properties of the composite compared to the base metal. The results from the study confirmed that a sound composite with higher hardness and strength was obtained. The microstructural characterization also confirmed that the grain structure is significantly refined, leading to property enhancement.

Keywords: Al 7075, Aluminium Metal Matrix Composites, Agro waste, Bamboo Leaf Ash, Rotary Stir Casting.

---
* **Corresponding author Murahari Kolli:** Lakireddy Bali Reddy College of Engineering, Mylavaram, Krishna District, Andhra Pradesh, India; E-mail: kmhari.nitw@gmail.com

Amar Patnaik, Albano Cavaleiro, Malay Kumar Banerjee, Ernst Kozeschnik & Vikas Kukshal (Eds.)

INTRODUCTION

Nowadays, the manufacturing scenario is changing in the usage of conventional materials as they have been replaced by advanced materials. Metal Matrix Composites (MMCs) are one such advanced materials that are playing a prominent role in the manufacturing sector due to their properties like high specific strength, lightweight, specific stiffness, wear resistance, corrosion resistance and elastic modulus, mostly applicable for aerospace, automobile, marine, mining and mechanical structures [1]. Metal matrix composites (MMC's) are metals and alloys combined with other materials that ultimately result in performance enhancement of the metal/alloy. Ever since their inception, copper, aluminium and magnesium-based MMC's were significantly researched [2]. In particular, Al-based MMC's can be broadly categorized as synthetic ceramic derivatives, agro-waste derivatives and industrial waste derivatives [3]. Aluminium based MMC's were manufactured with many synthetic ceramic reinforcements due to their strength and desirable properties, such as Silicon Nitride (Si_3N_4), Alumina (Al_2O_3), AluminiumTitanate (Al_2TiO_5), Zirconia (ZrO_2), Aluminium Nitride (AlN), Boron Carbide (B_4C), Silicon Dioxide (SiO_2), etc. due to their strength and desirable properties [4]. Expensiveness and inadequate availability of conventional ceramic reinforcements in many developing countries prompted a compulsory paradigm shift in the choice of selection for reinforcement particles [5]. Hence, the usage of agro-waste and industrial waste-derived MMC's has gained prominence in the recent past because such reinforcements mitigated the expensiveness of the reinforcements and were able to achieve properties on par with ceramic composites [6]. There have been numerous reports on Aluminium Industrial waste agro waste derived MMC's.

Balasubramani *et al.* investigated the mechanical properties of Al 7075/B4C/coconut shell fly ash composites fabricated through the friction stir processing technique. It was observed from the results maximum tensile strength was indicated for 9% B4C and 3% CSFA in the matrix material. The hardness values are also enhanced with the reinforcement particles being added to the matrix [7]. Bhasha and Balamurugan examined the Al 6061 hybrid composites reinforced by nanoTiC/Rise husk ash using an *ex-situ* process *via* ultrasonic probe sonicator stir casting. Microhardness, tensile, and flexural strength were considered as testing results of the composite. When 3-6% of TiC was added with 3% of RHA, a better result of the composite was found; further, increasing the addition of 6-9% TIC with 3% of RHA gave results with lower values because of the agglomerations and clusters formed on the composite [8]. Gireesh *et al.*, attempted to use aloe vera particles as reinforcement into the aluminum matrix to examine the mechanical properties of the composites. They concluded that the addition of aloe vera particles in the Al matrix significantly improved the tensile strength, impact strength, and hardness compared with the base and Al/FA composites [9]. Joseph and Babaremu studied different kinds of literature and identified the various agricultural wastes being added as reinforcement particles addition into aluminum matrix materials to enhance the mechanical properties of the composites. Some of the wastes include reinforcement materials like groundnut shells, coconut shells, rice husk, breadfruit seed ash, aloe vera, bean pod ash, cow horn, and so on [10].

Jose *et al.*, investigated the mechanical characterization of Al 6061 alloy by adding lemon grass ash as reinforcing particles in the MMC by compocasting. The results showed that tensile strength and microhardness values increase linearly because the adequate diffusion of LGA reinforcement particles hindered the movement of dislocations in the MMC [11]. Manikandan and Arjunan fabricated Al 7075 hybrid metal matrix composites with agro-waste, cow dung ash, and boron carbide particles as reinforcement in the matrix, and a two-stage stir casting method was also adopted in their study. The influence of reinforcement particles of the composites was examined for mechanical, microstructural and tribological properties. It was concluded that adding two or more reinforcements in the aluminum matrix enhanced the properties compared to single-reinforcement composite materials [12].

Kumar and Birru investigated the effect of BLA reinforcement on Al material by fabricating the composite using the stir casting method. Hardness and tensile strength properties were considered as measuring characteristics of the composite when the addition of BLA reinforcement's particles from 2% to 8% to the matrix material. They concluded that 2% to 4% of reinforcement added to material increase the hardness and tensile properties [13, 14]. Alanemeand andAdewuyi studied the mechanical properties of Aluminium magnesium silicon alloy composites filled with Al2O3 as well as BLA reinforcements. Particulates of Al_2O_3 compounded by 0%, 2%, 3%, and 4% wt BLA were added to fabricate 10 wt.% of reinforcement in the matrix material and adapted the dual stir casting technique. Tensile strength, yield strength, and elongation, hardness tests were examined in their study [15]. Singh *et al.* examined the mechanical properties of Al 6063 composites filled with groundnut shell ash using the liquid metallurgy technique. The result indicated increases in the tensile strength and compressive strength with groundnut shell ash added to Al 6063 material [16].

Many such instances are available in the literature. It was also notified in the literature that although numerous techniques are available for manufacturing MMC's and HMMC's, stir casting has always been a vastly economical and well-established process. Also, the structural properties of the composites prepared through the stir casting route were superior to composites prepared through other techniques due to the homogeneity of particle distribution achieved by continuous stirring action. In the present investigation, Al 7075 alloy was considered as it is to be a strong material with high strength compared to steels, also having good fatigue strength with average machinability. The elements of Al 7075 alloys are iron, magnesium, zinc, copper, chromium, manganese, titanium, silicon, and other metals. The bamboo leaf ash (BLA) has been selected as reinforcement for the matrix material. BLA particles are chosen to replace conventional ceramic particles. When BLA particles were added to the matrix material, they were uniformly distributed, and its effect enhanced the material properties of Al alloy.

Hence, in an attempt to improve the mechanical properties while retaining ductility and fracture toughness, the current investigation deals with the fabrication and mechanical and microstructural characterization of Al 7075/BLA reinforced through liquid metallurgy stir casting route with varying volume percentages of reinforcements by weight (Wt.). The degree of improvement is compared against the Al alloy and Al 7075/BLA MMC's developed through Stir Casting process with identical process parameters and volume percentages from 2% to 8% each.

MATERIALS AND METHODOLOGY

The base metal matrix considered for the current study is Al 7075 alloy, and its ingots are procured from commercial suppliers in Hyderabad, India. The chemical combinations and mechanical properties of the as-received material are given in Table **1**.

Table 1. Chemical composition of Al 7075 alloy.

Elements	Zn	Mg	Cu	Si	Mn	Fe	Cr	Ti	Al
Wt%	5.56	2.71	1.42	0.33	0.21	0.34	0.27	0.12	Remaining

The reinforcement particulates used as BLA are added at varying percentages in the metal matrix by weight. BLA is chosen as a potential reinforcement because it offers superior wettability thereby ensuring regularity in the dispersal of particles in the MMC. The average sizes of BLA particulates are 50-70 μm. The reinforcement material, bamboo leaves, was collected from the Mylavaram, Krishna district region, Andhra Pradesh, India. The bamboo leaves are converted to reinforcement particles as per the following steps and shown in Fig. (**1**)

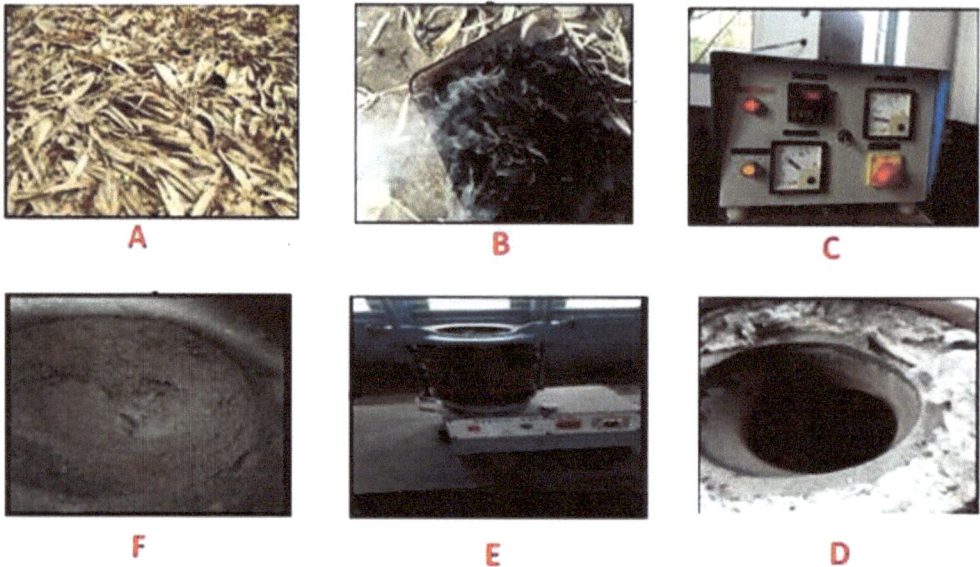

Fig. (1). Preparation of Bamboo leaf ash reinforcements.

Stage 1. Collection of dry bamboo leaves.

Stage 2. A specially designed metallic drum was used in firing the dry bamboo leaves.

Stage 3. Unwanted residuals are removed from the BLA, using a sintering machine.

Stage 4. By using a sieving machine, BLA particle sizes were found.

Stage 5. The shape and size along with the elemental composition of BLA particles, are using the SEM and EDS analysis.

EXPERIMENTS

The production of MMC is realized through the stir-casting process. A typical outline of the setup is shown in (Fig. **2**). Initially, Al 7075 ingots are charged, and the furnace is heated up to 780°C for 30 Minutes. The mixture of metal matrix is constantly stirred at 800 RPM for a duration of 20 minutes by simultaneously adding reinforcements that are preheated up to 300°C in order to improve wettability according to the weight percentages. Enhancement of the stirring time improved the homogeneous distribution of the particles and decreased grain structure size, and higher hardness values were achieved. However, excessive stirring caused a serious increase in oxidation and impurities in the metal matrix. Hence, the selection of parameters from the literature for the current study was carefully carried out to achieve optimum results. Further, the molten metal is poured into mould cavities preheated up to 450°C for 3 Hours, are charged into the metal matrix. The molten metal is then pressurized using a mechanized die into the mould cavity. The obtained specimens are of dimensions ø20x120mm observed in Fig. (3). The test specimens are cut from the base metal with CNC Turning and Wire EDM.

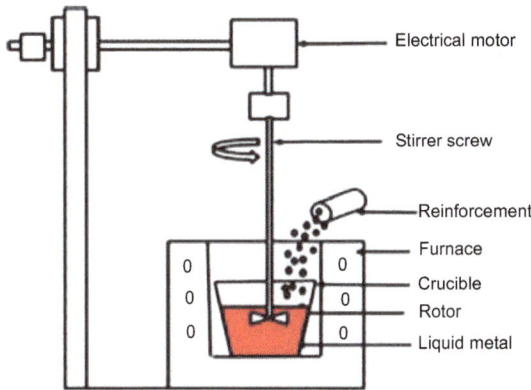

Fig. (2). Outline of Stir Casting Setup.

Fig. (3). Obtained Specimens after Stir Casting.

To characterize the mechanical properties of the specimen viz. Ultimate Yield Strength (UYS), Tensile Strength, Ductility, Hardness and Impact Energy, the specimens were cut, following the ASTM standards like ASTM B559 and ASTM E10-18 using the wire cut EDM and CNC machines. The tensile testing is done on Universal Tensile Testing Machine. The Hardness testes are conducted on the adopted Hardness Tester model B-3000H. The V- Notch Specimens are cut and Tested on Impact Energy Tester Model: IT 30 of FIE make. The grain refinement of the composites was studied using optical microscopy. The samples are first ground with successive meshed grits of 1200, 1600 and 2000, respectively, and the samples are then polished into a mirror-like finish using diamond slurry and alumina paste on a polisher.

RESULTS AND DISCUSSION

Metallographic Characteristics

The SEM and EDS profiles of BLA particulate matter are shown in Fig. (**4**) and (Fig. **5**). The EDS profile (Fig. **5**) of BLA has shown peaks of Silicon (Si), Oxygen (O), Calcium (Ca), Potassium (K), Chlorine (Cl) and Magnesium (Mg) in their oxide form and are consistent with reports of several other investigations. The silicon dioxide's (SiO_2) dominance in the chemical composition is confirmed by the highest peak in the EDS profile. This finding is well supported by the literature [13]. On the other hand, the SEM image (Fig. **4**) of BLA confirmed that the average size was almost 50-70 μm.

| 200 μm | EHT = 10.00 kV | Mag = 100 X | Date :7 Feb 2020 | ZEISS |
| | WD = 12.0 mm | Signal A = SE1 | Time :18:38:08 | |

Fig. (4). SEM Image of B4C Particles.

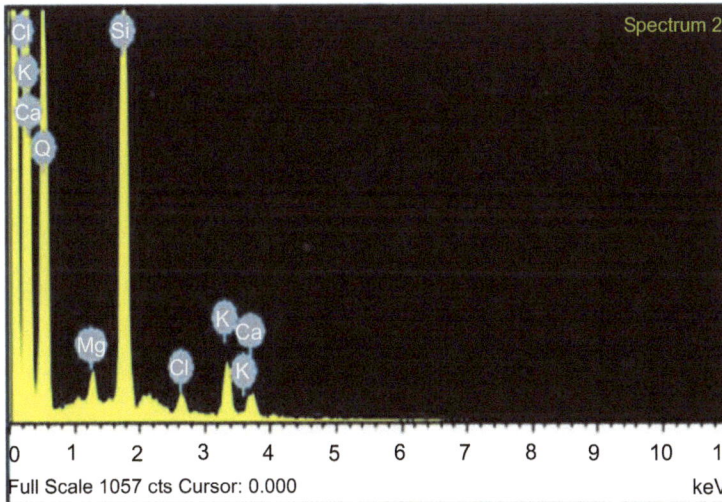

Fig. (5). EDS Profile of BLA Particles.

Microstructure

The interface characteristics and the microstructure of the composites dictate the properties demonstrated by the alloys at large. The results of the microstructural studies are shown in Fig. (**6**). All the microstructures of MMC's have revealed a substantial grain enhancement with a growing weight percentage of the reinforcements. This is due to the fact that the dispersal of the particles inside the matrix alloy inhibited the devolvement of the α- Grains during solidification and the BLA particles acted as effective grain refiners [14]. Variables such as the type of reinforcements and solidification rates of the composites are accountable for the distribution of the particles inside the matrix and prevent particle agglomerations [13].

The microstructures depicted in Fig. (**6**) for reinforcement compositions from 2 through 8 revealed that particle agglomerations and segregations are nominal, and successive refinement of grain in the matrix is evident with an accumulative weight percentage of reinforcements due to the restraint of free α- Grains development by BLA particles. This phenomenon is a consequence of the hard BLA particles acting as grain nucleation sites whilst solidification of matrix α- Grains. Amplified grain nucleation can be seen in Fig. (**6**) at higher concentrations of reinforcement particles inside the matrix.

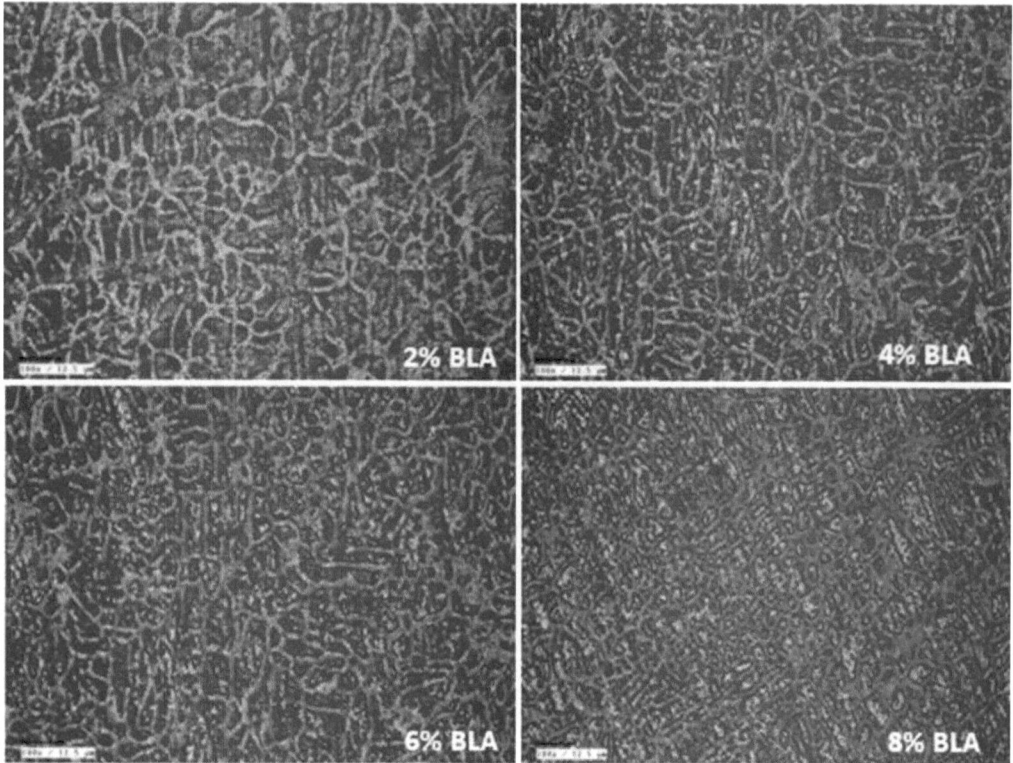

Fig. (6). Microstructural images of AL-7075/BLA MMC.

Tensile and Ultimate Yield Stress

The experimental tensile and UYS values can be observed in Fig. (7 and 8). The effect of BLA reinforcements in the matrix is investigated in the MMC. Tensile and UYS exhibited a proportional relationship with the percentage of reinforcements in MMC's. The trend was clearly specifying that a 2-8% percentage of BLA reinforcements in the matrix enhances the tensile and ultimate yield stress values. With the BLA addition to matrix material shown, increasing the trends due to the load transferring criteria signifies that the strengthening of the composite occurred due to the fact that the load applied on the composite gets transferred to the much harder reinforcement particles inside the MMC at the interface boundaries which then leads to improved restriction to plastic deformation, consequently work hardening the composite in itself and increasing the strength [14].

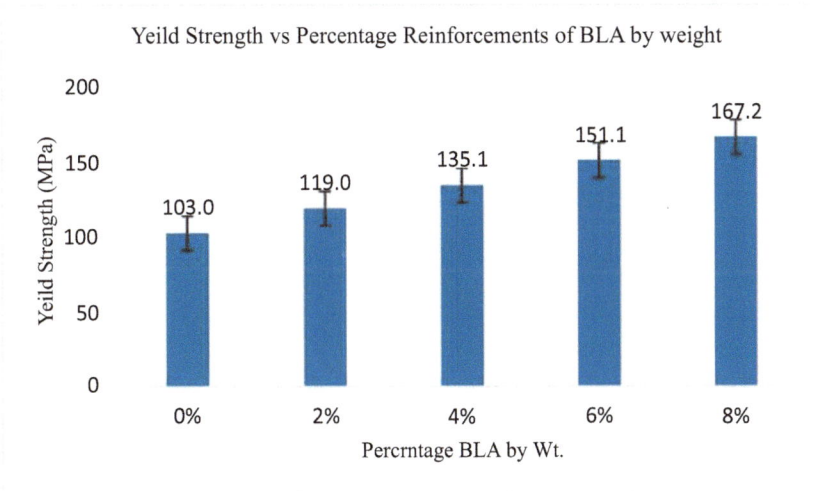

Fig. (7). Yield Strength Vs % of BLA reinforcements.

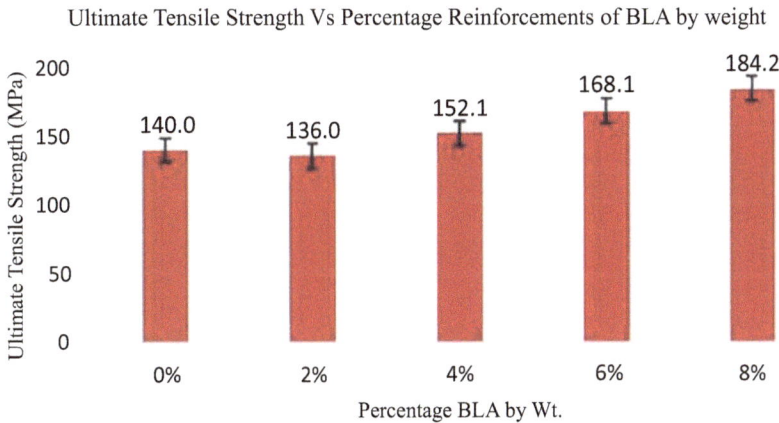

Fig. (8). Ultimate Tensile Strength Vs % of BLA reinforcements.

Hardness

Fig. (9) revealed that adding reinforcement particles to Al 7075 has increased the hardness values significantly. Also, as the composition by Wt. %of reinforcement particles progressed, the hardness has also increased substantially except for Al 7075- BLA MMC, which exhibited a slight decrease in hardness beyond 6%. This is due to the fact that as the volume % of the reinforcement particles increases, the dislocation density of the MMC increases, and the particulate matter tend to restrict the movements of these dislocations more effectively, which results in the increased Hardness. Also, the downward movement of the indenter is resisted by the particles in the matrix, which further advances the hardness of the composite [15]. The hardness value may have dropped for Al-7075-BLA MMC after 6% as the increased occurrence of ash particles may have served as a potential site of crack initiation [14]. These results were in good agreement with other investigations as well the wettability of the reinforcement particles plays a major role in the hardness.

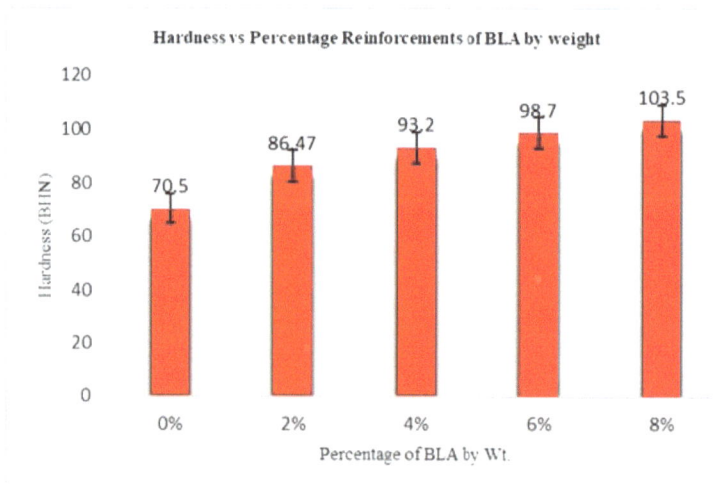

Fig. (9). Hardness Vs % of BLA reinforcements.

Ductility

The ductility of the composites is investigated and analyzed as a function of % elongation of the composites. In the current investigation, % elongation of the composites reduced with an increase in the volume percentage of reinforcements as observed in Fig. (**10**). These results are consistent with literature and reports that summarize that adding reinforcement particles to soft metal matrix decreases the ductility of the composites by various mechanisms. The development of particle clusters with increasing volume % of reinforcements which may contribute to the reduction of ductile phases in the composite, in turn, contributed to the reduction in elongation for MMC's with increasing BLA content [16].

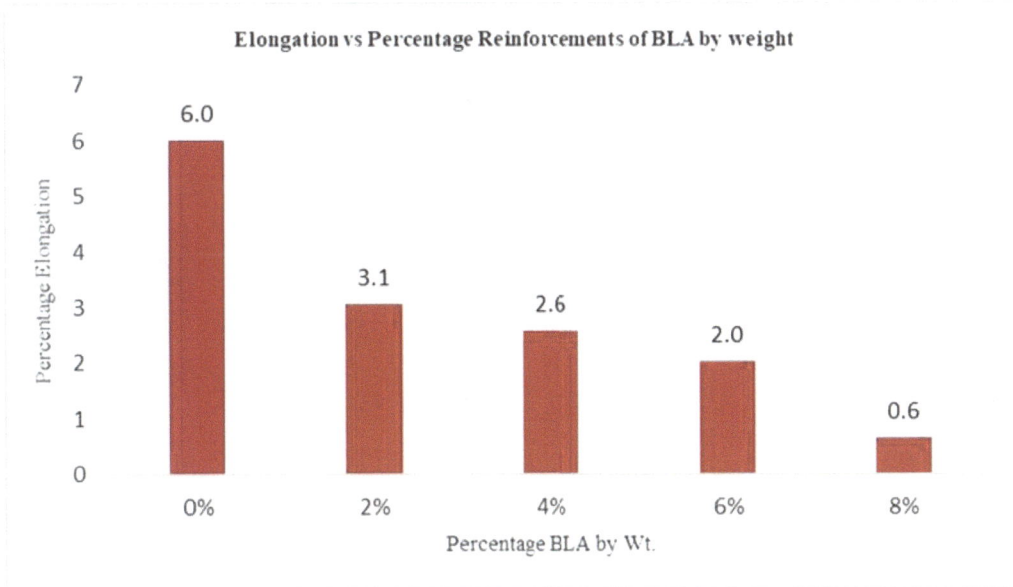

Fig. (10). Percentage Elongation Vs % of BLA reinforcements.

Impact Energy

The fracture toughness of a composite is determined by the amount of energy absorbed by the specimen of fracture. The impact energy responses are shown in Fig. (**11**). It is revealed that the cumulative increment in the volume percentage of the reinforcements resulted in a diminishing trend of impact energy values. Due to the fact that the barricading effect of reinforcements on the movement of dislocations causes the increase in the rate of work hardening, consequently decreasing the toughness values is the primary reason for the decrement in the impact energy of the composites [17, 18].

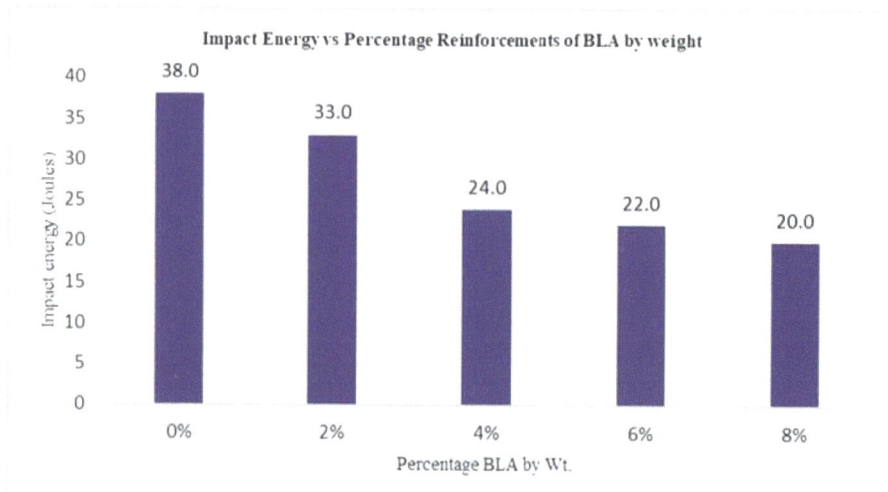

Fig. (11). Impact Energy Vs % of BLA reinforcements.

CONCLUSION

In the present study, Al 7075-BLA MMC's are successfully fabricated with reinforcements added to metal matrix in varying volume percentages by wt. The following conclusions are inferred from the investigations:

1. The microstructures of MMC's are fairly uniform, and there is a significant grain refinement inside the matrix.
2. The hardness of the MMC's increased significantly with an increase in the volume percentage of reinforcements. The highest hardness was achieved for Al 7075-BLA, *i.e.,* 103.50 BHN.
3. The Tensile and UYS values have increased significantly with an increase in reinforcement particulate composition. Al 7075 BLA has shown over 38% increment in strength to the base metal.
4. The elongation of Al 7075-BLA composites decreased drastically when compared to the base metal. However, the ductility retention was fairly minimum for Al 7075-BLA and Al 7075-B4C-BLA composites.
5. Fracture toughness has been improved significantly for Al 7075 BLA composite due to better interfacial reaction bonding.

REFERENCES

[1]   W.S. Lee, W.C. Sue, and C.F. Lin, "The effects of temperature and strain rate on the properties of carbon-fiber-reinforced 7075 aluminum alloy metal-matrix composite", *Compos. Sci. Technol.,* vol. 60, no. 10, pp. 1975-1983, 2000.

[2]   A. Mortensen, and J. Llorca, "Metal Matrix Composites", *Annu. Rev. Mater. Res.,* vol. 40, no. 4, pp.

243-270, 2010.

[3]   M.O. Bodunrin, K.K. Alaneme, and L.H. Chown, "Aluminium matrix hybrid composites: A review of reinforcement philosophies; mechanical, corrosion and tribological characteristics", *J. Mater. Res. Technol.,* vol. 4, no. 4, pp. 434-445, 2015.

[4]   X. Zhang, Y. Chen, and J. Hu, "Recent advances in the development of aerospace materials", *Prog. Aerosp. Sci.,* vol. 97, pp. 22-34, 2018.

[5]   K. Shirvanimoghaddam, H. Khayyam, H. Abdizadeh, M.K. Akbari, A.H. Pakseresht, F. Abdi, A. Abbasi, and M. Naebe, "Effect of B4C, TiB$_2$ and ZrSiO$_4$ ceramic particles on mechanical properties of aluminium matrix composites: Experimental investigation and predictive modelling", *Ceram. Int.,* vol. 42, no. 5, pp. 6206-6220, 2016.

[6]   N.E. Udoye, O.S.I. Fayomi, and A.O. Inegbenebor, "Realization of agro waste fiber-particulate for low cost aluminium based metal matrix composite: A review", *1st International Conference on Sustainable Infrastracture Development* Convant University, Canaan Land, Ota, Negeria pp.24-28, 2019.

[7]   B. Subramaniam, B. Natarajan, B. Kaliyaperumal, and S.J.S. Chelladurai, "Investigation on mechanical properties of aluminium 7075-boron carbide-coconut shell fly ash reinforced hybrid metal matrix composites", *China Foundry,* vol. 15, no. 6, pp. 449-456, 2018.

[8]   A. Chinnama hammad Bhasha, *Silicon,* vol. 14, no. 2, pp. 13-26, 2022.

[9]   C.H. Gireesh, K.D. Prasad, K. Ramji, and P.V. Vinay, "Mechanical characterization of aluminium metal matrix composite reinforced with aloe vera powder", *Mater. Today Proc.,* vol. 5, no. 2, pp. 3289-3297, 2018.

[10] O.O. Joseph, and K.O. Babaremu, "Agricultural waste as a reinforcement particulate for aluminum metal matrix composite (AMMCs): A review", *Fibers,* vol. 7, no. 4, pp. 33-41, 2019.

[11] J. Jose, T.V. Christy, P.E. Peter, J.A. Feby, A.J. George, J. Joseph, R.G. Chandra, and N.M. Benjie, "Manufacture and characterization of a novel agro-waste based low cost metal matrix composite (MMC) by compocasting", *Mater. Res. Express,* vol. 5, no. 6, p. 066530, 2018.

[12] R. Manikandan, and T.V. Arjunan, "Mechanical and tribological behaviours of aluminium hybrid composites reinforced by CDA-B4C", *Mater. Res. Express,* vol. 7, no. 1, p. 016584, 2020.

[13] B.P. Kumar, and A.K. Birru, "Characterization of Al-4.5% Cu alloy with the addition of silicon carbide and bamboo leaf ash", *Kovove Materialy-Metallic Materials,* vol. 56, no. 5, pp. 325-337, 2018.

[14] B.P. Kumar, and A.K. Birru, "Microstructure and mechanical properties of aluminium metal matrix composites with addition of bamboo leaf ash by stir casting method", *Trans. Nonferrous Met. Soc. China,* vol. 27, no. 12, pp. 2555-2572, 2017.

[15] K.K. Alaneme, and E.O. Adewuyi, "Mechanical behaviour of Al-Mg-Si matrix composites reinforced with alumina and bamboo leaf ash", *Metallurgical and materials engineering,* vol. 19, no. 3, pp. 177-188, 2013.

[16] J. Singh, N.M. Suri, and A. Verma, "Affect of mechanical properties on groundnut shell ash reinforced Al6063", *Int J Technol Res Eng,* vol. 2, no. 11, pp. 2619-2623, 2015.

[17] L. Poovazhagan, K. Kalaichelvan, and T. Sornakumar, "Processing and performance characteristics of aluminum-nano boron carbide metal matrix nanocomposites", *Mater. Manuf. Process.,* vol. 31, no. 10, pp. 1275-1285, 2016.

[18] C.U. Atuanya, A.T. Esione, and F.A. Anene, "Effects of Bamboo leaf stem ash on the microstructure and properties of cast Al–Si–Mg/Bamboo leaf stem ash particulates composites", *J. the Chin. Advan. Mat. Soc.,* vol. 6, no. 4, pp. 543-552, 2018.

CHAPTER 15

Experimental Investigation on the Joint Efficiency of Grit Blasted and Silica Particle Coated Adhesively Bonded Carbon and Glass Fibre Reinforced Polymer Composite Laminates

Mohammed Yusuf A. Yadwad[1,*], Vishwas G.[1] and **N. Rajesh Mathivanan[1]**

[1] Department of Mechanical Engineering, PES University Bangalore-560085, India

Abstract: Composite material is formed when one or more material is distributed or reinforced in a continuous second phase called a matrix. Composites have many superior properties, including low density, high strength-to-weight ratio, and good durability, which make them attractive in many industries. Composite materials have been used extensively in various applications. In any application where the strength-t--weight ratio plays a vital and important role, Fibre Re-inforced Polymer's (FRP) is the best material and offers the most efficient solution. Adhesive bonding is one of the most powerful joining techniques for FRP's because of its high mechanical properties. It has applications in all the fields like aerospace, marine technology, defence systems, and automotive industries, as well as structural applications and sports. However, the mechanical performance is biased undesirably by contaminants, like release agents, and also an excess of matrix in the top layer. In order to generate the most appropriate surface pre-treatment, their effect on adhesively bonded joints of carbon and glass fibre re-inforced polymer composite laminates have been investigated. The adhesively bonded surfaces are treated with grit blasting and silica particle coating and later tested in order to determine the failure modes. It was found that the mechanical properties of adhesively bonded joints depend on the surface characteristics of the substrate. The results indicate that it is possible to increase the bond strength of the joints to maximum by various surface treatments.

Keywords: Adhesive bonding, Carbon fibre re-inforced polymer, Grit blasting, Glass fibre re-inforced polymer, Surface treatment, Silica particle coating.

INTRODUCTION

Fibre re-inforced polymer composite laminates can be used to get net-shaped manufactured parts. However, there is a challenge in getting parts produced by

---

* **Corresponding author Mohammed Yusuf A. Yadwad:** Department of Mechanical Engineering PES University Bangalore-560085, INDIA; E-mail: mdyusufyadwad@gmail.com

Amar Patnaik, Albano Cavaleiro, Malay Kumar Banerjee, Ernst Kozeschnik & Vikas Kukshal (Eds.)

net-shape manufacturing, thereby it leads to products produced near net-shape manufacturing. Secondary manufacturing processes are inevitable to parts produced using FRP composite laminates. Machining and joining are some of the commonly performed secondary operations on composite laminates [1].

The annual demand for Carbon and Glass Fibre Reinforced Polymer (CFRP/GFRP) composite laminates has been steadily increasing, essentially in aircraft, defence, marine technology, sports goods, *etc.* and many other engineering applications. The automotive sector has made the proper utilization of composite laminate adhesive bonding hence the demand for the light weight. It has the application in automobiles like interior panels made up of aluminium frame; another example is they are used in the oil and gas industry where every single weight adds up to the platform, resulting in the outcome of the fuel and cost of the oil that has pumped and stored. The adhesive bonding is feasible for both the movement of the platform and the actual of the crude oil and hydrocarbon at very high temperatures. Hence, the adhesive joining of composite materials is very widely searching its place into many new things which exits the old methods by conventional fastening. Composites are retained to reduce the weight of the products and improve the aesthetic and environmental resistance, which permits the improved structure and strong design to weight criteria [2 - 4].

Adhesive bonding plays a vital role in joining composite laminates. Adhesive bonding is a material joining process in which an adhesive is placed between the adherend and surfaces, and solidifies to produce an adhesive bond. Recently, adhesively bonded joints have overtaken mechanical joints, such as mechanical fasteners, the conventional fasteners often result in the cutting of fibres which will give rise to stress concentration by reducing the structural integrity [5, 6]. Particularly with regard to composite in fields of engineering application by considering the advantages like higher strength-to-weight ratio, low cost of fabrication and higher damage tolerance. Four step bonding procedure proposed by Kyungtae Kim *et al.* enhanced greater damage protection as well as the accessibility of application to the conventional method [7].

Adhesives are used for bonding a wide variety of comparable and different metallic and non- metallic materials, composites, and apparatuses with various shape, size, and thickness. The merits of bonding with adhesive over conventional joining or bonding like mechanical closure, which is fastening, is now acknowledged. Specific adhesives deliver greater design flexibility, allocate load over a larger area hence decreasing the stress concentration as well as improving the fatigue and corrosion resistance of the bonded joints. The adhesive selection is preferred bond material for the given application. There is a need to choose the proper adhesive because various adhesives are available, and all the properties of adhesives match themselves. So, the selection requires knowledge properties of selected adhesives and where the adhesive is being utilised. The assortment of adhesive process is very problematic because no adhesives are used as a universal adhesive for all the methods and the selection of the adhesives are very complex manner because there are a number of adhesives available in the market. However, the selection of an adhesive is the main factor, such as the nature and type of the adherend surface to be joined and curing the adhesive for the application of expected manner and adhesive related to the environment. Hence the price of adhesive is a very important criterion for the selection in the situation of production [8].

The joining surface plays a dominant role in the process of adhesive bonding, which influences the quality of an adhesive joint. The pertinent surface pre-treatment can often confer additional properties to the surface of the bond. Davis et at. [9, 10] strongly insists to have surface pre-treatment prior to the application of adhesives is recommended to achieve maximum mechanical strength. The appropriate surface pre-treatment has a direct influence on the strength of the bond. Composite laminates usually suffer a cohesive kind of failure, perhaps the failure is largely associated with the surface preparation of bonded joints. The challenge in using adhesive for carbon fibre reinforced polymer bonded joints is because of the presence of a thick polymer layer of 2-10μm on the top of the surface of the laminates [11].

Based on the loading conditions such as shear and peel strength, the toughness of the material, ductile and fragile behaviour as well as high fatigue cycles, creep, strength-to-weight ratio and stiffness to weight ratio and wetting properties of the surface of the bond and environmental degradation and resistance, the strength of the adhesive bond strongly depends upon the type of the adhesive. The increased utilization of the resin matrix reinforcement for the composite material is very necessary to develop a feasible system of adhesive. The epoxy adhesive that is usually used for the composite matrix is applied for the bonding of composites based on the adhesives and due to the mixing of resin and adhesive [12].

Surface pre-treatment is followed by the important and crucial step in the adhesive joining process; proper adhesive bond testing is performed to improvise the surface pre-treatment. Choosing surfaces to be pre-treated mainly depends on the desired durability and strength of the bond, even if taking economy like time as well as the cost involved in the preparation, which plays an important role selection. Perfect pre-treatment is very necessary for better strength of the joint and maintaining the ever-lasting integrity of the structural joints bonded. Poor pre-treated surface results in the adhesive joint failure of the bond, which will occur in the adhesive and substrate interface. Surface pre-treatment is followed to provide the consistent and accuracy of the surface of the bond and also the quality of the substrate. To reduce the formation of the boundary layer, for example, the weak boundary layer, the material attached weakly comes out after the abrasion method which have moved towards the surface of the polymer composite and also the mould the agent released from peel-ply [13]. To achieve the proper and better bond, the molecular contact between the adhered and the adhesive is necessary. The proper selection of the surface pre-treatment will optimize the contact degree between which the chemical method and the change of substrate surface. To achieve the proper and better substrate surface topography also changes the profile of the surface and also improvising the area of the surface to be bonded and to be abraded surface [14].

It is mandatory to safeguard the substrate surface before the bonded joints as a developed surface is very highly tangible towards the adhesives and also the contamination [15]. To hold the integrity of the surface of the adherend, it is required to bond the surface with less amount of the time of pre-treatment and also to develop primer which is feasible with the substrate to be coated. The theory behind the adhesion is more complicated and be developed for the other collaborated to mechanical interlocking and Vander walls theory also the diffusion theory *etc* [16].

OBJECTIVES

1. To prepare the coupons of carbon and glass fibre reinforced polymer composite laminates according to ASTM D5868-01 standards.
2. To evaluate the influence of grit blasting as a type of mechanical surface pre-treatment for adhesively bonded joints.
3. To evaluate the influence of silica particle coating as a type of chemical surface pre-treatment for adhesively bonded joints.
4. To examine the failure characterisation and the suitable surface pre-treatment method for adhesively bonded joints.

EXPERIMENTAL INVESTIGATION

Material Properties

The materials used here are woven type carbon fibre, and the matrix materials are Bisphenol-A based epoxy resin LY556 (200 gms) and hardener HY955 mixed in appropriate ratio with room temperature. A catalyst is used to increase the rate of a chemical reaction without itself being changed at the end of a chemical reaction. The promoter is used to enhance the action of the catalyst without actually having a catalytic value itself. The adhesive used here is 3M scotch weld DP-420 with two parts for joining the specimens. Silica particles with a diameter 200-300nm and N, N-dimethyl formamide is DMF solution, were used for the coating of the surfaces of composite specimens. Acetone is utilised to clean the surfaces of the composite specimens. All the chemicals were utilized and performed as received from the chemist without any additional purification processes.

Specimen Preparations

The best suitable fabrication technique for GFRP laminates or any other fibre reinforced polymer composites is the hand layup process. Hand layup process is a simple process compared to other processes. In this process, initially release agent or releasing gel is applied to mould surface. Generally, metallic or ceramic moulds are used in the fabrication process. The matrix material is mixed in proper proportion as specified by the manufacturer. The woven fibre mates are cut as per the mould size, and many layers are placed one after the other to achieve the desired thickness. The matrix polymer material is applied evenly on the surface of each woven mat layer. Once the layers are stacked up, the breathing sheet is covered on the top surface and applied for a vacuum pressure of 6 bar at room temperature; force is applied on the top surface to get a compressive pressure. Fabrication of CFRP laminate is carried out as the same method as the fabrication of GFRP laminates as shown in Fig. (**1**). The woven type fabrics are used and cut in the shape of mould; the required number of layers are stacked up using the resin mixture. They are allowed to cure for 24-48hrs at normal room temperature. GFRP and CFRP samples are cut using CNC milling using ASTM D5868; the edges of the samples are filed accurately. The specimens are cleaned using acetone to remove the dust particles. The dimensions of the specimens are checked using a vernier caliper, and they are maintained for 101mm x 24.2mm. The prepared specimens are then joined using 3M scotch weld DP420 structural epoxy adhesive. The standards are referred to for the joining process, "ASTM D5859" gives the procedure for preparing single lap joints (SLJ) of FRP to FRP [17].

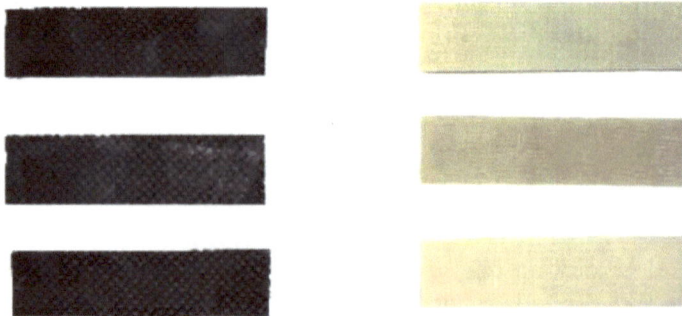

Fig. (1). Specimens prepared of CFRP and GFRP laminates [20].

Surface Preparations

Grit Blasting

The surface pre-treatments carried out in this work are aluminium grit blasting and silica particle coating as shown in Fig. (**2**). The specimens were thoroughly cleaned with acetone, and grit blasted using an Aluminium oxide blast having a grit size of G40, maintaining the air pressure of 3 bar through an 8mm diameter spray nozzle. The nozzle was held perpendicular to the surface and moved back and forth at a distance of around 300mm. To remove alumina particles produced during grit blasting as shown in Fig. (**3**), the specimens were washed with acetone [18].

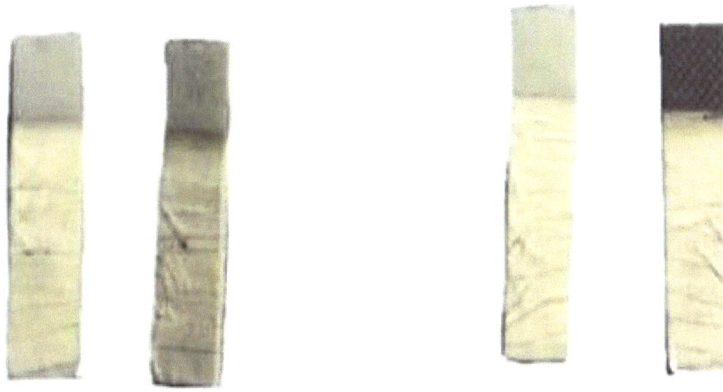

Fig. (2). Grit Blasted CFRP and GFRP specimens [16].

Fig. (3). Aluminium oxide grit blasting machine, *source: internet* [16].

Silica Particle Coating

The areas of adhesion of the carbon fibre-reinforced polymer coupons are thoroughly roughed by using sandpaper to increase the physical surface area, which will adsorb the silica particles in a very improved manner (Fig. **4**). The pre-treated coupons were thoroughly cleaned with acetone and then dried in the oven. Simultaneously, the silica particles are submerged in a dimethyl formamide-based solvent and then sonified to get a homogenous colloidal solution. Ultrasonic pulverization of 5-10min is conducted for a solution to be made properly without any chemical imbalance and an optimal silica particle concentration is prepared. The prepared solution is applied on the adhesion areas of the carbon fibre where the 8ml of the silica solution is applied into the adhesion area according to the standards ASTM 5868 which is the area of 25mm x 25mm in size to make the silica particle solution with $0.25mg/cm^2$ concentration. The DMF solvent is later allowed to evaporate at room temperature slowly for a duration of 30min [7].

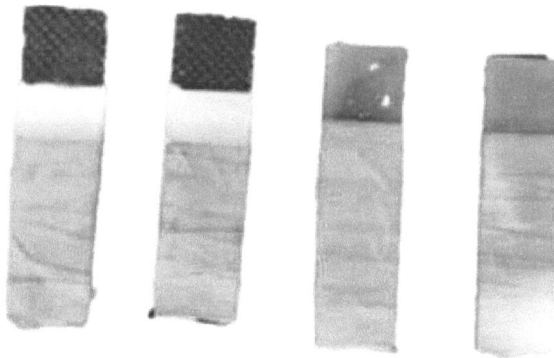

Fig. (**4**). Specimens coated with silica particle coating [7].

Adhesive Bonding

3M DP420 structural epoxy adhesive is used for this research work; it is a two-part adhesive generally used for joining of FRP's and also for metals. It exhibits properties like high strength, excellent tensile properties, resistance to any environment extreme, and easy mixing with a work-life of 20min. The manufacturer represents that this adhesive is a low halogen content adhesive [19].

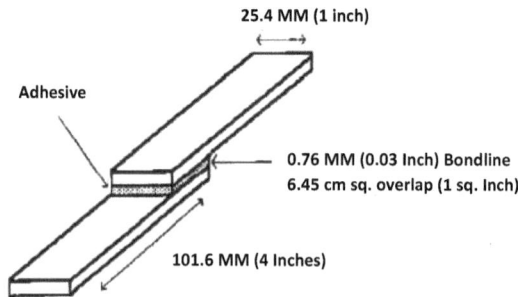

Fig. (**5**). Lap shear overlay according to ASTM D5868-01 standard [13].

There is a method of joining process for the various application. The main method is that the method of joining is to be selected based on cost, weight, and also performance (Fig. **5**). Apply the adhesive at the adhesion area and prepare a single lap joint according to the ASTM D 5868-01 by mixing the two parts of adhesive according to supplier procedure, cure the adhesive for at least 6 hours, control the joint geometry by appropriate fixturing, using bench vice and clamps and maintain the bondline thickness, allow the bonded parts to be cooled and cured completely before testing the specimen (Figs. **6** and **7**).

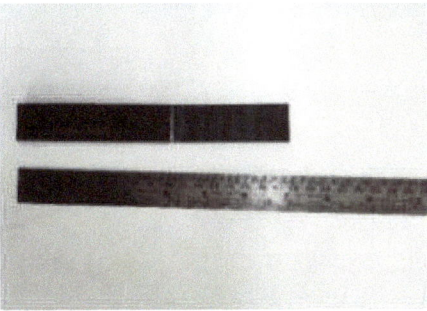

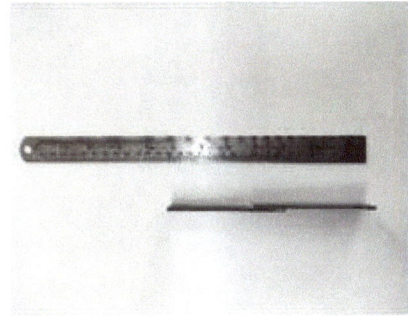

Fig. (6). Adhesively bonded Single Lap Joint of CFRP [13].

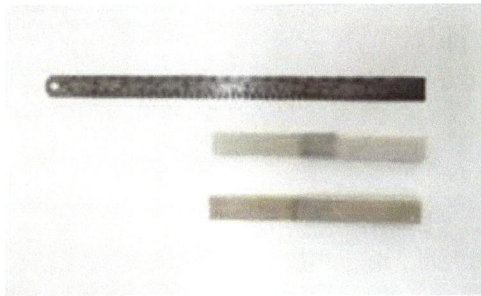

Fig. (7). Adhesively bonded Single Lap Joint of GFRP [13].

Overlap Shear Test

The determination of the efficiency of adhesively bonded single lap joint is done by overlap shear strength; the pull test was performed using the UTM machine in FIE Make Universal Testing Machine (Fig. **8**). Alignment tabs of 20mm x 20mm are used to maintain proper alignment during the pull test. UNITEK-9550 under the cross-head speed of 13mm/min, proper instructions are followed as per ASTM D5868 standard.

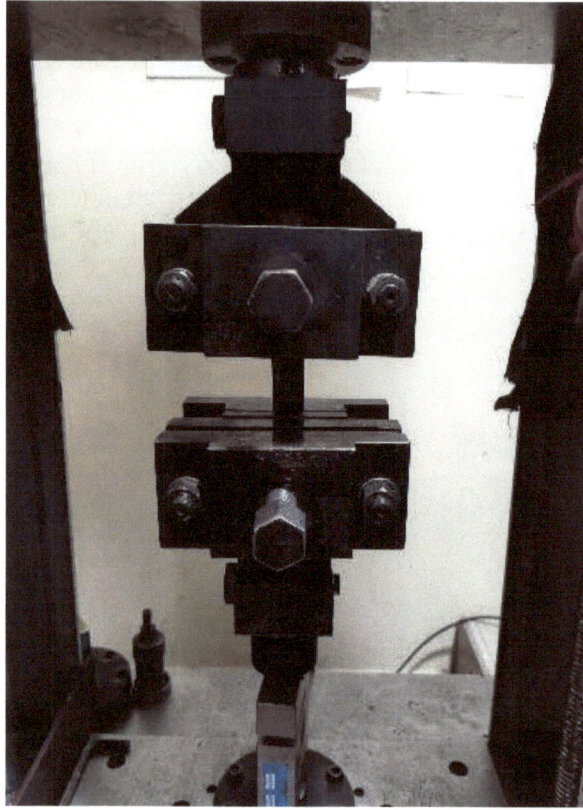

Fig. (8). Testing of specimens in Universal testing machine [14].

RESULTS AND DISCUSSION

The test results are tabulated in Table 15.1, where we observe that the maximum force measured is 5.363 kN at 0.700 mm displacement for CFRP to CFRP and the fracture occurred at 0.810 mm displacement. For GFRP to GFRP, the maximum force measured is 6.198 kN and the maximum displacement measured is 1.140 mm, and the specimen fractured at 1.250 mm. displacement. The test results for silica particle coating are tabulated below in Table. **1** where the maximum force measured is 6.145 kN at a maximum displacement of 1.270 mm. for CFRP-CFRP bond and for the GFRP-GFRP the measured maximum force is 2.335 kN at 0.980 mm. displacement.

Table 1. Test results for adhesively bonded specimens [12]

Materials	CFRP		GFRP	
Parameters	Grit Blasting	Silica Coating	Grit Blasting	Silica Coating
Thickness (mm)	2	2	2	2
Maximum force (kN)	5.363	6.145	6.198	1.127
Displacement (mm)	0.700	1.160	1.140	0.200
Maximum Displacement(mm)	0.810	1.270	1.250	0.260

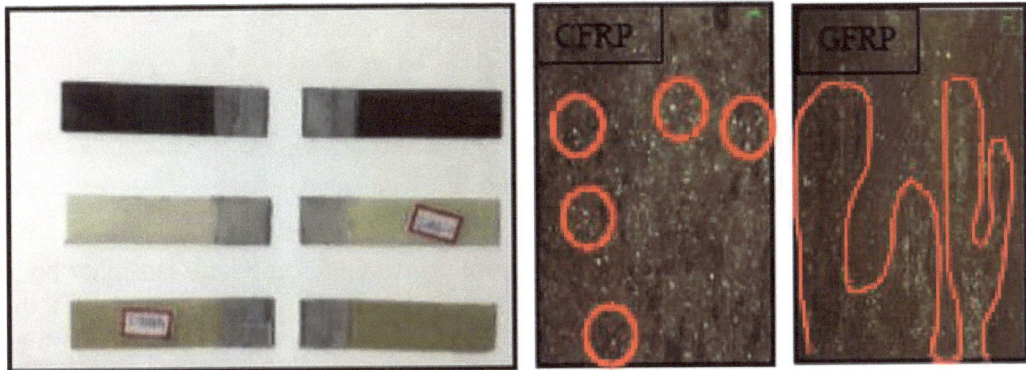

Fig. (9). Failure mode characterisation of grit blasted specimens [12].

The surface pre-treated with grit blasting on CFRP resulted in adhesive failure and exhibited the maximum fracture load value of 5.363kN (Fig. **9**). But in the case of GFRP, it was found that the fibres of GFRP were exposed towards the surface of adhesion during the grit blasting process and the adhesive was totally submerged with the substrate, resulting in a cohesive failure and yielded the maximum fracture load value of 6.198kN. Due to the fabric material of the CFRP, the adhesive was unable to adhere to the substrate, which did not result in a micro ploughing of the substrate, unlike GFRP which was bearing brittle properties that expelled micro ploughing and tend to the maximum load carrying capacity.

The CFRP bond yielded 6.145kN as full fracture load when Silica particle coating was used as a surface pre-treatment, but the GFRP bond yielded 1.127kN. The scale, form, and dispersion of the silica particle treated onto the surface of CFRP and GFRP exteriors were studied. With silica particle coating, uniform distribution of spherically formed silica particles on the CFRP bond's surface was observed. Furthermore, the silica particles are evenly distributed on the CFRP surface and do not form a single coat. Due to the crosslinking density, this was not found in the case of GFRP bonds. The CFRP surface responds as a filler material due to the silica particle coating, increasing cross-linking density and providing a strengthened effect, consequential in strong crack development inhibition. The intensity at which the silica particles adsorbed on top of the surface of the CFRP was such that they did not originate off in any circumstances was important (Fig. **10**). Clearly indicates the uniform distribution of silica particle coating.

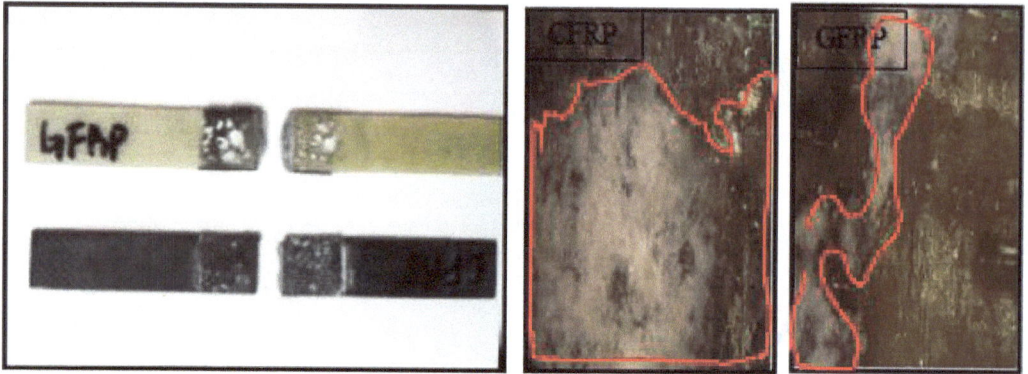

Fig. (10). Failure mode characterisation of silica particle coated specimens [12].

Fig. (**11**) indicates the graph for carbon fibre reinforced composite laminate bond using surface pre-treatment as grit blasting, which clearly indicates that the maximum force achieved is 5.363 kN at a displacement of 0.700 mm and steep fall leading to fracture at 0.810 mm maximum displacement, whereas in silica particle coating using surface pre-treatment the maximum force achieved is 6.145 kN at 1.160 mm displacement and steep fall leading towards fracture at a maximum displacement of 1.270 mm.

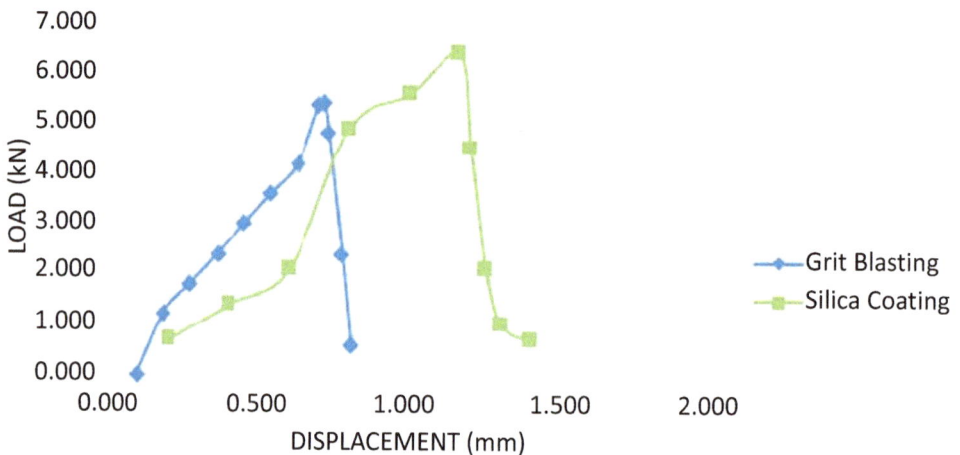

Fig. (11). Load vs Displacement graph for surface treated CFRP laminates [4].

Fig. (**12**) indicates the graph for the GFRP bond using grit blasting as surface treatment, the maximum force achieved is 6.198 kN at 1.140 mm displacement, where as for silica particle coating, the maximum force achieved is 1.127 kN at 0.200mm at a maximum displacement of 0.260mm.

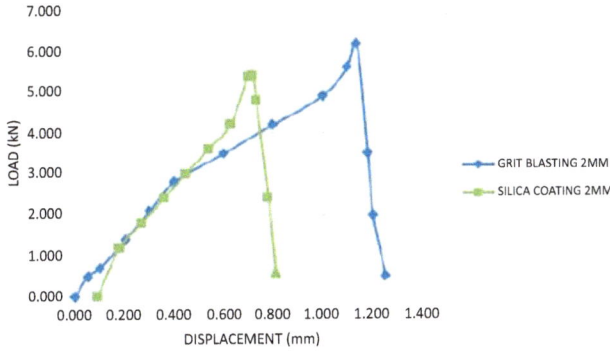

Fig. (12). Load *vs.* Displacement graph for surface-treated GFRP laminates [15].

The graph is plotted in Fig. (**13**), for CFRP bond using surface pre-treatment as grit blasting, which clearly indicates that the maximum force achieved is 5.363 kN at a displacement of 0.700 mm and steep fall leading to fracture at 0.810 mm maximum displacement, whereas in GFRP bond the maximum force achieved is 6.198 kN at 1.140 mm displacement and steep fall leading towards fracture at a maximum displacement of 1.250 mm.

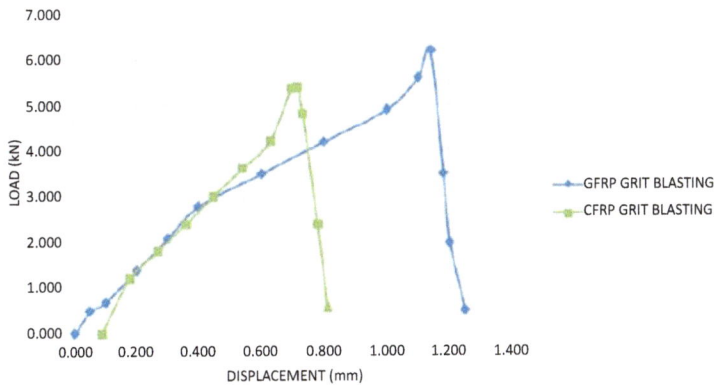

Fig. (13). CFRP and GFRP surface treated with grit blasting [17].

Fig. (**14**) represents the graph for CFRP bond using surface pre-treatment as silica particle coating, which clearly indicates that the maximum force achieved is 6.145 kN at a displacement of 1.160 mm and steep fall leading to fracture at 1.270 mm maximum displacement, whereas in GFRP bond, the maximum force achieved is 1.127 kN at 0.200 mm displacement and steep fall leading towards fracture at a maximum displacement of 0.260 mm.

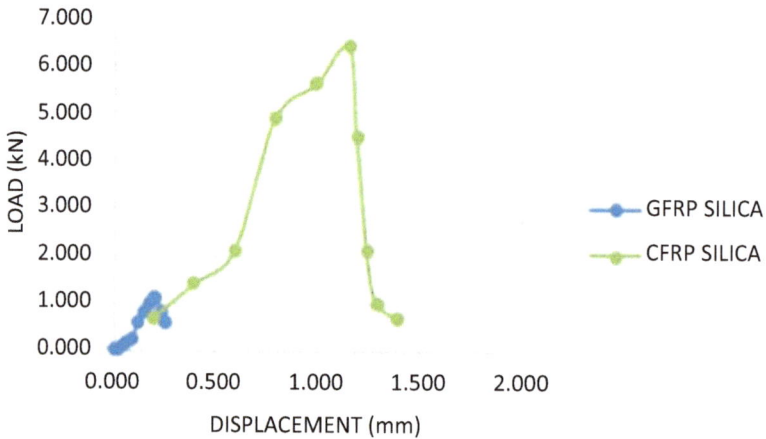

Fig. (14). CFRP and GFRP laminates surface treated with silica particle coating [10].

CONCLUSION

Experimental Investigation on the Joint Efficiency of Grit Blasted, and Silica Particle Coated Adhesively Bonded Carbon and Glass Fibre Composite Laminates led to the following conclusions:

- The surface pre-treated with silica particle coating on CFRP-CFRP adhesively bonded joints exhibited the highest tensile strength yielding a value of 6.145 kN, which is 12.72% more when compared to the surface pre-treated with grit blasting which exhibited the lowest value of 5.363 kN
- GFRP-GFRP adhesively bonded joints treated with grit blasting exhibited the highest tensile strength yielding a value of 6.198 kN, which is 81.81% more when compared to surface pre-treated with silica particle coating; the tensile strength exhibited the lowest value of 1.127 kN
- When compared with GFRP-GFRP and CFRP-CFRP, the surface pre-treated using grit blasting resulted in improved tensile strength in GFRP-GFRP bonds

of about 13.47%

- When compared with GFRP-GFRP and CFRP-CFRP, the surface pre-treated using silica particle coating resulted in improved tensile strength in CFRP-CFRP bonds about 81.65%.

CONSENT OF PUBLICATION

We, give our consent for the publication of identifiable details, which can include photograph(s) and/or videos and/or case history and/or details within the text ("Material") to be published in the above Journal and Article.

ACKNOWLEDGEMENT

We would like to thank PES University Bangalore, for all the support, from the preparation of specimens to testing and microscopic surface analysis facilities.

REFERENCES

[1]   C. Worrall, E. Kellar, and C. Vacogne, *Joining of fibre-reinforced polymer composites a good practice guide.*, 2020, pp. 1-77.

[2]   G. Jeevi, S.K. Nayak, and M. Abdul Kader, "Review on adhesive joints and their application in hybrid composite structures", *J. Adhes. Sci. Technol.,* vol. 33, no. 14, pp. 1497-1520, 2019.
[http://dx.doi.org/10.1080/01694243.2018.1543528]

[3]   J. Lisiecki, M. Łukasik, P. Synaszko, D. Nowakowski, and P. Orzechowski, "Influence of surface preparation in composite bonded joints", *Compos. Theory Pract,* pp. 234-237, 2016.

[4]   K.B. Sim, "Enhanced surface properties of carbon fibre reinforced plastic by epoxy modified primer with plasma for automotive applications", *Polymers,* vol. 12, no. 3, 2020.
[http://dx.doi.org/10.3390/polym12030556]

[5]   W. Wang, J.A. Poulis, S. Teixeira De Freitas, and D. Zarouchas, "Surface pre-treatments on CFRP and titanium for manufacturing adhesively bonded biomaterial joints", 18th Eur. Conf. Compos June, 2020.

[6]   J. Bardis, and K. Kedward, *Effects of surface preparation on long-term durability of composite adhesive bonds. report No. DOT/FAA/AR-01/08,* p. 23, 2001.

[7]   S. B. Kumar, *Adhesive Bonding in Composite Structures for Aerospace Applications,* Doctoral thesis, Nanyang Technological University: Singapore, 2006.

[8]   D.K. Rajak, D.D. Pagar, P.L. Menezes, and E. Linul, "fibre-reinforced polymer composites: Manufacturing, properties, and applications", *Polymers,* vol. 11, no. 10, 2019.
[http://dx.doi.org/10.3390/polym11101667]

[9]   A.V. Pocius, "The Surface Preparation of Adherends for Adhesive Bonding", *Adhes. Technol,* pp. 181-218, 2012.

[10] K. Kim, Y. C. Jung, S. Y. Kim, B. J. Yang, and J. Kim, "Adhesion enhancement and damage protection for carbon fibre-reinforced polymer (CFRP) composites via silica particle coating," *Compos. Part A Appl. Sci. Manuf.,* vol. 109, no. March, pp. 105–114, 2018
[http://dx.doi.org/10.1016/j.compositesa.2018.02.042]

[11] V.V. Prasad, and P.D. Maganti, "Mechanical characterization of carbon fibre reinforced composite - Variety of single lap joint", *Mater. Today Proc.,* vol. 21, pp. 1149-1154, 2020.
[http://dx.doi.org/10.1016/j.matpr.2020.01.064]

[12]   M. Schweizer, D. Meinhard, S. Ruck, H. Riegel, and V. Knoblauch, "Adhesive bonding of CFRP: A comparison of different surface pre-treatment strategies and their effect on the bonding shear strength", *J. Adhes. Sci. Technol.,* vol. 31, no. 23, pp. 2581-2591, 2017.
[http://dx.doi.org/10.1080/01694243.2017.1310695]

[13]   J. Specimen, and T. Results, "Results, ASTM D5868-01.pdf", pp. 5-6.

[14]   H.Y. Kebede, T.M. Kim, and D.H. Bae, "Tensile strength assessments of CFRP adhesive bonded joint", *Int. J. Adv. Technol.,* vol. 08, no. 02, 2017.
[http://dx.doi.org/10.4172/0976-4860.1000180]

[15]   R.J. Zaldivar, H.I. Kim, G.L. Steckel, J.P. Nokes, and D.N. Patel, "The effect of abrasion surface treatment on the bonding behaviour of various carbon fibre reinforced composites", *J. Adhes. Sci. Technol.,* vol. 26, no. 10–11, pp. 1573-1590, 2012.
[http://dx.doi.org/10.1163/156856111X618425]

[16]   G. Kelly, Joining of Carbon Fibre Reinforced Plastics for Automotive Applications. 2004.

[17]   L.F.M. da Silva, T.N.S.S. Rodrigues, M.A.V. Figueiredo, M.F.S.F. de Moura, and J.A.G. Chousal, "Effect of adhesive type and thickness on the lap shear strength", *J. Adhes.,* vol. 82, no. 11, pp. 1091-1115, 2006.
[http://dx.doi.org/10.1080/00218460600948511]

[18]   G. Di Bella, G. Galtieri, E. Pollicino, and C. Borsellino, *Joining of GFRP in marine applications,* no. December, p. 2013, 2012.

[19]   M.A. Wahab, "Joining composites with adhesives theory and applications", 2016, p. 316.

[20]   P. Banakar, "Preparation and characterization of the carbon fibre reinforced epoxy resin composites", *IOSR J. Mech. Civ. Eng.,* vol. 1, no. 3, pp. 15-18, 2012.
[http://dx.doi.org/10.9790/1684-0131518]

CHAPTER 16

Colossal Dielectric Properties Of $(Ta_{0.1}Sm_{0.9})_{0.04}Ti_{0.96}O_2$/PVDF Composites For Energy Storage Applications

Dileep Chekkaramkodi[1,*], Muhammed Hunize Chuttam Veettil[1] and **Murali Kodakkattu Purushothaman[1]**

[1] *Department of Mechanical Engineering, National Institute of Technology Calicut, Calicut-673601, India*

Abstract: In this study, $(Ta_{0.1}Sm_{0.9})_{0.04}Ti_{0.96}O_2$/Polyvinylidene fluoride composites were synthesized and analyzed for colossal dielectric properties. The ceramic powder was prepared by solid-state ceramic route, confirmed its phase purity through an X-ray diffractometer, and composites with different volume fractions were synthesized by finely dispersing the filler in the Polyvinylidene fluoride (PVDF) matrix followed by compression molding. Dielectric properties (dielectric constant and loss) up to 1 MHz were studied using an impedance analyzer. A high dielectric constant of 45 along with an acceptable loss of 0.089 was obtained for an optimum filler volume of 50% at 1 kHz. Hence the composites can be effectively used for energy storage applications.

Keywords: Ceramics, Colossal permittivity, Dielectrics, PVDF, Impedance analyzer.

INTRODUCTION

Polymers having excellent dielectric performance are used in various fields such as electrical engineering, microelectronics, high energy density storage, etc [1]. In electronic circuits, polymer dielectric materials are suitable owing to their mechanical flexibility, easiness of processing and low cost. However, for applications requiring good dielectric permittivity (ε') and low loss ($\tan\delta$), polymer materials are not desired, as it has a very low dielectric constant compared to ceramic materials. However, ceramic materials have drawbacks of difficulty in fabrication into larger size and brittle nature. Ceramic/polymer composites can surmount these disadvantages and are generally used for these purposes.

Corresponding author Dileep Chekkaramkodi: Department of Mechanical Engineering PES University Bangalore-560085, India; E-mail: mdyusufyadwad@gmail.com

Amar Patnaik, Albano Cavaleiro, Malay Kumar Banerjee, Ernst Kozeschnik & Vikas Kukshal (Eds.)

$CaCu_3Ti_3O_{12}$(CCTO) and its derivatives are non-ferroelectric oxide ceramics exhibiting good dielectric permittivity in the range of 10^4-10^6 [2]. Hence, they are widely used to enhance the dielectric performance of various polymer materials. However, they possess comparatively a high dielectric loss (> 0.1). Co-doped TiO_2 ceramics are other groups of dielectric materials that show good dielectric permittivity and low loss [3 - 6]. Compared to the CCTO family, the high activation energy of the co-doped TiO_2 is responsible for this low dielectric loss [7].

Recently, various elements have been co-doped in TiO_2 for tuning its dielectric properties. Among these, Ta and Sm co-doped TiO_2 show good dielectric properties [8]. TiO_2 shows a dielectric constant of 140 at 1 kHz. Co-doping of Ta would increase the dielectric constant and loss, whereas the presence of Sm reduces the dielectric loss. In this family, $((Ta_{0.1}Sm_{0.9})_{0.04}Ti_{0.96}O_2)$ is a promising ceramic material, which is having ε' of 703.71 and tanδ of 0.08 at 1 kHz at room temperature. In this study, to exploit the advantages of a heterogeneous system, a ceramic/polymer composite of Ta and Sm co-doped TiO_2 (TSTO)/PVDF has been developed. TSTO powders prepared in two different particle sizes have been used as filler material to study the effect on packing density, water absorption, and dielectric properties of the composite material.

Experimental Part

Preparation of TSTO

TSTO ceramic powder was prepared via conventional solid-state reaction using TiO_2 (99.9% purity, Merk), Ta2O5 (99.98% purity, Sigma Aldrich) and Sm2O3 (99.9% purity, Sigma Aldrich). The raw materials were weighed according to stoichiometric ratio and mixed for an hour using agate mortar in de-ionized water medium to ensure uniform mixing. It was oven dried to remove water content. The powder thus obtained was calcined at 1100°C for 2 hours. Then the obtained powder was checked for its phase purity and carefully ground. Half of the powder thus prepared was ball-milled using 3mm diameter zirconia balls for 12h at 250 rpm to reduce particle size. Ball: powder ratio was 8:1.

Preparation of Polymer/Matrix Composites (PMCs)

TSTO powders with two different particle sizes were used as the filler materials. PVDF polymer (Alfa Aesar) was chosen as a polymer matrix. Finely dispersed TSTO in the PVDF matrix was prepared at various filler fraction (fTSTO = 0.1 to 0.5) by ball milling in an ethanol medium for 2h. Then the mixture was heated to evaporate ethanol to form a composite powder. Polymer composite samples were prepared by hot pressing the composite powder at 220 °C for 20 min, at a pressure of 10 MPa, in a cylindrical mould of 10 mm size.

Characterization

An X-ray diffractometer (Panalytical, Aeris) was employed to analyse the crystal structure and phase purity of the TSTO powder. A particle size analyser (Malvern Panalytical's master sizer 3000) was used to analyse the size and distribution of the fillers. Scanning electron microscopy (SEM) (Jeol 6390LA/OXFORD XMXN) was used to examine microstructure of the composite. Moisture absorption characteristics were studied as per the standard IPC-TM-650 2.6.2.1 method.

Dielectric Measurement

The composite samples were polished and silver coated for dielectric measurements. Impedance analyser (Agilent 4294A) was used to measure the capacitance and dissipation factor of the filler and composite using a parallel capacitance technique in the frequency range of 40 Hz to 1MHz at a 500mv oscillation voltage.

RESULTS AND DISCUSSIONS

Particle size analysis curves of the powders before (PA) and after (PB) ball milling are shown in Fig. (**1**). Before ball milling, the average particle size and specific surface area of the powder have been found at 5.15µm and1759 m^2/kg respectively, whereas after ball milling specific surface area has been increased to 10280 m^2/kg and average particle size reduced to 1.06µm.

The XRD pattern of PA and PB powders is illustrated in Fig. (**2**), which matches standard JCPDS pattern 21-1276 and clearly indicates the phase purity of the synthesized TSTO ceramic fillers. Compared to PA powder, the peak intensity of PB powder is found to be less. It is attributed to internal stress resulting from the lattice defects due to the ball milling. During the milling time, this lattice stress decreases the crystallinity of the material and the intensity of peaks gets reduced. The morphologies of PA and PB filler materials observed using SEM images are depicted in the inset (1) and (2), respectively.

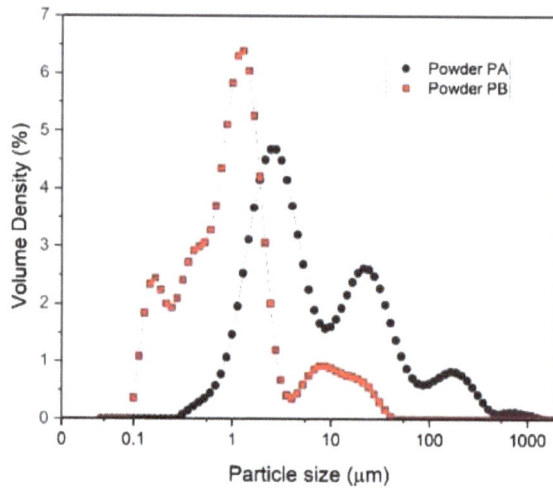

Fig. (1). Particle size distribution of calcined and milled TSTO.

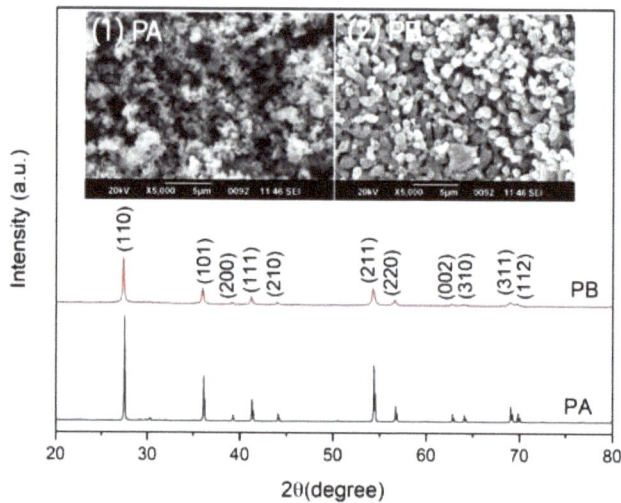

Fig. (2). Powder XRD patterns of TSTO with different particle sizes. Inset (1) and (2) shows SEM micrographs of PA and PB powders, respectively.

To get the dielectric properties of ceramic filler material, the compacted ceramic powder was sintered at 1500^0C for 6h. An optimal relative density of 98% is obtained for sintered TSTO. The dielectric properties of sintered samples are depicted in Fig. (3). It exhibits good dielectric performances at a wide frequency range. The ε' values of 703 and 582 have been obtained at 1 kHz and 1 MHz, and corresponding tanδ values are 0.08 and 0.008, respectively.

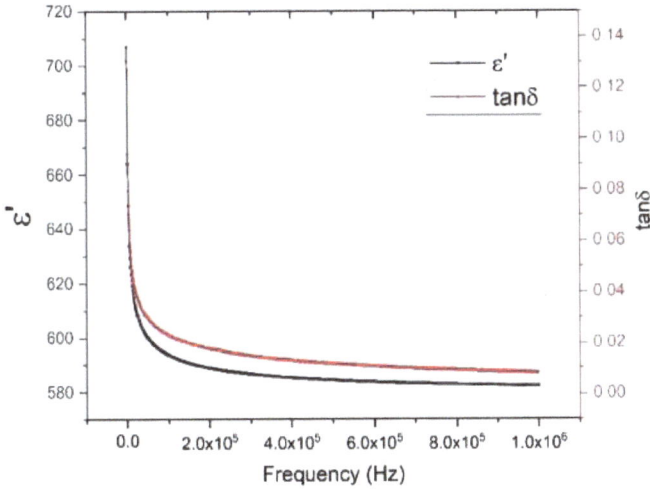

Fig. (3). ε′ and tanδ of TSTO filler sintered at 1500°C as a function of frequency at 25°C.

The cross-sectional SEM images of the brittle fractured (PA)- TSTO/PVDF and (PB)-TSTO/PVDF composite samples with fTSTO= 0.4 are illustrated in Fig. **(4a)** and Fig. **(4b)**, respectively. In the case of PB-filled PVDF, uniform dispersion and better packing of the filler particle in the PVDF matrix are observed than in the PA-filled PVDF because of the wider distribution and smaller particle size of the former.

Fig. (4). SEM micrographs of cross-section of (a) PA-TSTO/PVDF and (b) PB-TSTO/PVDF composites with fTSTO =0.4.

Fig. **(5)** shows a relation between filler fraction and relative density of the TSTO/PVDF composites. The densities of composites were measured by the

Archimedes method. Even relative density of (PA)-TSTO/PVDF and (PB)-TSTO/PVDF at 0.5 filler fraction were obtained as 90.37% and 91.06%, respectively. Generally, smaller-sized filler particles with multimodal distribution enhance better dispersion in the matrix compared to larger-sized ones, which can be justified by the obtained higher density of composites prepared using PB-TSTO fillers. At higher filler fractions, a reduction of density was observed, which is attributed to the induced porosity in the composite material.

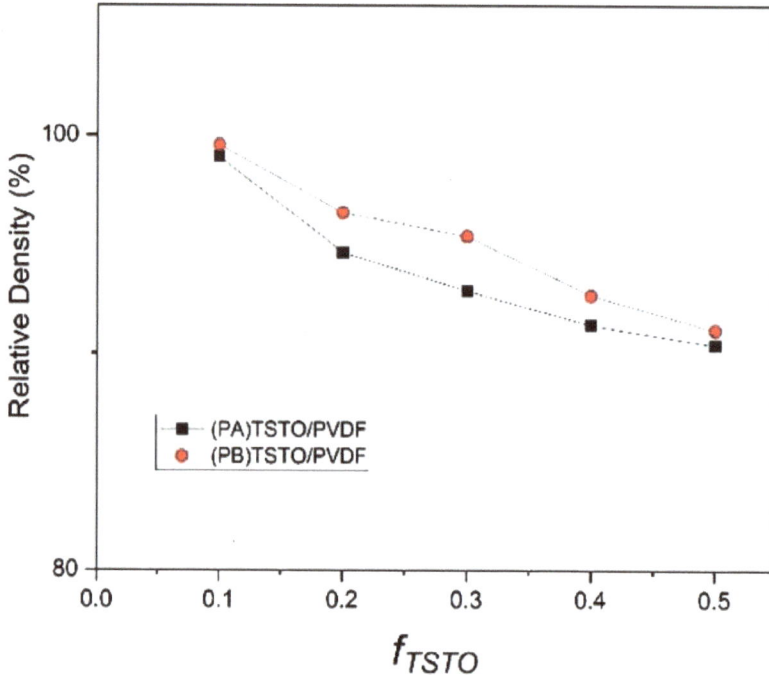

Fig. (5). Relative density of composites at various filler fractions.

The permittivity and loss of the TSTO/PVDF composites at different filler loading as a function of frequency are given in Fig. (**6a** to **d**). It has been observed that the ε' value of composites slightly varies with frequency. The ε' values of composites are found to increase with an increase in filler fraction due to the high ε' of TSTO ceramic particles and resulting enhancement in the TSTO/PVDF interfacial area. Under the application of an electric field, polarization developed in the TSTO/PVDF interface due to the blocking effect of free charges as explained in the Maxwell-Wagner-Sillars (MWS) effect is attributed to the increase in the dielectric permittivity with filler loading. As illustrated in Fig. 6. (b) and (d), tanδ of composites significantly varies with filler size. A reduction in tanδ was observed in a low-frequency range of 50 Hz to 1000 Hz, followed by an increase in dielectric loss with frequency. The DC conduction of free charge carriers in the

TSTO/PVDF interface is responsible for this effect. In the case of composites, lower tanδ values have been obtained at higher filler fractions due to low tanδ values of ceramic filler material compared to the PVDF polymer used.

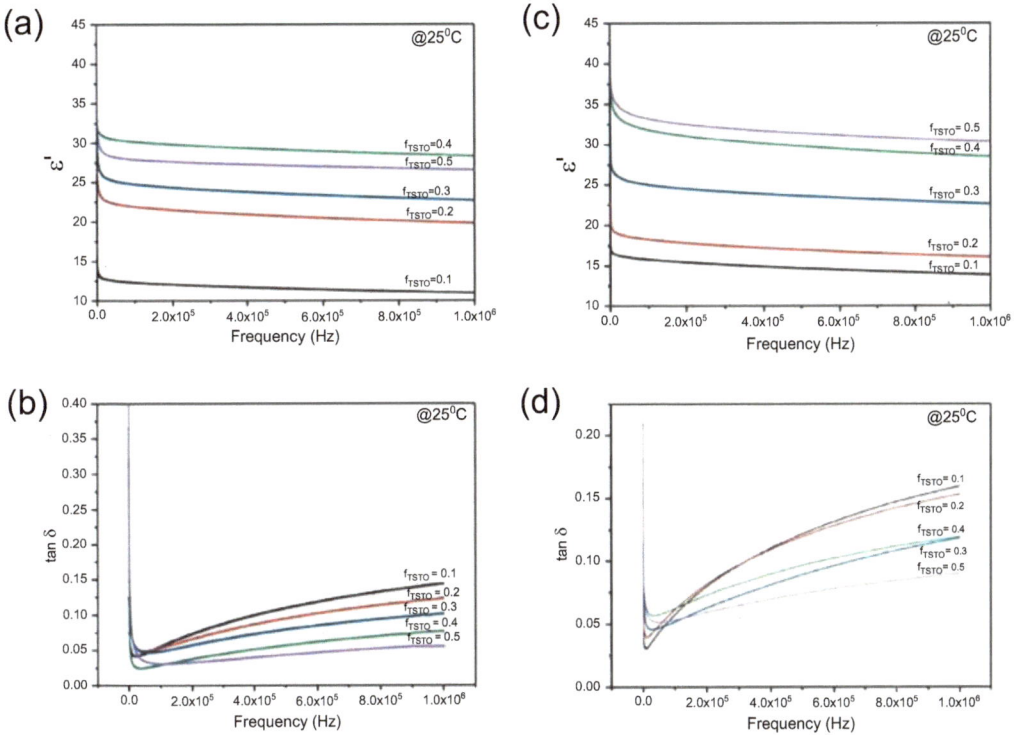

Fig. (6). Variation of ε' and tanδ with frequency at 25°C for TSTO/PVDF composites made using (a, b) PA and (c, d) PB powders, respectively.

Fig. (7) shows the conductivity-frequency relation of TSTO/PVDF composites. Conductivity increases with filler fraction, concurrently giving rise to increased tanδ values at 1 kHz. In 1 MHz, low dielectric loss (0.008) of filler material is responsible for reduced loss of composites. The quick increase in tanδ resulted from the effect of Debye relaxation, caused by C-F dipole polarization in the γ- and β- phases of the PVDF matrix [9]. The loss tangent values of all samples were less than 0.1 in the frequency range of 10^3 to 10^5 Hz.

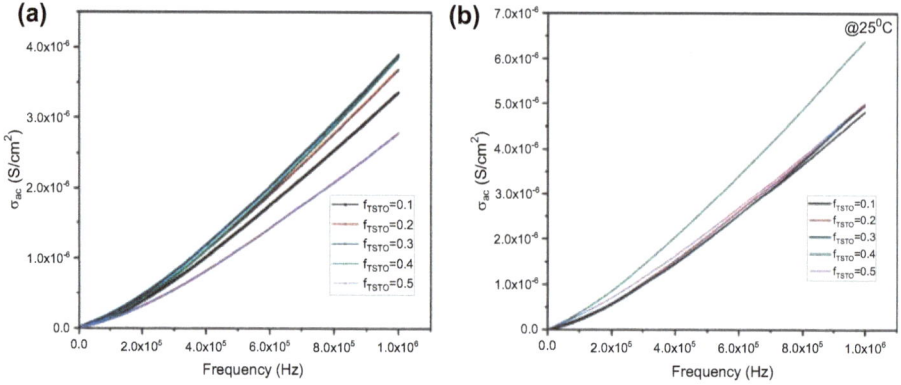

Fig. (7). Variation of conductivity with frequency at 25°C for TSTO/PVDF composites using **(a)** PA and **(b)** PB powders, respectively.

The dielectric permittivity and dielectric loss of TSTO/PVDF composites at 1 kHz and 1 MHz for various filler loadings are illustrated in Fig. (**8a** - **8d**) respectively. It is observed that the ε′ values of composites continuously increase with filler fraction. In PB-filled PVDF, ε′ is reduced from filler fraction 0.4 to 0.5. Loading more ceramic fillers leads to agglomeration and reduction in the interfacial area. At 1 MHz, a small increase is observed. The tanδ value of (PA)-TSTO/PVDF and (PB)-TSTO/PVDF composites at 1 kHz is increased continuously. A reduction in tanδ is observed in both types of filler/PVDF composites at 1MHz. PVDF has a tanδ value of 0.02 in 1 kHz and 0.17 in 1 MHz is responsible for these curves. Also, the ceramic filler has a tanδ value of 0.08 in 1 kHz and 0.008 in 1 MHz. An optimum ε′ and tanδ of 43.97 and 0.086 (fTSTO=0.5) are observed in (PA)-TSTO/PVDF at 1 kHz, whereas in (PB)- TSTO/PVDF, the values are 45.47 and 0.089 (fTSTO =0.5), respectively.

To validate the enhanced ε′ value of TSTO/PVDF composite with various fTSTO values, ε′ values have been theoretically calculated using the logarithmic model, Effective Medium Theory (EMT) model, and the Maxwell-Garnet model, the equations of these models, were given in (1) to (3), respectively [9].

$$ln\varepsilon' = (1 - f_{TSTO})ln\varepsilon_{PVDF} + f_{TSTO}ln\varepsilon_{TSTO} \tag{1}$$

$$\varepsilon' = \varepsilon_{PVDF}\left[1 + \frac{3f_{TSTO}(\varepsilon_{TSTO}-\varepsilon_{PVDF})}{2\varepsilon_{PVDF}+\varepsilon_{TSTO}+(\varepsilon_{TSTO}-\varepsilon_{PVDF})f_{TSTO}}\right] \tag{2}$$

$$\varepsilon' = \varepsilon_{PVDF}\left[1 + \frac{f_{TSTO}(\varepsilon_{TSTO}-\varepsilon_{PVDF})}{\varepsilon_{PVDF}+n(\varepsilon_{TSTO}-\varepsilon_{PVDF})(1-f_{TSTO})}\right] \tag{3}$$

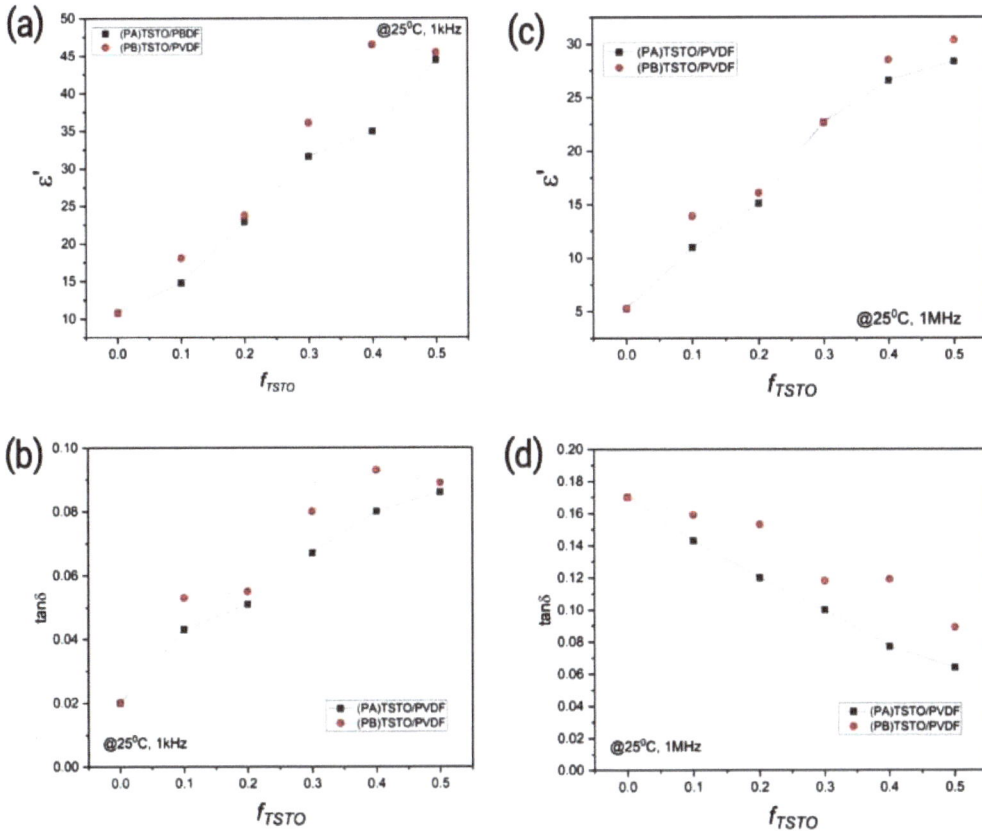

Fig. (8). Variation of ε′ and tanδ with different (PA) and (PB) TSTO filler fraction in the PVDF matrix at 1 KHz (Figs (a) and (b)) and 1 MHz (Figs (b) and (d)).

Where ε_{PVDF} is the dielectric constant of PVDF (10.78 at 1 kHz and 5.26 at 1 MHz), ε_{TSTO} is the dielectric constant of TSTO filler (693 at 1 kHz and 582 at 1 MHz) and n is the morphological fitting factor of the filler. The value of n was 0.23 and 0.18 for PA and PB powder, respectively. Comparison of obtained

ε′ with the theoretical values at 1 kHz and 1 MHz illustrated in Fig. (**9a** and **9b**), respectively. At lower filler volume, ε′ of all models are nearly matching with experimental values. However, at higher filler volume, it shows a significant difference between values. Among the various models, the logarithmic model shows a higher value compared to experimental values and the Maxwell-Garnet model shows lower values. EMT models show nearly accurate ε′ compared to obtained value. For lower filler volume, the enhancement of ε′ is due to the high dielectric permittivity of TSTO ceramics. In higher filler volume, interfacial polarization contributes a major rise in ε′ values. It is responsible for the deviation between theoretical and experimental.

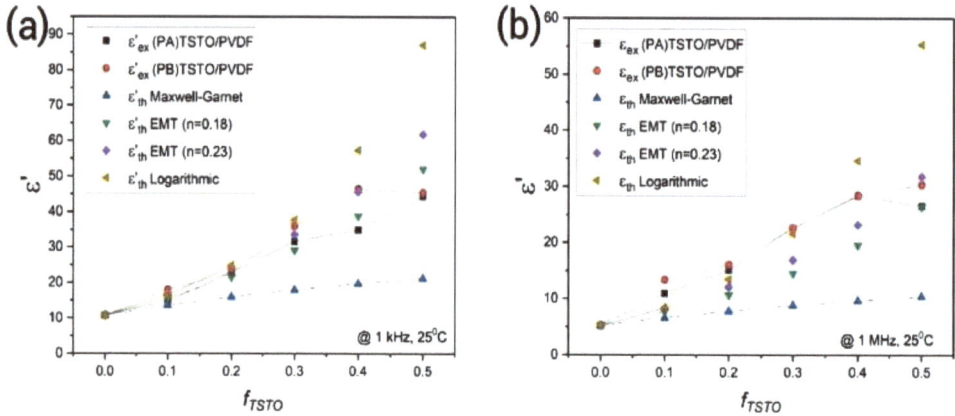

Fig. (9). Theoretically predicted and experimental ε′ values of TSTO/PVDF composites at **(a)** 1 kHz and ε′ values. **(b)** 1 MHz at 25⁰C.

The moisture absorption properties of TSTO/PVDF composites have been investigated using the IPC-TM-650 2.6.2.1 method and the findings are presented in Fig. (**10**). Because of the increasing porosity (lowering density), moisture absorption in composites is found to increasing with filler loading. PA- filled PVDF has a higher porosity than PB-filled PVDF, as given in Fig. (**6**). Hence moisture absorption of PA-filled PVDF is relatively high compared to the PB-filled composites.

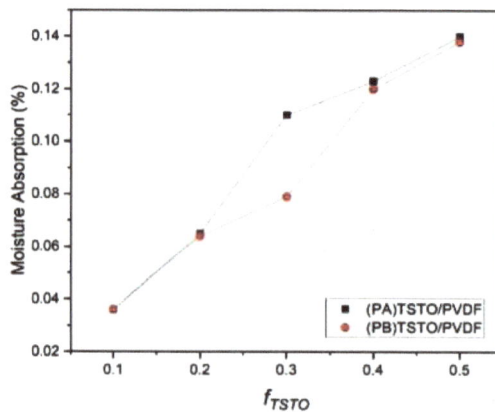

Fig. (10). Moisture absorption curve of TSTO/PVDF composites in various fTSTO.

CONCLUSION

$^{(Ta}0.1^{Sm}0.9^){}0.04^{Ti}0.96^{O}2^{/PVDF}$ composites were successfully fabricated with various filler fractions. The TSTO (ε' of 703.71 and tanδ of 0.08 at 1 kHz) filler material was synthesized via the conventional solid-state reaction method by calcining at 1100°C. Two different particle sizes (5.15μm and 1.06μm) were made using a ball mill. The phase purity of the TSTO phase was confirmed using XRD. Composite samples of 1 cm diameter 1.5 mm height were prepared by fine dispersing the filler in the PVDF (ε' of 10.78 and tanδ of 0.02 at 1 kHz) matrix followed by compression molding at 220°C, 9.8MPa. SEM microstructural analysis showed that the filler particles were homogenously distributed in the matrix. A high ε' of 45.47 and tanδ of 0.089 was obtained at filler volume 0.5 at 1 kHz. The higher values of ε' were theoretically validated using models and low tanδ obtained was due to the low dielectric loss of the filler used. The high dielectric constant with low loss of the composite can be effectively used for energy storage applications.

ACKNOWLEDGEMENTS

The authors are thankful to HOD, MED, NIT Calicut, Director, CMET Thrissur, M/s Carborandom Universal Limited Kochi for the support extended to carry out the various phases of the work.

REFERENCES

[1] Y. Fan, Z. Wang, Y. Huan, T. Wei, and X. Wang, "Enhanced thermal and cycling reliabilities in (K,Na)(Nb,Sb)O3-CaZrO3-(Bi,Na)HfO3 ceramics", *J. Adv. Ceram.,* vol. 9, no. 3, pp. 349-359, 2020.

[2] L. Jiao, P. Guo, D. Kong, X. Huang, and H. Li, "Permittivity and low dielectric loss", *J. Mater. Sci. Mater. Electron.,* vol. 31, no. 4, pp. 3654-3661, 2020.

[3] J. Li, F. Li, Y. Zhuang, L. Jin, L. Wang, X. Wei, Z. Xu, and S. Zhang, "Microstructure and dielectric properties of (Nb + In) co-doped rutile TiO2 ceramics", *J. Appl. Phys.,* vol. 116, no. 7, pp. 2-11, 2014.

[4] W. Dong, W. Hu, A. Berlie, K. Lau, and H. Chen, "Colossal dielectric behavior of ga+nb co-doped rutile TiO2", *ACS Appl. Mater. Interfaces,* vol. 7, no. 45, pp. 25321-25325, 2015.

[5] B. Guo, P. Liu, X. Cui, and Y. Song, "Colossal permittivity and dielectric relaxations in Tl + Nb co-doped TiO2 ceramics", *Ceram. Int.,* vol. 44, no. 11, pp. 12137-12143, 2018.

[6] Z. Wang, H. Chen, T. Wang, Y. Xiao, W. Nian, and J. Fan, "Enhanced relative permittivity in niobium and europium co-doped TiO2 ceramics", *J. Eur. Ceram. Soc.,* vol. 38, no. 11, pp. 3847-3852, 2018.

[7] W. Hu, Y. Liu, F. Brink, and J.W. Leung, "Electron-pinned defect-dipoles for high-performance colossal permittivity materials", *Nat. Mater.,* vol. 12, no. 9, pp. 821-826, 2013.

[8] X. Wang, B. Zhang, G. Shen, L. Sun, Y. Hu, L. Shi, X. Wang, C. Jie, and L. Zhang, "Colossal permittivity and impedance analysis of tantalum and samarium co- doped TiO2 ceramics", *Ceram. Int.,* vol. 43, no. 16, pp. 13349-13355, 2017.

[9] Z. Wang, M. Fang, H. Li, Y. Wen, C. Wang, and Y. Pu, "Enhanced dielectric properties in poly(vinylidene fluoride) composites by nanosized Ba(Fe0.5Nb0.5)O3 powders", *Compos. Sci. Technol.,* vol. 117, pp. 410-416, 2015.

CHAPTER 17

Modelling the Effects of Carbon Nanotube (CNT) and Interphase Parameters on Mechanical Properties of CNT-Reinforced Nanocomposites

Saurabh Mishra[1], Surendra Kumar[1,*] and Amit Kumar[1]

[1] CSIR-Central Mechanical Engineering Research Institute, Durgapur-713209, India

Abstract: CNT-reinforced polymer nanocomposites are emerging as a pioneer material for structural applications because of their enhanced mechanical properties as compared with neat polymers. The load transfer mechanisms and effective mechanical properties of these nanocomposites are strongly influenced by CNT parameters (volume fraction, length, aspect ratio, etc.), and thickness and mechanical properties of the interfacial region between the embedded CNT and the matrix. In this paper, modelling studies have been carried out to analyze the effects of these parameters on the effective elastic properties of a polymethyl methacrylate matrix embedded with single-walled CNTs. A three-phase continuum mechanics-based 3-D model of the nanocomposite is analyzed using the finite element method to predict the effect of an interphase on the elastic properties (elastic modulus and Poisson's ratio) of the nanocomposite in longitudinal and transverse directions. The effect of the interphase having a varied modulus (ranging from that of CNT to that of matrix) through its thickness is also investigated. The Mori-Tanaka homogenization method is also applied to the three-phase and multi-phase micromechanical models to determine its feasibility in estimating the influence of the interphase on the elastic properties of the nanocomposite.

Keywords: Carbon nanotube (CNT), Finite element analysis, Interphase, Mori-Tanaka method, Nanocomposite.

INTRODUCTION

Nanostructured composites have garnered major research interest for structural applications in recent years because of their unprecedented mechanical properties. Carbon nanotubes (CNTs) are perhaps the most favourable and extensively used nanofillers possessing high elastic modulus, multi-functionality, physical characteristics, and other mechanical properties [1]. In general, different theor-

*** Corresponding author Surendra Kumar:** CSIR-Central Mechanical Engineering Research Institute, Durgapur-713209, India; E-mail: surend_kr@yahoo.com

Amar Patnaik, Albano Cavaleiro, Malay Kumar Banerjee, Ernst Kozeschnik & Vikas Kukshal (Eds.)

etical and experimental studies have indicated that the insertion of CNTs into the polymers has effectively improved the elastic properties of nanocomposites [2 - 4]. However, theoretical predictions have some discrepancies with experimental results. These discrepancies are attributed to the following reasons: (1) Poor dispersion and aggregation of CNTs into bundles, (2) Misaligned and entangled CNTs, and (3) Insufficient knowledge of interfacial characteristics. Out of these concerns, the properties of interphase are an influential issue and need to be addressed properly in order to obtain optimistic effective properties of polymers embedded with CNTs [5]. The effect of surface to volume ratio of nanotubes in poly(vinyl alcohol) was experimentally studied by Cadek *et al.* [6]. Their findings revealed maximized surface area for small diameter multi-walled nanotubes compared to single walled nanotubes, resulting in more crystallinity of polymer on the nanotube surface. Namilae and Chandra [7] investigated the role of interface on the elastic properties of nanocomposites using a hierarchical multi-scale model comprising molecular dynamics (MD) simulation and finite element simulation. It was reported that chemical functionalization has effects on the interfacial strength, and the higher the attachments more will be the interfacial strength. Montazeri and Naghdabadi [8] used a multi-scale model comprising molecular mechanics and finite element method (FEM) to investigate the effect of interphase on (10,10) SWCNTs reinforced polymethyl methacrylate (PMMA) polymer composite. A comparative study revealed a 1.8% reduction in effective young's modulus obtained from the three phases representative volume element (RVE) model as compared to MD simulations and found it a more reliable model. Arash *et al.* [9] estimated the elastic properties of interphase by examining the fracture behaviour of CNT/PMMA composites. The study revealed that interfacial bonding relies on the aspect ratio of CNTs, and that by increasing the aspect ratio, the properties of the interface were significantly improved. Amraei *et al.* [10] developed a closed-form model to compute transverse isotropic elastic properties and thickness of interfacial region between polymer matrix and CNTs. MD simulations performed on cylindrical RVE for different SWCNTs revealed the dependency of the thickness of interphase on the diameter of CNTs. Malekimoghadam and Icardi [11] examined a multi-scale hybrid model comprising CNTs, carbon fiber, interphase and matrix. They discovered that adding a small amount of CNTs into the matrix (about 2 wt%) reduced the interfacial radial stress, which is responsible for debonding. It was also reported that non-bonded interphase merely affected the interfacial properties of hybrid composites, but no impact was observed on Young's moduli. Studies using continuum mechanics have obtained some observations in the context of the interface in the CNT-reinforced composites but are sporadic in nature. In order to bridge the gap between the analytical model and atomic level analysis, Guru *et al.* [12] parametrically studied the effect of the interphase with MD simulation in conjunction with FEM and reported variation in longitudinal elastic modulus of a composite at different thicknesses and stiffnesses of the interphase. However, the stiffness of the interphase was kept constant through its thickness.

In order to further reduce ambiguity in terms of different CNT and interphase parameters, this paper presents a detailed study of SWCNT-reinforced PMMA polymer composite with and without interphase. Continuum -based finite element (FE) models of two-phase and three-phase RVEs are developed, and parametric analyses are done to investigate the influence of CNT at different volume fractions and aspect ratio. Also, the effects of interfacial properties, such as thickness and stiffness, on the elastic properties of the nanocomposite have been studied comprehensively. The novelty of this paper is reflected in the study of the effect of interphase having a varied modulus (ranging from that of CNT to that of the polymer) through its thickness. This study is also aimed at providing a comparative analytical model to validate the results obtained from the FE analysis. Accordingly, the Mori-Tanaka homogenization method is applied to three-phase and multi-phase micromechanics models for estimating the effects of interphase parameters on the properties of the nanocomposite, and the predicted results are verified with those obtained by FE modelling.

MODELLING AND COMPUTATIONAL METHODOLOGY

Finite Element Modelling of RVE

3-D FE modelling is undertaken for the RVE of a nanocomposite of CNTs embedded in a homogeneous isotropic polymer matrix. The modelling is performed for the two-phase as well as the three-phase composites. The RVE of square cross-section has been chosen in such a way that the assumed CNT and interphase are densely packed inside the surrounding matrix, and computational efforts are minimized. The analyses of RVEs give the same elastic properties as those of the composites while also reducing the efforts required in the time-consuming analyses of complex nanostructures. The material system taken is (10,10) SWCNT embedded in a poly methyl methacrylate (PMMA) polymer matrix. Since, end caps of CNT do not contribute to the effective elastic properties of RVE, and they are discarded from the FE models. The diametral dimension of CNT is calculated from its chiral vectors. In order to obtain dimensions of the square RVE, a basic approach of volume fractions is applied. Both long and short CNTs are considered. The type of reinforcement in which a CNT spans the full length of the RVE is termed a long CNT, while if the CNT lies inside the RVE, it is denoted as a short CNT. Fig. (**1**) illustrates the RVEs having fully-embedded long and short CNTs inside the polymer matrix.

(a)

(b)

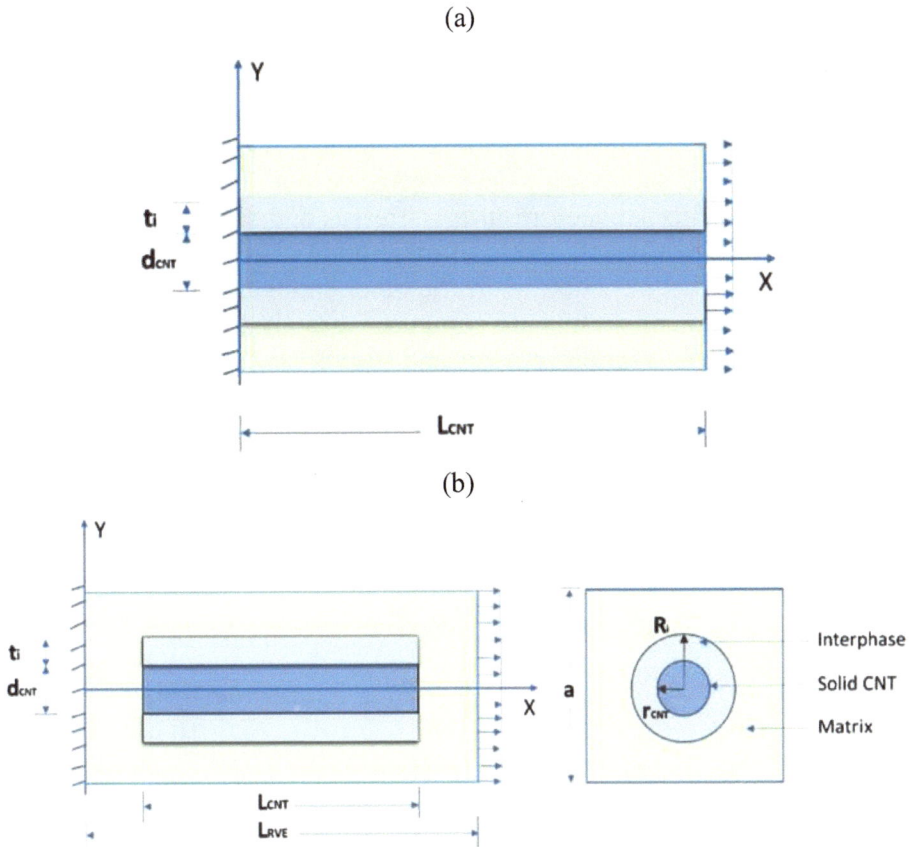

Fig. (1). Schematic of three phase RVE models with **(a)** long CNT **(b)** short CNT.

The two-phase RVE has CNT and polymer matrix perfectly bonded, and no interphase is considered between them. The three-phase RVE consists of an interphase also such that the outer surface of the embedded CNT coincides with the inner surface of the interphase, and both are assumed to be perfectly bonded. In order to predict different elastic constants of CNT-reinforced nanocomposites, FE analyses are performed at different boundary conditions and parameters. CAD models of RVEs are prepared using geometrical tools and then discretized into finite elements using ANSYS software. In order to get the most feasible solution, discretization and its order play a prominent role. On the basis of different mesh featuring parameters, the appropriate size of the mesh is selected in such a way that the solution deviates the least and gets converged. In this work, hex mesh elements are generated in quadratic order. FE models of the three-phase RVEs having long and short CNTs are depicted in Fig. (**2**). Once the RVE is meshed, different boundary constraints and loads are applied in appropriate directions. The structural response of the RVE is recorded in terms of stresses and strains. In the FE simulations to evaluate the five independent elastic properties, RVE is subjected to three different loadings (uniaxial tension, lateral uniform tension and shear loading) using appropriate periodic boundary conditions. These loading conditions are selected according to the type of response required to establish a constitutive relationship between resulting stresses and strains in the RVE. As a first loading condition, a tensile strain of magnitude 1% is applied on one of the ends of the RVE in the x direction while the other end is kept fixed (Fig. **1a**). The corresponding stresses and strains are computed directly from the postprocessing of the FEA results. Only average stresses and strains calculated at the mid-section of the RVE are chosen in the constitutive equations. The constitutive relationship between stresses and strains in the RVE is as given by Eq. (1).

$$
\begin{Bmatrix} \epsilon \\ \epsilon \\ \epsilon \\ yz \\ xz \\ xy \end{Bmatrix} =
\begin{bmatrix}
\dfrac{1}{E_x} & -\dfrac{v_{xy}}{E_x} & -\dfrac{v_{xy}}{E_x} & 0 & 0 & 0 \\[2ex]
-\dfrac{v_{xy}}{E_x} & \dfrac{1}{E_y} & -\dfrac{v_{yz}}{E_y} & 0 & 0 & 0 \\[2ex]
-\dfrac{v_{xy}}{E_x} & -\dfrac{v_{yz}}{E_y} & \dfrac{1}{E_y} & 0 & 0 & 0 \\[2ex]
0 & 0 & 0 & \dfrac{2\left(1+v_{yz}\right)}{E_y} & 0 & 0 \\[2ex]
0 & 0 & 0 & 0 & \dfrac{1}{G_{xy}} & 0 \\[2ex]
0 & 0 & 0 & 0 & 0 & \dfrac{1}{G_{xy}}
\end{bmatrix}
\begin{Bmatrix} \sigma \\ \sigma \\ \sigma \\ yz \\ xz \\ xy \end{Bmatrix}
\qquad (1)
$$

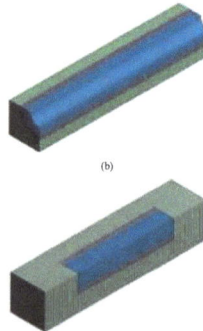

Fig. (2). FE models of three-phase RVE with (**a**) unidirectional long CNT and (**b**) unidirectional short CNT.

Mori-Tanaka Model

The evaluation of the elastic properties of unidirectional composites can be easily done by implementing appropriate micromechanical models. These models are based on the empirical relations between the elastic constants and volume fractions of the fiber and matrix material, and do not require any other experimental data. Some of the elastic constants are evaluated accurately by these models, and some show discrepancies due to certain orientation and distribution issues. Mori-Tanaka method [13] is a reliable semi-analytical homogenization approach used to compute a composite material's mechanical properties. The method was later modified by Benveniste [14] and based on which more feasible models were developed. The expression of effective elastic tensor reformulated by Benveniste [14] is

$$[C] = [C_m] + V_f \left([C_f] - [C_m] \right) [A], \tag{2}$$

in which [Cm] and [Cf]are the elastic tensors of the polymer matrix and fiber, respectively. Vf is the volume fraction of the fiber. [A] is termed as concentration tensor and is given by Eq. (3).

$$[A] = \left[A^{(dil)} \right] \left[V_m [I] + V_f \left[A^{(dil)} \right] \right]^{-1}, \tag{3}$$

where [A(dil)]is the dilute strain concentration tensor given by Eq. (4).

$$\left[A^{(dil)} \right] = \left[[I] + [S][C_m]^{-1} \left([C_f] - [C_m] \right) \right]^{-1} \tag{4}$$

In Eqs. (3) and (4), Vm is the volume fractions of the matrix, [I] is a 4th order unity tensor, and [S] is Eshelby tensor [15] whose computation is based on the Poisson's ratio of the matrix and shape of the fiber.

A simple and convenient approach to the Mori-Tanaka model in closed-form expressions was proposed by Abaimov *et al.* [16]. With the assistance of these expressions, the Mori-Tanaka method can be directly implemented to calculate the elastic properties of two-phase composites. The closed-form equations as proposed by Abaimov *et al.* [16] for the elastic constants of unidirectional composites, are depicted in Eqs. (5) to (10), in which different terms have their usual meanings; and subscripts 'f' and 'm' in these terms stand for fiber (inclusion) and matrix, respectively.

$$E_x^c = V_f E_{f,x} + V_m E_m + 2 V_f V_m Z_1^c \left(v_{f,xy} - v_m \right)^2 \tag{5}$$

$$E_y^c = \cfrac{E_x^c / \left(1 - v_m^2\right)}{\cfrac{1}{1-v_m^2} + 2V_f \cfrac{E_x^c}{Z_2^c}\left[1 + v_{f,yz} - \cfrac{E_{f,y}}{E_m}\left(1+v_m\right)\right] + V_f Z_1^c \cfrac{E_{f,x}}{E_m}\left(\cfrac{1+v_m}{E_m} - \cfrac{2}{E_{f,x}} + \cfrac{1-v_{f,yz}}{E_{f,y}}\right)} \tag{6}$$

$$\tag{7}$$

$$v_{xy}^c = v_{xz}^c = v_m + 2V_f \frac{Z_1^c}{E_m}\left(v_{f,xy} - v_m\right)\left(1 - v_m^2\right)$$

$$v_{yz}^c = 1 - \frac{E_y^c}{E_x^c}\left\{ 2 + 2\frac{V_f}{V_m}\frac{E_{f,x}}{E_m}\left(1-v_m^2\right)\left[1 - 2\frac{Z_1^c\left(V_f E_{f,x}\left(1-v_m^2\right) + V_m E_m\left(1-v_{xy}^2\right)\right)}{E_{f,x}E_m}\right] \right. \tag{8}$$

$$\left. - \frac{E_x^c}{E_m}\left(1+v_m\right)\right\}$$

$$\tag{9}$$

$$G_{xy}^c = \frac{E_m}{2V_m\left(1+v_m\right)}\left\{ 1 + V_f - 4V_f\left[1 + V_f + 2V_m\frac{G_{f,xy}}{E_m}\left(1+v_m\right)\right]^{-1}\right\}$$

$$G_{yz}^c = \cfrac{E_m}{2\left(1+v_m\right) + V_f\left[\cfrac{V_m}{8\left(1-v_m^2\right)} + \cfrac{1}{E_m/G_{f,yz} - 2\left(1+v_m\right)}\right]^{-1}}, \tag{10}$$

where

$$Z_1^c = \left[-2V_m\frac{v_{f,xy}^2}{E_{f,x}} + V_m\frac{1-v_{f,yz}}{E_{f,y}} + \frac{\left(1+v_m\right)\left(1+V_f\left(1-2v_m\right)\right)}{E_m}\right]^{-1}$$

$$Z_2^c = E_{f,y}\left(3 + V_f - 4v_m\right)\left(1+v_m\right) + V_m E_m\left(1+v_{f,yz}\right)$$

These closed-form expressions of the Mori-Tanaka approach have been derived for the two-phase composite model comprising transverse isotropic fiber and isotropic matrix material. Since the Mori-Tanaka model is an inclusion-based model, for composites with multiple inclusions, a multi-step homogenization approach can be used. In our work, two-step, as well as multi-step approaches, have been used to apply these expressions for the three-phase and multi-phase composite models. As a first step, interphase and CNT are considered as the matrix and inclusion, respectively. Later, in the second step, the interphase and CNT are replaced by an equivalent fiber, and elastic constants obtained from the first step are assigned to the equivalent fibre embedded in the polymer matrix. The same procedure is applied for the multi-phase composite model having interphase with varied stiffness through its thickness.

NUMERICAL RESULTS

To calculate the effective elastic constants of the CNT/PMMA composites, finite element analyses have been performed on different square RVEs. In this work, 12, 17 and 28 vol% of CNTs are considered for the two-phase model in order to analyze the influence of embedded CNT on the elastic properties of nanocomposites. While in the three-phase model, the volume fraction of embedded CNT is kept constant at 12%, and an interphase of varied thickness is introduced by replacing an equal volume of polymer matrix. The dimensions of RVEs are calculated based on the CNT volume fraction and thus the side of square RVE is taken as 4.34, 3.65 and 2.84 nm at 12, 17 and 28 vol% of CNT, respectively. The material properties of the equivalent SWCNT and PMMA matrix are taken as reported in [17], and the interphase is assigned varied elastic properties. The elastic properties and dimensions of each phase are listed in Table 1. An interphase having an elastic modulus less than that of the polymer is referred to as a soft interphase and accounts for voids, debonding, or weak interfacial interaction. Whereas stiff interphase having an elastic modulus greater than that of the polymer represents strong interfacial interaction.

Table 1. Material properties and dimensions for nanocomposite constituents.

Property	Constituents		
	CNT	*Polymer Matrix*	*Interphase*
Young's modulus (GPa)	$E_{fx} = 600; E_{fy} = 10$	$E_m = 2.5$	$E_i = 0.25, 25$
Shear modulus (GPa)	$G_{fxy} = 255; G_{fyz} = 4.3$	0.93	0.093, 9.3
Poisson's ratio	$v_{fxy} = v_{fyz} = 0.175$	0.34	0.34
Dimensions (nm)	$d_{CNT} = 1.696; t_{CNT} = 0.34$	$L_{RVE} = 10$	$t_i = 0.17$ to 0.68

As an important step in performing finite element analysis, mesh sensitivity analysis was also performed. For a typical case of the three-phase FE model having soft interphase, the convergence of results (in the form of the longitudinal modulus) with mesh refinement has been shown in Table 2, in which N_S and N_L represent a number of elements along each side and along the length of the RVE.

Table 2. Mesh sensitivity analysis for the prediction of elastic properties of the CNTRC having soft interphase (ti=tCNT).

Mesh size	Longitudinal modulus (in GPa)
6 × 6×14	73.48
26×26×62	73.905
32×32×76	73.92
46×46×110	73.93
68×68×162	73.935
90×90×210	73.935

Effect of CNT Volume Fraction

To predict the effect of the CNT volume fraction (VCNT) on the basic functional characteristics of the nanocomposite, its effective mechanical properties are predicted at three different volume fractions of (10,10) CNTs. At first, 12 vol% of long and uniformly aligned CNT embedded inside the PMMA polymer matrix is analyzed under different loading and boundary conditions. On the basis of different stresses and strains, the elastic properties are computed using Eq. (1). Similar approaches have been adopted for 17 and 28 vol% of embedded CNTs. The effect of volume fractions of CNTs on effective moduli (longitudinal as well as transverse) of the nanocomposite is depicted in Fig. (3).

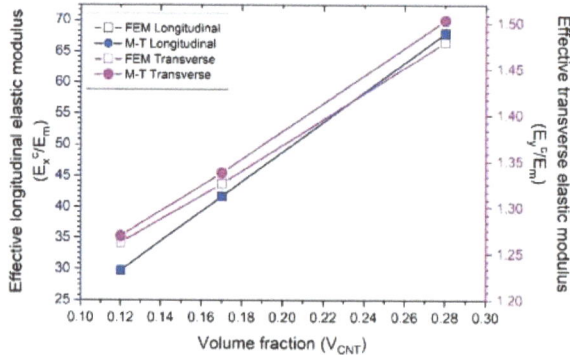

Fig. (3). Variation of nanocomposite's effective elastic moduli with vol% of SWCNT.

It can be inferred from Fig. (3) that reinforcement of CNT has significantly improved the elastic modulus of neat PMMA polymer. On increasing the volume fraction of CNTs from 12 to 28%, the effective longitudinal modulus has increased from 30 to 68 times that of PMMA. However, the effective transverse elastic modulus increased only from 1.2 to 1.45 times that of the matrix. The significant increase in longitudinal modulus occurred due to the perfectly aligned long CNT, which offered maximum stiffness longitudinally. The predicted values of different elastic moduli and Poisson's ratios of the nanocomposite are listed in Table **3**.

In this table, properties predicted using closed-form expressions of the Mori-Tanaka (M-T) model are also shown and are observed to be matching with the FEM results. It can be noted that the M-T model has predicted transverse elastic modulus with very little deviation. Table **3** also compares the predicted results for the three CNT volume fractions with MD simulation results reported in [17], and it is found that the two-phase RVE models underestimate the longitudinal modulus of the nanocomposite by about 22 to 25%. It can thus be inferred that a three-phase RVE model with strong interphase is necessary to account for the strong interfacial energy between the CNT and the polymer, as observed in the MD simulation.

Table 3. Predicted elastic properties of CNT/PMMA composite at different volume fractions of long (10,10) SWCNTs.

Material constants	Volume fraction of CNT (V_{CNT})								
	0.12			*0.17*			*0.28*		
	FEM	Mori-Tanaka model	MD Result [17]	FEM	Mori-Tanaka model	MD Result [17]	FEM	Mori-Tanaka model	MD Result [17]
E_x (GPa)	74.2	74.2	94.6	104.1	104.1	138.9	169.8	169.8	224.2
E_y(GPa)	3.13	3.17	2.9	3.27	3.34	4.9	3.60	3.76	5.5
v_{xy}	0.32	0.32	--	0.31	0.31	--	0.29	0.29	--
v_{yz}	0.49	0.49	--	0.49	0.48	--	0.48	0.45	--
G_{xy} (GPa)	1.40	1.19	--	1.97	1.32	--	2.11	1.65	--

Effect of Interphase Parameters

The interfacial region developed between the matrix and CNT plays a major contribution to the effective elastic properties of the nanocomposite. In order to investigate the influence of interphase properties on the elastic properties of the nanocomposite, FE studies were conducted for interfacial regions (spanning throughout the length of the CNT) of different elastic and geometrical properties. In this analysis, the volume fraction of CNT is kept constant at 12 vol%, and the thickness of the interphase is varied from 0.5tCNT to 2tCNT. Parametric studies are performed with two types of interphase (a) soft interphase (Ei=0.1Em) and (b) stiff interphase (Ei=10Em), both having thicknesses of 0.17, 0.34, 0.51 and 0.68

nm. The comparative results of variations in effective elastic moduli of CNT/PMMA with varied thickness and modulus of interphase are shown in Fig. (**4**).

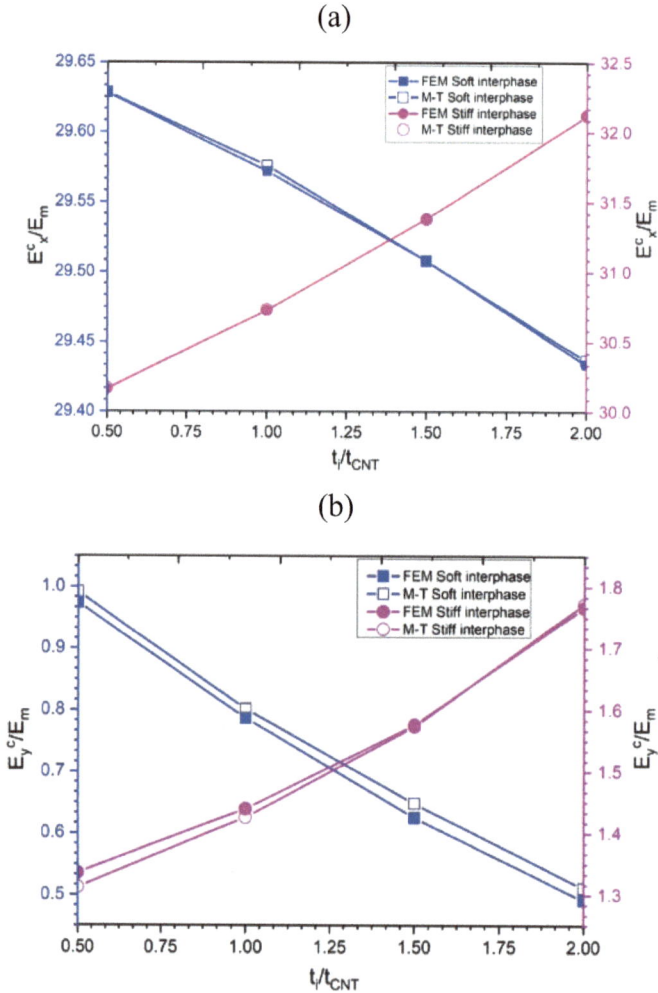

(a)

(b)

Fig. (4). Variation of CNT/PMMA nanocomposite's moduli (a. Longitudinal and b. Transverse) with interphase thickness for two different interphase stiffnesses (VCNT=0.12).

It is observed from the figure that the presence of interphase has a substantial influence on the elastic properties of CNT/PMMA nanocomposites. Simulation results demonstrate that the soft interphase has a very small influence on the longitudinal modulus as compared to the stiff interphase. In the case of the stiff interphase, the effective longitudinal modulus has increased by 1.6% at ti= 0.17 nm and by 8.2% at ti= 0.68 nm, as shown in Fig. (**4a**). On the other hand, the soft interphase regime has relatively weakening effect and an increase in its thickness causes a decrease in the elastic modulus of the composite. So far as the effective transverse modulus of the composite is concerned, a significant decrease of about 22% to 61% in its value was observed in the case of the soft interphase when its thickness varied from 0.17 to 0.68 nm; whereas a rise of about 6.6% to 41% happened in the case of the stiff interphase (Fig. **4b**). Also shown in Fig. (**4**) are the results predicted using the two-step closed-form M-T model applied to the three-phase composite RVE. A comparison of these analytical results with the FE predictions shows almost negligible deviation for both longitudinal and transverse elastic moduli. In the case of the stiff interphase having $t_i = 0.68$ nm, the longitudinal modulus of the CNT/PMMA was calculated to be 80.3 GPa (about 8.2% larger than that predicted using a two-phase model) and thus, its deviation from the MD prediction decreased from 22% to 15%.

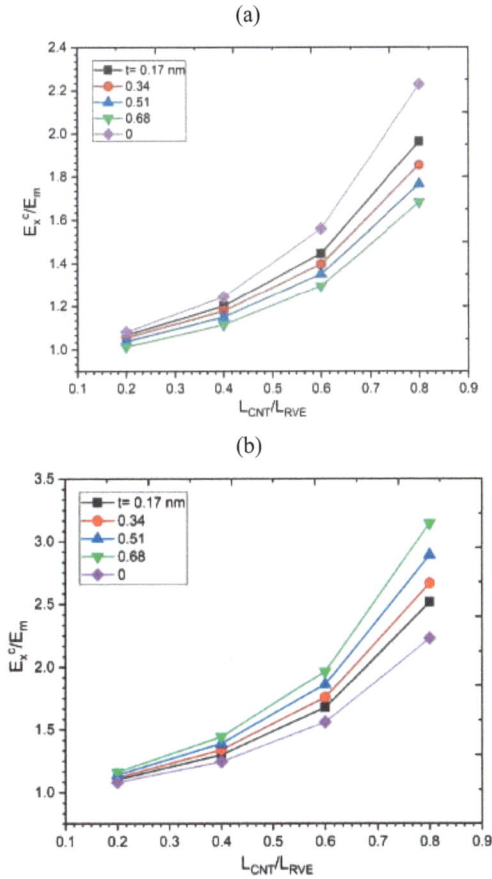

(a)

(b)

Fig. (5). Variation of longitudinal elastic modulus of CNT/PMMA nanocomposite with LCNT/LRVE ratio: (a) soft interphase and (b) stiff interphase of different thicknesses.

Interphase with Varied Modulus

In another approach, the interphase of different thicknesses was considered, and its modulus (Ei) was linearly varied throughout its thickness in the analysis; Ei being equal to Ef,x at the CNT-interphase junction and Em at the interphase-matrix junction. Mori-Tanaka prediction for the longitudinal modulus of the CNT/PMMA was found to be 89.5 GPa, 97.8 GPa, and 106.7 GPa for interphase thicknesses ti= 0.17 nm, ti= 0.25 nm and ti= 0.34 nm, respectively. Thus, the deviation of the result from the MD prediction further decreased to 5%, 3%, and 12%, respectively, for the three thicknesses of the interphase.

Effect of Short CNTs

To analyze the effect of short (10,10) SWCNTs on elastic modulus of CNT/PMMA polymer nanocomposites, four different LCNT/LRVE ratios varying from 0.2 to 0.8 have been studied for both two- and three-phase RVEs. This variation in LCNT/LRVE ratio results in CNT aspect ratio (LCNT/dCNT) varying from 1.2 to 4.7. The effect of LCNT/LRVE ratio is parametrically analyzed for different thicknesses of soft and stiff interfacial regions, as shown in Fig. (5).

It can be noted from the results that the embedded short CNTs have drastically reduced the longitudinal elastic modulus of the nanocomposite as compared to long-embedded CNTs. However, an increase in LCNT/LRVE ratio from 0.2 to 0.8 has slightly enhanced the elastic modulus in both cases. Moreover, stiff interphase has a beneficial effect on the longitudinal modulus of the nanocomposite in the case of short CNT also, and this gain in the modulus improves with an increase in the interphase thickness. On the other hand, soft interphase has further diminished the longitudinal modulus of the nanocomposite embedded with short CNTs.

CONCLUDING REMARKS

The effective elastic properties of PMMA polymer nanocomposite reinforced with (10,10) SWCNTs were studied using RVE-based FEM analysis and the closed-form Mori-Tanaka model, by including the effect of the interfacial region between the embedded CNT and the matrix. Parametric studies were conducted for different stiffnesses and thicknesses of the interphase, and the predicted elastic properties were compared with those calculated for the two-phase RVE model consisting of only CNT and the matrix. Interphase is also characterised by having a varied modulus throughout its thickness. Both long (fully-embedded) and short

CNTs were taken, and the effects of volume fraction and aspect ratio of CNTs were analyzed.

Simulation analyses revealed that the addition of 12 and 28 vol% of long (10,10) CNTs inside the PMMA polymer matrix remarkably enhanced Young's modulus of CNT/PMMA nanocomposites to 30 times and 68 times that of PMMA, respectively. In addition, the interphase has indicated the potential to affect the stiffness of the nanocomposites depending upon their thickness, and the interfacial region should be stronger for enhanced load transfer ability and superior effective properties of such composites. It was demonstrated that the two-step Mori-Tanaka model could appropriately predict the effects of interphase parameters on the effective elastic constants of the nanocomposites having long CNTs. A significant finding was observed by the incorporation of interphase having varied modulus through its thickness and having a thickness in the range of the CNT thickness. This type of interphase is observed to have a stiffening effect on the nanocomposite and resulted in a significant decrease in the deviation of the FEM prediction for the effective elastic properties when compared with the MD simulation result.

ACKNOWLEDGEMENT

This research work was a part of a research project funded by the Aeronautics R&D Board, New Delhi. The financial support is acknowledged.

REFERENCES

[1]   J.N. Coleman, U. Khan, W.J. Blau, and Y.K. Gun'ko, "Small but strong: A review of the mechanical properties of carbon nanotube-polymer composites", *Carbon,* vol. 44, no. 9, pp. 1624-1652, 2006.

[2]   D. Qian, E.C. Dickey, R. Andrews, and T. Rantell, "Load transfer and deformation mechanisms in carbon nanotube-polystyrene composites", *Appl. Phys. Lett.,* vol. 76, no. 20, pp. 2868-2870, 2000.

[3]   X.L. Chen, and Y.J. Liu, "Square representative volume elements for evaluating the effective material properties of carbon nanotube-based composites", *Comput. Mater. Sci.,* vol. 29, no. 1, pp. 1-11, 2004.

[4]   M. Chwał, and A. Muc, "Transversely isotropic properties of carbon nanotube/polymer composites", *Compos B Eng.,* vol. 88, pp. 295-300, 2016.

[5]   F. Zhang, and Y. Liu, "Interphase structures and properties of carbon nanotube-reinforced polymer nanocomposite fibers". *Carbon nanotube fibers and yarns.,* M. Miao, Ed., Woodhead Publishing, 2020, pp. 71-102.

[6]   M. Cadek, J.N. Coleman, and K.P. Ryan, "Reinforcement of polymers with carbon nanotubes: The role of nanotube surface area", *Nano Lett.,* vol. 4, no. 2, pp. 353-356, 2004.

[7]   S. Namilae, and N. Chandra, "Multiscale model to study the effect of interfaces in carbon nanotube-based composites", *J Eng Mater Technol Trans ASME.,* vol. 127, no. 2, pp. 222-232, 2005.

[8]   A. Montazeri, and R. Naghdabadi, "Investigation of the interphase effects on the mechanical behavior of carbon nanotube polymer composites by multi-scale modeling", *J. Appl. Polym. Sci.,* vol. 117, no. 3, pp. 361-367, 2010.

[9]   B. Arash, Q. Wang, and V.K. Varadan, "Mechanical properties of carbon nanotube/polymer

composites", *Sci. Rep.,* vol. 4, pp. 1-8, 2014.

[10]   J. Amraei, J.E. Jam, B. Arab, and R.D. Firouz-Abadi, "Modeling the interphase region in carbon nanotube-reinforced polymer nanocomposites", *Polym. Compos.,* vol. 40, no. S2, pp. E1219-E1234, 2019.

[11]   R. Malekimoghadam, and U. Icardi, "Prediction of mechanical properties of carbon nanotube-carbon fiber reinforced hybrid composites using multi-scale finite element modelling", *Compos B Eng.,* vol. 177, no. 7, p. 107405, 2019.

[12]   K. Guru, T. Sharma, K.K. Shukla, and S.B. Mishra, "Effect of interface on the elastic modulus of CNT nanocomposites", *J. Nanomech. Micromech.,* vol. 6, no. 3, p. 04016004, 2016.

[13]   T. Mori, and K. Tanaka, "Average stress in matrix and average elastic energy of materials with misfitting inclusions", *Acta Mater.,* vol. 21, no. 5, pp. 571-574, 1973.

[14]   Y. Benveniste, "A new approach to the application of Mori-Tanaka's theory in composite materials", *Mech. Mater.,* vol. 6, no. 2, pp. 147-157, 1987.

[15]   T. Mura, *Micromechanics of Defects in Solids.* 2nd ed. Springer, 1987.

[16]   S.G. Abaimov, A.A. Khudyakova, and S.V. Lomov, "On the closed form expression of the Mori-Tanaka theory prediction for the engineering constants of a unidirectional fiber-reinforced ply", *Compos. Struct.,* vol. 142, pp. 1-6, 2016.

[17]   Y. Han, and J. Elliott, "Molecular dynamics simulations of the elastic properties of polymer/carbon nanotube composites", *Comput. Mater. Sci.,* vol. 39, no. 2, pp. 315-323, 2007.

CHAPTER 18

Vibration and Deflection Analysis of Quadrilateral Sandwich Plate with Functionally Graded Core

Shivnandan Bind[1], Manish Kumar[1] and Saroj Kumar Sarangi[1,*]

[1] Mechanical Engineering Department, National Institute of Technology Patna, India

Abstract: This chapter presents the analysis of the vibration and deflection of a quadrilateral sandwich plate with a functionally graded core for different temperature conditions. The plate is made of functionally graded carbon nanotube-reinforced composite (FG-CNTRC), and results are obtained with the help of ANSYS software. The uniform distribution (UD), functionally grading V type distribution (FG-V), functionally grading X type distribution, and functionally grading O type distribution are the four different distributions of the reinforcements grading that are taken into consideration as a core in the direction of thickness (FG-O). The uniformly distributed grading is applied to the face plate for all the cases. Young's modulus, mass density, and Poisson's ratio are all important material properties calculated by the extended rule of mixture with the CNT efficiency parameter, accounting for size dependence. Detailed analysis is done in this paper to reveal the effect of volume fraction and temperature on the natural frequency and central deflection of the quadrilateral sandwich plate, and compared it with the results of a normal quadrilateral plate. Numerical results are obtained and presented using Ansys R17.2 and MATLAB. The results suggested that UD-type grading has the highest natural frequency and lowest central deflection compared to other types of functionally grading material.

Keywords: CNT reinforcement, Functionally graded core, Sandwich plate, Volume fraction.

INTRODUCTION

Carbon nanotubes (CNTs) were discovered by Iijima in 1991 [1]; from then to now, much research have been conducted on carbon nanotubes. The carbon nanotubes possess superior characteristics like high mechanical strength as compared to regular steel and elastic modulus as well as excellent thermal and electrical conductivities [2]. In a recent development, researchers are combining

* **Corresponding author Saroj Kumar Sarangi:** Department of Mechanical Engineering, National Institute of Technology Patna, India; E-mail: sarojksarangi@yahoo.com

Amar Patnaik, Albano Cavaleiro, Malay Kumar Banerjee, Ernst Kozeschnik & Vikas Kukshal (Eds.)

their research of carbon nanotubes by grading them functionally to form multiphase composites with smooth property variations along the thickness [3]. In functionally grading, there are mainly three types of distribution *i.e.* functionally grading V type distribution (FG-V), functionally grading X (FG-X) type distribution, and functionally grading O (FG-O) type distribution [4]. These functionally grading properties vary linearly in the thickness direction as per the distribution of CNTs volume fraction. These are superior to conventional laminated composite structures. The laminated composite structures have delamination-related problems. High temperatures are frequently encountered during the entire life cycle of composite structures when employed in industrial or automotive applications. These temperatures impact the stiffness and strength of the material and can cause dynamic instability in these structures [5]. Plate structure of irregular shape plays a noteworthy role in engineering applications like aircraft wings, naval structures, and a bridge, so when these structures undergo with due to thermal and mechanical load, stress concentration or in-plane compressible forces occur. The structure failure can be accountable by buckling, which reduces the load-carrying capacity of the structure [6]. These days, sandwich composite is a very good option all over the world because the sandwich structure has a good strength-to-weight ratio of material and high bending stiffness as compared to traditional composite structures. Therefore, the sandwich structure is an excellent solution in a number of applications.

Zhu *et al.* used the theory of shear deformation by infinite element analysis to analyse the static deflection as well as free vibration of a composite plate reinforced by carbon nanotubes. They also examined the impact of various boundary conditions and found that the results obtained using this method agreed with those obtained using the ANSYS [2]. It was found that the plate width-t--thickness ratio and its boundary conditions have an impact on the bending deflection, whereas the central axis stress and the CNT volume fraction can be neglected for the vibration effect of VCNT.

Aicha *et al.* analysed the sandwich plates reinforced by carbon nanotubes for static as well as dynamic analysis with the help of FSDT theory. Two conditions were observed for this analysis: (1) Functionally graded core with uniformly distributed face sheets and (2) Functionally graded core with uniformly distributed face sheets. The findings suggested that the functionally graded face sheet sandwiched reinforced plate has a strong level of resistance to deflection in comparison to other types of reinforcement due to the concentration of carbon nanotube at the bottom and upper face sheet sandwich plate. However, for core-reinforced sandwich plate decrease in dimensionless natural frequency is the most noticeable. Due to the small dimensionless frequency, increasing aspect ratio (a/h) is caused by fundamental modes [7]. Wang *et al.* meshless approach is used to examine the irregular quadrilateral plate where they obtained results for a different side-to-thickness ratio with varying side angles. It was observed that the quadrilateral plate for a composite structure reinforced by carbon nanotubes was thermally affected in terms of vibration and buckling. It was concluded that the volume fraction of carbon nanotube, functionally grading distribution, geometry parameters, and temperature affect the frequency and deflection of skew symmetric plates. Initially the three modes of natural frequencies decrease as the temperature rises. Near the value of temperature at which critical buckling starts, the natural frequency mode of the first order approaches zero [8]. The value of frequency of the first order vibration mode approaches zero near critical buckling temperature. Rasool *et al.* studied the nanocomposite cylindrical structure reinforced with single wall carbon nanotube subjected to impact load to analyse the effects by mesh free process [9]. It was observed that the different grading of carbon nano tube, carbon nano tube volume fraction and structure thickness affects the stress flow and frequency in cylindrical structure. As the thickness is reduced, the first parameters of frequency are reduced and the stress propagation intensity is increased. The value of frequency and the wave of stress intensity propagation both increase as the volume fraction of carbon nanotubes increases. Kumar *et al.* studied the composite beam reinforced with the carbon nanotube and analysed its vibrational effects. It was concluded that the vibration frequency of the reinforced beam is dependent on the volume fraction of carbon nanotube & distribution pattern of reinforcement [10].

The chapter emphasises the natural frequency and central deflection of a quadrilateral sandwich plate with a functionally graded core, which is not very commonly studied in the literature. The core material is assumed to be functionally graded in the thickness direction, while the face plates are uniformly distributed in all the cases. The plate is assumed to be orthotropic, and four types of grading are considered, *i.e.*, F.GO, F.GX, F.GV, and uniformly distributed. To obtain the material properties of each layer, the volume fraction of the carbon nanotube (CNT) is varied from 0.11 to 0.17, and MATLAB is used to calculate the corresponding material properties. ANSYS R17.2 software is used to validate the obtained results and compare them with previously published papers.

The main objective of the study is to find the superior grading type for natural frequency and deflection by comparing the obtained results. The study will contribute to the understanding of the behavior of functionally graded sandwich plates and will help in the design of more efficient and cost-effective structures.

MATHEMATICAL FORMULATION

The quadrilateral sandwich plate is made of 8 isotropic laminas in which 4 layers (two layers for top face sheets and two layers for bottom face sheets) are considered for the core. The thickness of the face plates and core is taken as the same. The face sheets (top and bottom) are considered to be uniformly distributed whereas for core functionally F.GV, F.GX, F.GO and uniformly distributed grading is taken. The properties of the carbon nanotube and isotropic polymer matrix of each layer are evaluated by the extended rule of mixture with consideration of CNT efficiency parameter $r\eta_1, \eta_2, \eta_3$. The required modulus of elasticity and rigidity of composite layer is obtained by using the following formula [11] (Fig. **1**).

$$E_{11} = \eta_1 V_{CNT}(Z) E_{11}^{CNT} + V_m E^m \tag{1}$$

$$\frac{\eta_2}{E_{22}} = \frac{V_{CNT}(Z)}{E_{22}^{CNT}} + \frac{V_m}{E_m} \tag{2}$$

$$\frac{\eta_3}{G_{12}} = \frac{V_{CNT}(Z)}{G_{12}^{CNT}} + \frac{V_m}{G_m} \tag{3}$$

$$V_{CNT}(Z) + V_m = 1 \tag{4}$$

Here $E_{11}^{CNT}, E_{22}^{CNT}$ and G_{12}^{CNT} are the modulus of elasticity and rigidity of single wall carbon nanotube. E^m, E_m are the matrix modulus of elasticity and rigidity. η_1, η_2 and η_3 are the CNT/matrix efficiency parameter. V_{CNT} and V_m are carbon nanotube volume fraction and matrix volume fraction, respectively.

The uniformly distributed (U.D) functionally grading, F.GO, F.GV and F.GX graded distribution of the carbon nanotube along the thickness of the plate is evaluated by the formula given below [8].

(a)

(b)

(c)

(d)

(e)

(f)

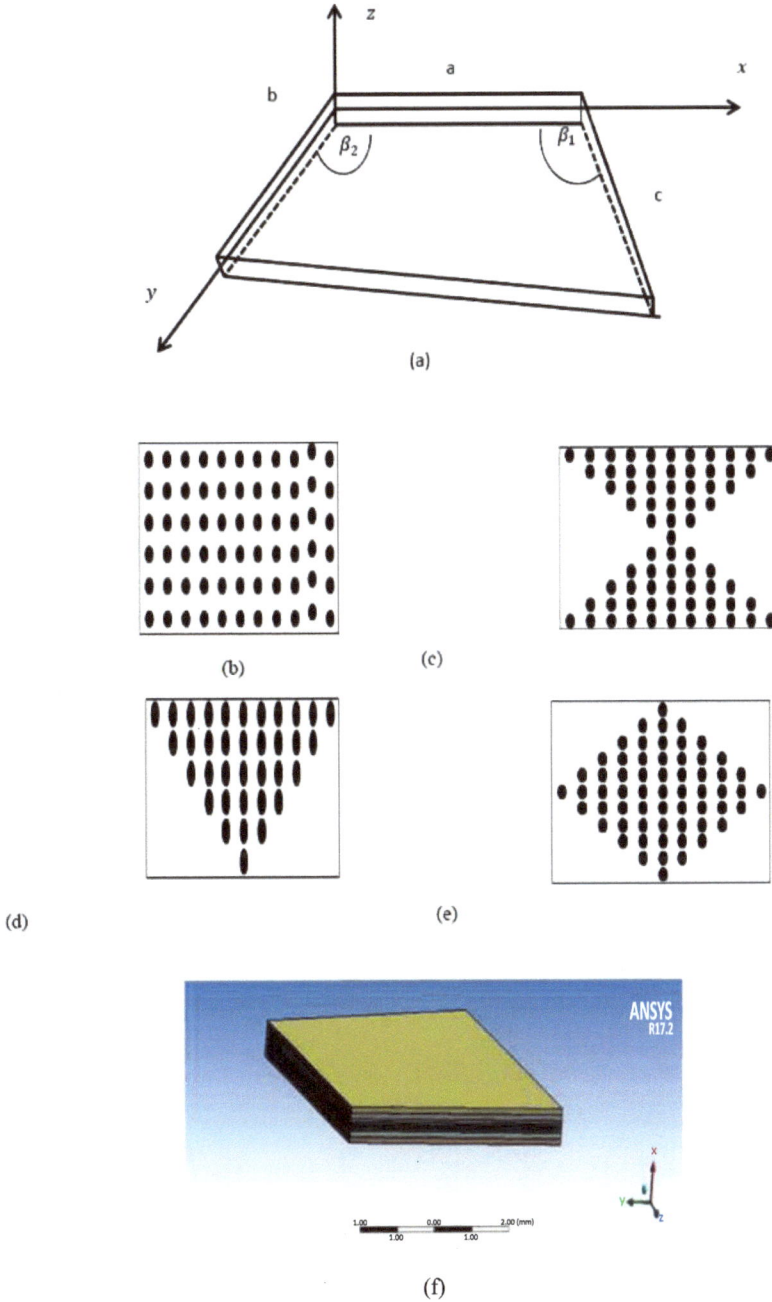

Fig. (1). (a) Schematic diagram of quadrilateral lamina for quadrilateral sandwich plate here b/a=0.8, a/h=10, c/a=0.7,β1=105°,β2=75° and taken h=3mm for sandwich quadrilateral plate **(b)** uniformly distribution, **(c)** X type FG, **(d)** V type FG, **(e)** O type FG and **(f)** modelled figure in ANSYS 17.2 for the quadrilateral sandwich plate.

$$V_{CNT}(Z) = V^*_{CNT} \qquad \text{[U.D CNTRC]} \qquad (5)$$

$$V_{CNT}(Z) = 2\left(1 - \frac{2|z|}{h}\right)V^*_{CNT} \qquad \text{[F.GO CNTRC]} \qquad (6)$$

$$V_{CNT}(Z) = \left(1 + \frac{2z}{h}\right)V^*_{CNT} \qquad \text{[F.GV CNTRC]} \qquad (7)$$

$$V_{CNT}(Z) = 2\left(\frac{2|z|}{h}\right)V^*_{CNT} \qquad \text{[F.GX CNTRC]} \qquad (8)$$

Where,

$$V^*_{CNT} = \frac{w_{CNT}}{w_{CNT} + \left(\frac{\rho^{CNT}}{\rho^m}\right) - \left(\frac{\rho^{CNT}}{\rho^m}\right)w_{CNT}} \qquad (9)$$

Where w_{CNT} is mass fraction of the CNTRC plate. ρ^m & ρ^{CNT} are matrix & carbon nanotube density respectively. Similarly, density, thermal expansion coefficient in the transverse and longitudinal direction (a_{11}, & a_{22}) poison ratio v_{12} of composite quadrilateral sandwich plate is determined by

$$\rho = V_{CNT}(Z)\rho^{CNT} + V_m\rho^m \qquad (10)$$

$$v_{12} = V_{CNT}(Z)v^{CNT}_{12} + V_m v_m \qquad (11)$$

$$a_{11} = V_{CNT}(Z)a^{CNT}_{11} + V_m a^m \qquad (12)$$

$$a_{22} = (1 + v^{CNT}_{12})V_{CNT}(Z)a^{CNT}_{22} + (1 + v_m)V_m a^m - v_{12}a_{11} \qquad (13)$$

MATERIALS PROPERTIES

SWCNT properties change according to temperature, so in this study, CNTs material property is used at different temperatures, with data from Zhu *et al.* [2] (Tables **1** and **2**).

Table 1. Properties of (10,10) SWCNT at different temperatures.

T(k)	E_{11} (TPa)	E_{22} (TPa)	G_{12} (TPa)	$\alpha_{11}^{CNT}(\frac{10}{K})$	$\alpha_{22}^{CNT}(\frac{10}{K})$
300K	5.6462	7.0801	1.9445	3.4582	5.1681
500K	5.5301	6.9343	1.9643	4.5361	5.0182
700K	5.4743	6.8642	1.9641	4.6672	4.8943

Table 2. CNT efficiency parameters.

V_{CNT}^*	η_1	η_2
0.11	0.149	0.934
0.14	0.150	0.941
0.17	0.149	1.381

RESULTS AND DISCUSSION

In this analysis, matrix material taken is PMPV, and its property values considered are $\alpha^m = 73.5 * 10^{-6}K^{-1}$, $\rho^m = 1150^{kg}/_m^3$, $E^m = 2.1G$ pa, $\rho_m = 0.34$ taken at room temperature and assumed constant at given temperature for complete analysis. For carbon nanotube reinforcement, the armchair (10,10) SWCNTs is chosen, and effective material properties like modulus of elasticity, poison ratio, density, and thermal expansion coefficient are evaluated by a formula for each layer. The plate dimensions are taken as a/h=10, c/a=0.7, b/a=0.8, β1=105°,β2=75°. For calculation, the thickness of plate is considered 3 mm (1 mm for each face plates and 1 mm for core). Four types of functionally distribution U.D, F.GX, F.GO, F.GV, CNTRCs are considered with 2 types of B.C SSSS & CCCC (all four bottom edges are in simply supported condition and fixed condition).

In view of validation of mathematical approach to find out various material properties and result obtained by ANSYS R17.2, the result is prepared and compared to P. Zhu *et al.*'s previously published result [2]. Table **3** presents a square FG plate with V_{CNT}^* = 0.11 a/h =10, for UD grading with SSSS and CCCC boundary condition. While comparing, it is visible from Table **3** that the value of frequency of F.G square plate is more in CCCC boundary condition than SSSS boundary condition. On increasing the volume fraction of the carbon nanotube from 0.11 to 0.14 and 0.14 to 0.17, the natural frequency of the square plate is increasing for the same type of grading and dimension. And obtained results show good agreement with the previously published paper by P. Zhu *et al.* So, based on this validation, natural frequency and central deflection of the quadrilateral sandwich plate are proceeded.

The obtained value of natural frequency is mentioned in Table **4** for CCCC boundary condition and in Table **5** for the SSSS boundary condition. It is found that at a constant temperature of 300K and MODE 1, the value of frequency is higher for UD type grading & minimum for F.G-V type grading at all respective temperatures. The value of natural frequency increases for the respective type grading distribution towards higher value when we fix the temperature (for example, 300K) and volume fraction of carbon nanotube (for example, 0.11) on increasing MODE from 1 to 2 & 2 to 3, but overall increased natural frequency of UD is maximum and V is minimum. Similar operations are carried out with carbon nanotubes with volume fractions at 0.14 and 0.17 at temperatures of 500K and 700K for MODE 1, 2, and 3. To further comprehend these shifting natural frequency patterns using carbon nanotube volume fractions, graphs are displayed in Figs. (**3a** and **3b**). For a natural frequency with a maximum value of UD type and a minimum value of V type at temperatures of 300K, 500K, and 700K and a volume percentage of 0.11 of carbon nanotube with MODE 1, 2, and 3 of the sandwich plate. The acquired value of natural frequency is shown in Table **5** when all other parameters are held constant and only the boundary condition is changed from CCCC to SSSS.

Table 3. Comparison of dimensional less natural frequency.

V^*_{CNT}	a/h	F.G type	P. Zhu *et al.* (2012) (SSSS)	2021 (SSSS)	P. Zhu *et al.* (2012) (CCCC)	2021 (CCCC)
0.11	10	UD	17.700	17.8362	23.041	23.0695
0.14	10	UD	28.230	28.1008	37.972	37.6179
0.17	10	UD	34.448	35.006	46.250	46.925

$$\bar{\omega} = \omega (a^2/h)\sqrt{\rho^m/E^m}$$

Table 4. Non dimensional frequency

CCCC Boundary Condition						
V^*_{CNT}	Temperature	MODE	UD	F.G-V	F.G-X	F.G-O
0.11	300K	1	10.9761	10.6727	10.8818	10.8245
		2	16.9601	16.6107	16.7976	16.7097
		3	19.1839	18.9251	19.0265	18.9306

(Table 4) cont.....

-	500K	1	10.8653	10.5652	10.7719	10.7153
		2	16.7888	16.4435	16.6280	16.5412
		3	18.9901	18.7344	18.8345	18.7397
-	700K	1	10.8110	10.5124	10.7182	10.6616
		2	16.7048	16.3616	16.5450	16.4584
		3	18.8951	18.6409	18.7404	18.6460
0.14	300K	1	13.3164	12.7876	13.0403	12.9839
		2	20.5961	19.8901	20.1394	20.0528
		3	23.3011	22.6840	22.8460	22.7506
-	500K	1	12.2053	11.8526	12.0986	12.0348
		2	18.8773	18.4644	18.6937	18.5961
		3	21.3571	21.0434	21.1797	21.0729
-	700K	1	12.1438	11.7931	12.0377	11.9742
		2	18.7823	18.3718	18.5998	18.5024
		3	21.2496	20.9377	21.0731	20.9668
0.17	300K	1	13.4534	13.0738	13.3413	13.2643
		2	20.8272	20.3856	20.6338	20.5148
		3	23.5675	23.2390	23.3810	23.2523
-	500K	1	13.3164	12.9410	13.2054	13.1293
		2	20.6151	20.1785	20.4240	20.3061
		3	23.3278	23.0036	23.1435	23.0170
-	700K	1	13.2494	12.8760	13.1388	13.0631
		2	20.5110	20.0770	20.3208	20.2038
		3	23.2101	22.8882	23.0281	22.8993

The obtained value of natural frequency is changed to dimensionless by using formula $\bar{\omega} = \omega (a^2/h)\sqrt{\rho^m/E^m}$.

$$\bar{\omega} = \omega \left(\frac{a^2}{h} \right) \sqrt{\frac{\rho^m}{E^m}}$$

Table 5. Non dimensional frequency

			SSSS Boundary Condition			
V^*_{CNT}	Temp.	MODE	UD	F.G-V	F.G-X	F.G-O
0.11	300K	1	7.9776	7.7394	7.9068	7.8690
		2	10.5395	10.2742	10.4786	10.4220
		3	11.6142	11.3877	11.5402	11.4494
-	500K	1	7.8970	7.6648	7.8268	7.7895
		2	10.4327	10.1740	10.3730	10.3170
		3	11.4947	11.2765	11.4239	11.3338
-	700K	1	7.8575	7.6230	7.7840	7.7507
		2	10.3803	10.1199	10.3161	10.2653
		3	11.4343	11.2168	11.3613	11.2772
0.14	300K	1	9.6996	9.2674	9.4534	9.4152
		2	12.8094	12.3166	12.5497	12.4937
		3	14.1343	13.6947	13.8759	13.7855
-	500K	1	8.8902	8.6225	8.8107	8.7686
		2	11.7400	11.4410	11.6714	11.6084
		3	12.9506	12.6837	12.8565	12.7546
-	700K	1	8.8456	8.5792	8.7666	8.7244
		2	11.6812	11.3835	11.6128	11.5500
		3	12.8880	12.6203	12.7919	12.6906
0.17	300K	1	9.7913	9.5227	9.7391	9.6852
		2	12.9823	12.6298	12.8946	12.8167
		3	14.2762	14.0044	14.2038	14.0846
-	500K	1	9.6916	9.4259	9.6399	9.5866
		2	12.8502	12.5015	12.7634	12.6862
		3	14.1307	13.8621	14.0593	13.9412
-	700K	1	9.6428	9.3784	9.5913	9.5382
		2	12.7854	12.4384	12.6991	12.6223
		3	14.0595	13.7922	13.9884	13.8708

Tables **4** and **5** make it abundantly clear that the dimensionless frequency of U.D. is higher than that of other distributions, and graph 18.2 (a) illustrates the variational change of this frequency, showing that it is higher for MODE 3 and lowest for MODE 1, and that it decreases as temperature rises from 300 K to 500 K and from 500 K to 700 K.

From Tables **4 & 5**, it is clearly visible that the dimensionless frequency of F.G-V is lowest among other distributions so in graph 18.2 (b), the variational change of dimensionless frequency of F.GV is plotted in which the frequency is higher for MODE 3 and minimum for MODE 1 and on increasing temperature from 300K to 500K & from 500K to 700K, the value of natural frequency decreases.

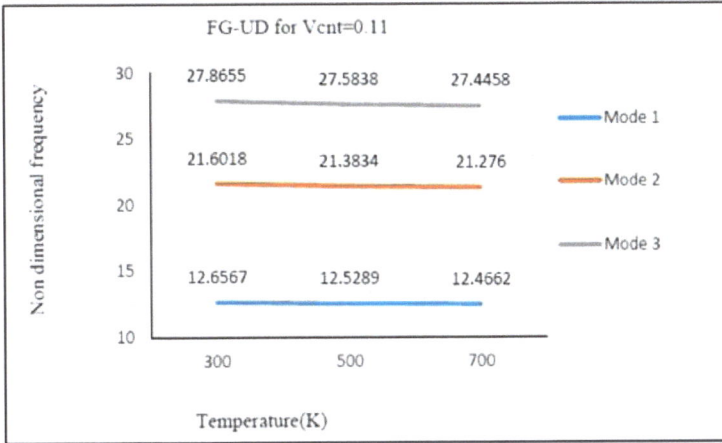

(a)

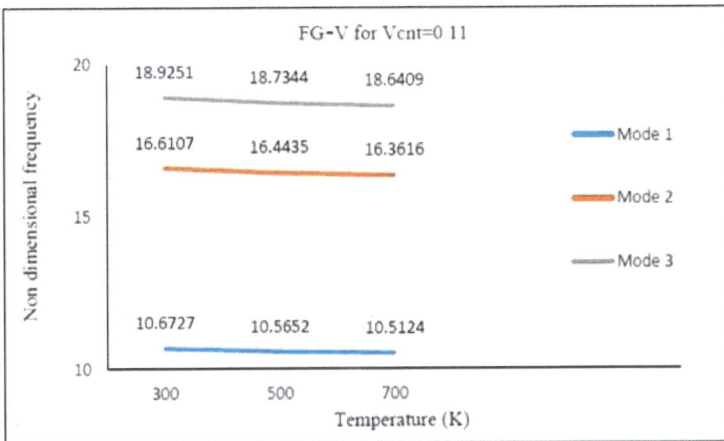

Fig. (2). The variation of dimensionless frequency with respect to different temperatures at a constant volume fraction of CNTs=0.11 in Figure (**a**) & (**b**) for UD & FG-V distribution.

The value of the natural frequency of the UD and FG-V types was chosen from Tables **4** and **5**, and a graph was generated to examine the variational change of the natural frequency on the volume fraction of the carbon nanotube while altering the MODE, as shown in Figs. (**3a** and **3b**).

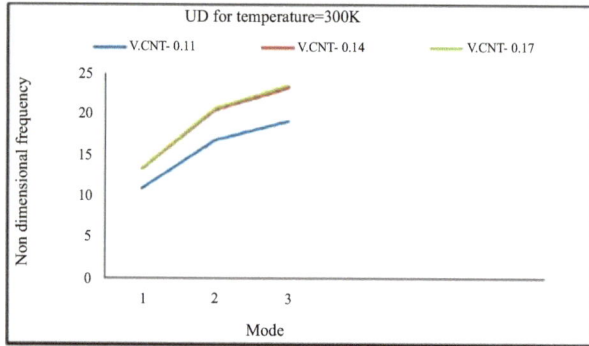

(a)

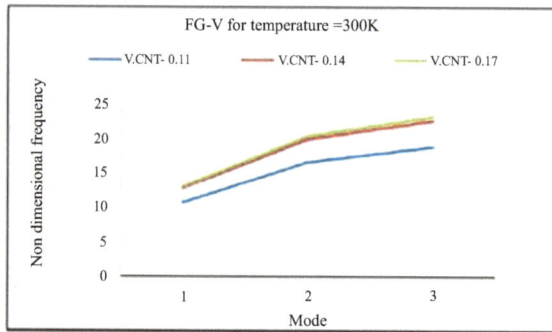

(b)

Fig. (3). The variation of dimensionless frequency at constant temperature of 300K for UD & FG-V distribution in Figure (a) & (b), respectively.

$$w_1 = {}^{w_0}/_h$$

Table 6. Dimensionless central deflection

CCCC Boundary Condition					
V^*_{CNT}	Temperature	U. D	F.GV	F.GX	F.GO
0.11	300K	14.7310	15.1690	14.9740	15.1350
-	500K	15.0340	15.4790	15.2800	15.4450
-	700K	15.1850	15.6350	15.4340	15.6010
0.14	300K	9.9480	10.5320	10.3530	10.4430
-	500K	11.8420	12.2030	12.0400	12.1690

(Table 6) cont.....

-	700K	11.9620	12.3270	12.1620	12.2930
0.17	300K	9.6870	9.9890	9.8420	9.9570
-	500K	9.8880	10.1950	10.0450	10.1630
-	700K	9.9880	10.2990	10.1470	10.2660

Now compute the central deflection of the plate using a uniformly distributed force of 1 MPa on the face plate for two distinct boundary conditions (SSSS and CCCC) and three different temperatures (300 K, 500 K, and 700 K) for the entire sandwich plate. Obtained results were changed to dimensionless by formula w1=w0h . With regard to the central deflection values listed in Tables **6** and **7** for CCCC and SSSS, it is evident that the deflection value for FG-V is at its highest level and the lowest level for UD. Additionally, as the temperature rises from 300 K to 500 K and 500 K to 700 K, the values of the corresponding central deflections increase with respect to all boundary conditions (SSSS and CCCC). When the volume fraction of carbon nanotube increases from 0.11 to 0.14 & 0.14 to 0.17, the value of central deflection decreases with all boundary conditions (SSSS and CCCC). The sandwich plate's stiffness, which depends on the core's grade and the volume percentage of the carbon nanotube, is the physical element causing this kind of result. The stiffness value may be deduced from the results shown in Tables **4, 5, 6,** and **7**.

$$w_1 = {}^{w_0}/_h$$

Table 7. Dimensionless central deflection

SSSS Boundary Condition					
V^*_{CNT}	Temperature	U. D	F.G-V	F.G-X	F.G-O
0.11	300K	18.9480	19.9663	19.1810	19.3810
-	500K	19.3380	20.3747	19.5760	19.7800
-	700K	19.5330	20.5797	19.7730	19.9790
0.14	300K	12.7890	13.8317	13.3180	13.4330
-	500K	15.2240	16.0600	15.4150	15.5760
-	700K	15.3790	16.2143	15.5720	15.7340
0.17	300K	12.4530	13.1323	12.5930	12.7400
-	500K	12.7100	13.4033	12.8540	13.0040
-	700K	12.8400	13.5393	12.9840	13.1360

CONCLUSION

Vibration and deflection analysis of the CNTRC quadrilateral sandwich plate is observed in three different temperatures, here face plate is analysed with UD grading & core with functionally grading O, V, X for every case. The plate's effective material properties were evaluated with the help of the extended mixture rule & parameter of CNTs efficiency. The conclusions are as follows.

- FG-V shows maximum central deflection & UD shows minimum central deflection at all boundary conditions (CCCC & SSSS) and on increasing temperature, the value of central deflection increases.
- UD shows maximum natural frequency & FG-V shows minimum natural frequency at all boundary conditions (CCCC & SSSS) & on increasing temperature, natural frequency decreases.
- Volume fraction of CNTRC has a significant effect as volume fraction increases, natural frequency increases & central deflection decreases for all boundary conditions (CCCC & SSSS) at the same temperature.

It is therefore deduced that stiffness is highest for UD type and lowest for V type FG, and that stiffness of quadrilateral sandwich plate increases with increasing carbon nanotube volume fraction and vice versa.

CONFLICT OF INTEREST

The authors declare that they have no known competing financial interests or personal relationships that could have appeared to influence the work reported in this chapter.

REFERENCES

[1] S. Iijima, "Helical microtubules of graphitic carbon", *Nature,* vol. 354, no. 6348, pp. 56-58, 1991. [http://dx.doi.org/10.1038/354056a0]

[2] P. Zhu, Z.X. Lei, and K.M. Liew, "Static and free vibration analyses of carbon nanotube-reinforced composite plates using finite element method with first order shear deformation plate theory", *Compos. Struct.,* vol. 94, no. 4, pp. 1450-1460, 2012. [http://dx.doi.org/10.1016/j.compstruct.2011.11.010]

[3] D. Li, Z. Deng, G. Chen, and T. Ma, "Mechanical and thermal buckling of exponentially graded sandwich plates", *J. Therm. Stresses,* vol. 41, no. 7, pp. 883-902, 2018. [http://dx.doi.org/10.1080/01495739.2018.1443407]

[4] J. Fan, J. Huang, J. Ding, and J. Zhang, "Free vibration of functionally graded carbon nanotube-reinforced conical panels integrated with piezoelectric layers subjected to elastically restrained boundary conditions", *Adv. Mech. Eng.,* vol. 9, no. 7, 2017. [http://dx.doi.org/10.1177/1687814017711811]

[5] G. Dereli, and B. Süngü, "Temperature dependence of the tensile properties of single-walled carbon nanotubes: O (N) tight-binding molecular-dynamics simulations", *Phys. Rev. B Condens. Matter*

Mater. Phys., vol. 75, no. 18, p. 184104, 2007.
[http://dx.doi.org/10.1103/PhysRevB.75.184104]

[6]   Y. Kiani, "Thermal post-buckling of FG-CNT reinforced composite plates", *Compos. Struct.,* vol. 159, pp. 299-306, 2017.
[http://dx.doi.org/10.1016/j.compstruct.2016.09.084]

[7]   A. Draoui, M. Zidour, A. Tounsi, and B. Adim, "Static and dynamic behavior of nanotubes-reinforced sandwich plates using (FSDT)", *J. Nano Res.,* vol. 57, pp. 117-135, 2019.
[http://dx.doi.org/10.4028/www.scientific.net/JNanoR.57.117]

[8]   J.F. Wang, S.H. Cao, and W. Zhang, "Thermal vibration and buckling analysis of functionally graded carbon nanotube reinforced composite quadrilateral plate", *Eur. J. Mech. A, Solids,* vol. 85, no. 104105, p. 104105, 2021.
[http://dx.doi.org/10.1016/j.euromechsol.2020.104105]

[9]   R. Moradi-Dastjerdi, M. Foroutan, and A. Pourasghar, "Dynamic analysis of functionally graded nanocomposite cylinders reinforced by carbon nanotube by a mesh-free method", *Mater. Des.,* vol. 44, pp. 256-266, 2013.
[http://dx.doi.org/10.1016/j.matdes.2012.07.069]

[10]   M. Kumar, and S.K. Sarangi, "Harmonic response of carbon nanotube reinforced functionally graded beam by finite element method", *Mater. Today,* vol. 44, pp. 4531-4536, 2021.

[11]   A. Alibeigloo, and A. Emtehani, "Static and free vibration analyses of carbon nanotube-reinforced composite plate using differential quadrature method", *Meccanica,* vol. 50, no. 1, pp. 61-76, 2015.
[http://dx.doi.org/10.1007/s11012-014-0050-7]

CHAPTER 19

Polymer Nanocomposites with Improved Electrical and Thermal Properties for Smart Electronic Material Applications

Tapan Kumar Patnaik[1,*]**, Simadri Priyanka Achary**[1]**, Jyoti Behera**[1] **and Sanjukta Mishra**[1]

[1] *Department of Physics, GIET University, Gunupur, Rayagada, Odisha, India*

Abstract: We know that in today's scenario, smart materials are in fame and are an important part of our daily life. Polymer nanocomposites (PNC) have attracted significant research and industrial interests due to their promising potential for versatile application. This nanocomposite technology has emerged from the field of engineering plastics and potentially expanded its application to structural materials, coatings, and packaging for medical products and electronic and photonic devices. The possibility of electrical and thermal conduction in a polymer matrix with low amounts of nanoparticles brings opportunities for highly demanding applications such as electrical conductors, heat exchangers, sensors, and actuators. The development of smart polymer nanocomposite (SPN) has been an area of high scientific and industrial interest in recent years, due to fantastic improvements achieved in these materials. SPN found potential applications in shape memory, self-healing, self-healing, self-cleaning, and energy harvesting. This paper mainly focuses on the most recent advances in polymer nanocomposites for everyday life applications, which are practically important and extremely useful. The applications of PNC are endless and still increasing rapidly due to their below-average cost and ease of manufacture. They make our lives more convenient and enjoyable. The main target is to study the applications of electrical and thermal properties of PNC for smart electronics materials.

Keywords: Actuators, Smart polymer nanocomposite, Self-healing polymer, Sensors.

INTRODUCTION

A polymer nanocomposite is a composite material comprising a polymer matrix and an inorganic dispersive phase that has at least one dimension, which is of nanometric scale [1]. Another highly potential application of polymer nanocomp-

* **Corresponding author Tapan Kumar Patnaik:** Department of Physics, GIET University, Gunupur, Rayagada, India; Email: tapanpatnaik@giet.edu

Amar Patnaik, Albano Cavaleiro, Malay Kumar Banerjee, Ernst Kozeschnik & Vikas Kukshal (Eds.)

osite is for energy, which includes energy generation and energy storage. Among the types of mostly applied nanofiller for nanocomposite energy applications are metal oxides, Nano clays, carbon nanotubes, and graphenes. Some applications of polymer nanocomposites in biomedical applications are shown in Fig. (**1**).

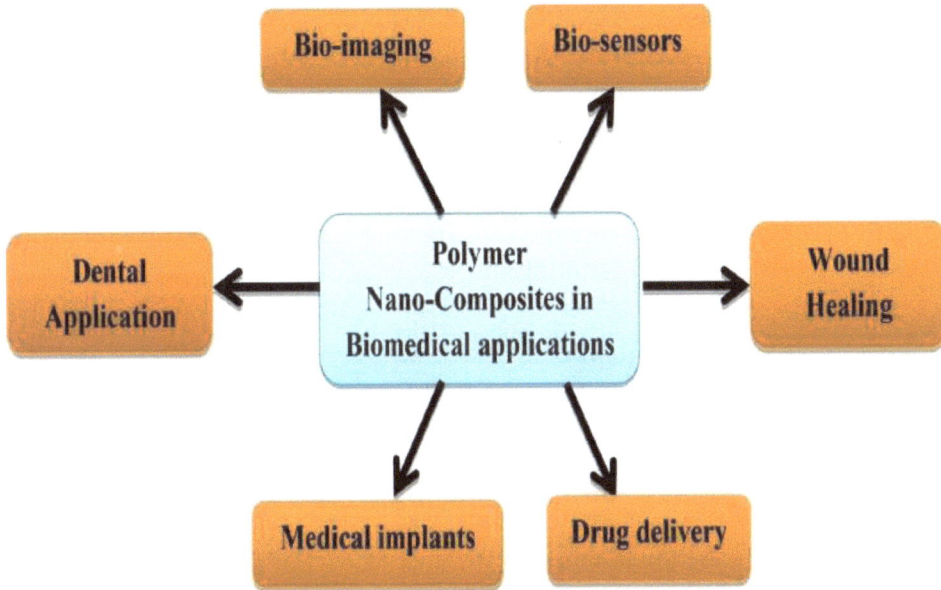

Fig. (1). Applications of polymer nanocomposites in biomedical applications [1].

The smartness of material development always comes from the inspiration and intelligence of nature. Smart materials can be defined as materials that incorporate the functions of sensing, actuation, and control. Smart materials are attracting increasing interest, especially in the era of the fourth industrial revolution and circular economy. Smart polymer nanocomposites (SPN) can be derived from shape memory polymers, stimuli-active polymers, smart electrorheological (ER) and magnetorheological (MR) polymers; self-healing polymer, self-cleaning polymer, self-healing polymer, self-sensing polymer, energy-harvesting and energy storage polymer are the latest hot research topics [2]. The addition of Nano filler can increase the performance of the SPN (*e.g.,* shape fixity, shape recovery, self-healing ability) due to their high specific surface area, nucleating effects, reinforcing effects, and inherent functionalities (*e.g.,* thermal conductivity, electrical conductivity). SPNs are widely used in various applications, for example, sensors and actuators, stre*tch*able electronics, wearable electronics, smart textiles, drug carriers and delivery, aircraft and aerospace applications. Ten years ago, Ratna and Karger-Kocsis documented a review on shape memory polymers and composites. When revisiting the 'future outlook' from Ratna and

Karger-Kocsis, they 'forecasted' that the blending and interpenetrating of network-based polymers in SMP could have led to the development of newer and novel SMPs. They also 'predicted' that SMPs are going to be the future materials for deployable structures for aircraft and spacecraft applications. And, nowadays, numerous research publications deal with SMP *via* blending and interpenetrating network strategies. This mini-research on smart polymer nanocomposite is mainly focused on the shape memory polymer (SMP) and self-healing polymer (SHP) [3], which is part of the 'heritage' of Professor Karger-Kocsis. This article provides a basic fundamental idea of how to fabricate SPN with different types of processing techniques. This could benefit researchers (especially 'beginners') looking for a feasible method to synthesize and prepare SPN, using as simple as solvent casting and melt blending, to some sophisticated methods (*e.g.,* electrospinning, 3D and 4D printing).

Smart Polymer Nanocomposite

Shape Memory Polymer

External stimuli can cause Shape Memory Polymer (SMP) to change shape temporarily, but it can then return to its original shape. The stimuli for shape memory alloy (SMA) is always limited to heat (*e.g.,* Joule heating) and magnetic field, whereas there are more 'triggering choice' for SMP, *e.g.,* light, water, solvent, pH, temperature, and electricity. Furthermore, compared with SMA, most of the SMP has more advantages, for example, low density, high elastic deformation, good biocompatibility, and bio-degradability, as well as fantastic tailor-ability. In general, SMP consists of permanent net points (a stable polymer network) and molecular switches of reversible nature [4]. The former memorizes the original shape, while the latter is used for setting the temporary shape. When molecules entangle, crystallize, cross-link, or intertwine in SMP, they create the permanent net points of SMP. There is a variety of thermosensitive SMP with physical (or chemical) network points that determine their original shape, while each temporary shape is fixed by switching domains associated with a thermal transition temperature (Ttrans), such as the crystallization/melting transition, the vitrification/glass transition, the liquid crystal anisotropic/isotropic transition, reversible molecule cross-linking, and supra molecular association/disassociation. The essential performance parameters for an SMP include Ttrans, shape fixity ratio (Rf), and shape recovery ratio (Rr) [4]. The viscoelastic and relaxation behaviors of SMP are attributed to the cooperative relaxation of each component. The working mechanism in the thermoresponsive SMP is governed by both the internal recovery stress and stored mechanical energy [5].

Self-healing Polymer

Self-healing polymer (SHP) is one type of smart material that can demonstrate healing-ability, in-situ damage repairing, and properties restoring. For composite constructions, this means longer service life, improved product reliability, less frequent inspections, and cheaper maintenance costs [3]. Extrinsic self-healing and intrinsic self-healing are the two main healing mechanisms. The former can heal the damage (*e.g.,* cracks) by the polymer itself, where the inherently thermally reversible bonds of the material create a healing effect under external heating, while the latter is pre-embedded with the healing agent (*e.g.,* microcapsules, micro-vascular networks, hollow fibers). Self-healing technology can be applied to thermoplastic and thermoset systems. Vijayan and AlMaadeed material, shell wall thickness, storage stability, and its response to external stimulations (*e.g.,* mechanical impact) are of great importance in determining self-healing efficiency [6].

To demonstrate the extrinsic self-healing technology and the importance of Nano filler for SHP nanocomposite, here we give an example of SHP nanocomposite for outer space application. One of the major challenges for space missions is that most of the materials degrade over time under extreme environments. Self-healing materials on composite spacecraft components were studied by Madara *et al.* [7]. A ruthenium Grubbs' catalyst was used to make microcapsules containing 5-ethylidene-2-norbornene and dicyclopentadiene monomers with self-healing properties. An epoxy resin (epoxy prepolymer Epon 828 and curing agent Epicure 3046) and single-wall carbon nanotubes were then combined in with the self-healing materials utilizing a vacuum centrifuging process. The SHP nanocomposite was infiltrated into carbon fiber-reinforced polymer layers (CFRP). Hypervelocity impact on CFRP specimens was then applied using a homemade implosion-driven hypervelocity launcher, which is common in the space environment [8]. The thickness distribution of the CFRP with SHP nanocomposite is much more homogenized compared to that of CFRP (with pure Epon 828); it denotes a positive healing impact (i.e as a result, there is reduced delamination propagation following treatment). The healing efficiency of the CFRP (4 layers of woven fiber) with SHP nanocomposite is about 83%. There are many ways in which intrinsic self-healing materials might repair themselves, including hydrogen bonding, $\pi-\pi$ stacking, and Diels–Alder bonds (*e.g.,* urea and Diels–Alder bonds).

In comparison to self-healing materials based on non-covalent interactions, dynamic covalent bonding-based materials have advantages such as high strength, higher temperature stability, and solvent resistance. Based on Diels–Alder (DA) chemistry, Zhao *et al.* [9] created a self-healing polysiloxane nanocomposite reinforced with furan-functionalized graphene. More than 90% of its tensile strength and 98% of its elongation at break were regained by the healable polysiloxane nanocomposite. Self-healing properties are facilitated by the reversible DA bonds between the polysiloxane chains and graphene functionalized with the furan group. The above-mentioned intrinsic self-healing technology is the potential to fabricate SHP nanocomposite with better healing performance and enhanced mechanical properties. The self-healing efficiency is one of the performance indicators for the SHP nanocomposite. The effectiveness and efficiency of self-healing for SHP nanocomposite are controlled by several factors, *e.g.,* the polymer matrix, the selection of nanofiller, the choices of self-healing mechanism (intrinsic or extrinsic), and the environment. The concept of self-healing for other materials, *e.g.,* cement and concrete in the industry (*e.g.,* hollow tubes extruded from cementitious material can be used to hold and release healing ingredients) and nature creatures (*i.e.,* self-healing of starfish) can be used as an additional 'reference' for the development of SHP nanocomposite [6].

Classification of Nanocomposites

Intercalated nanocomposites, flocculated nanocomposites, and exfoliated nanocomposites are three subcategories of nanocomposites based on the interfacial contact between the polymer matrix and layered silicates and are discussed below [10].

Intercalated Nanocomposites

In this class of nanocomposites, the polymer matrix is introduced into the layered silicate structure in a crystallographically uniform manner to swell the spacing between the platelets. A few molecular layers of polymers are often used to build these nanocomposites, which have characteristics that are strikingly comparable to those of ceramics [10].

Floccultaed Nano Composites

These nanocomposites are very similar to intercalated nanocomposites, sometimes, in this class of composite, the silicate layers are flocculated because of the hydroxylated edge-edge interaction of the silicate layers [10].

Exfoliated Nanocomposites

In exfoliated nanocomposites, depending upon the clay loading, the individual clay layers are separated in the continuous polymer matrix. The clay content of this class of nanocomposite is much lower than intercalated nanocomposite.

Fillers

Depending on their size, the fillers commonly employed to reinforce polymer materials can be categorized into three categories: coarse, medium, and fine. Particles smaller than one micron in diameter, or "conventional-sized filler," make up the first type of filler. The size of these particles can vary widely, although they tend to fall within the range of 1 μm to 100 μm. Submicron-sized particles make up the second group [11]. They range in size from 100 nm to 1 m in length. The term "microparticles" is used to refer to a single unit of the third type of particle. The nanometer-sized particles are the third group of particles. At least one dimension of these particles should be smaller than 100nm in diameter. A nanofiller can be one-dimensional, two-dimensional, or even three-dimensional (nanotubes, fibers, rods) (spherical particle) [12].

Thermal Properties

The thermal properties of nanocomposites can be analyzed by DSC from the weight loss on heating the nanocomposites; the thermal stability can be calculated from the heat resistance of nanocomposites using heat distortion temperature (HDT). The dependence of HDT on clay content has been investigated by several researchers. The improvement of thermal stability of nanocomposite by the addition of nanofillers has been studied by many researchers. The addition of laser oven SWNT to the epoxy nanocomposites. A 300% increased thermal conductivity by 10% aligned magnetically [13]. According to Kumanek *et al.* [14], the reduced interfacial area and the presence of shielded internal layers in MWNT make them the best candidates for improving the thermal conductivity of nanocomposites. They can be used in printed circuit boards, thermal interface materials, heat sinks, connectors, and high-performance thermal management systems because of their high thermal conductivity.

Electrical Properties

The electrical properties of nanocomposites depend on several factors, such as aspect ratio, dispersion, and alignment of the conductive nanofillers in the structure. Adding CNTs to nanocomposites improves their electrical conductivity. The ionic conductivity of nanocomposites is several orders of magnitude greater than that of the comparable material. Electroactive polymers can be incorporated into clay minerals to increase conductivity. The electrical conductivity increased by several orders of magnitudes with a very small loading (0.1 wt % or less) of nanotubes to the nanocomposites without altering the other properties. As shown in Fig. (**2**), conductive nanocomposites are used in many applications such as electrostatic dissipation, electrostatics painting, electromagnetic Interference shielding, *etc* [13].

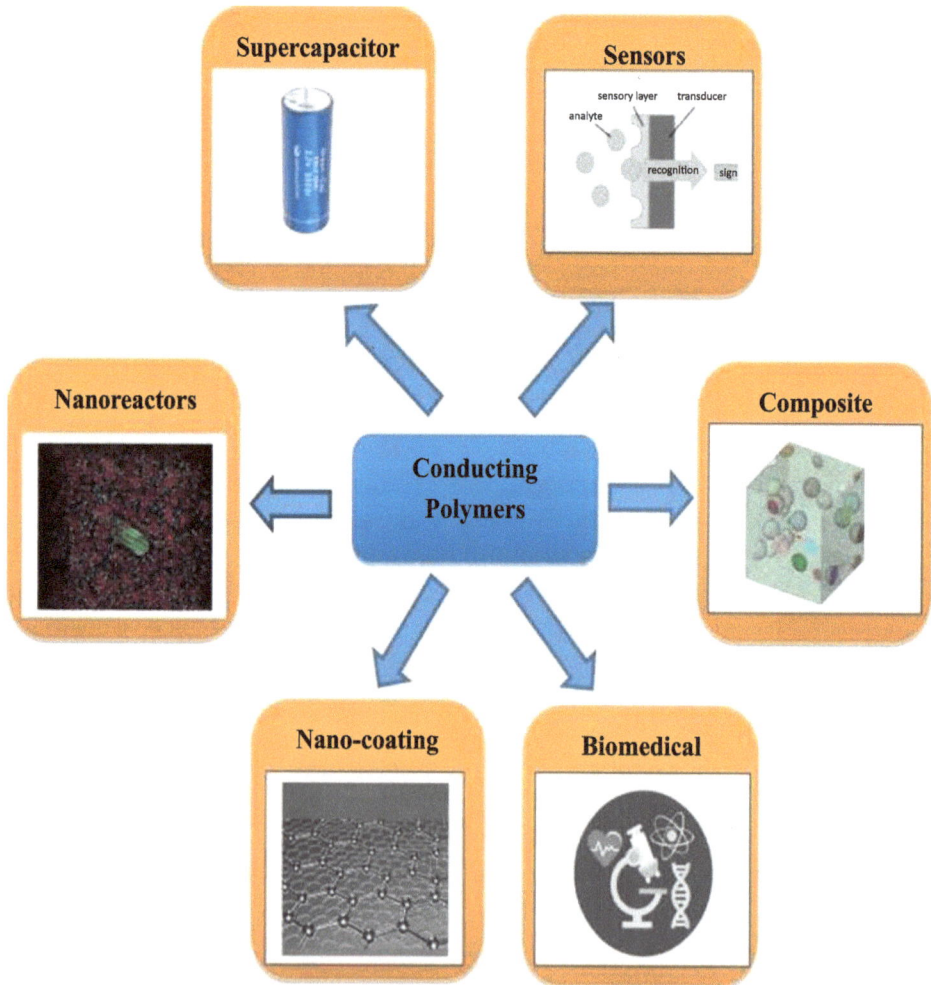

Fig. (2). Conducting polymers in the field of energy.

Methodology

There are various methods to synthesize polymer nanocomposites. The different methods are discussed below:

Melt Intercalation

The most common method for making thermoplastic polymer nanocomposites is to use melt intercalation. The polymer matrix is annealed at high temperatures, the filler is added, and the composite is then kneaded. Because no solvents are used, it is less harmful to the environment. Additional advantages include its compatibility with industrial procedures like extrusion or injection molding, which makes it more cost-effective. However, the filler's surface modification can be degraded by the procedure's high temperature. To achieve adequate dispersion and exfoliation, it is critical to optimize the processing conditions. Nanocomposites can be formed by weak electrostatic forces between the filler interlayers and compatibility with the polymer matrix.

Template Synthesis

Utilizing a nanopore membrane as a template, conductive polymers, metals, semiconductors, and carbon can all be made into nanofibers using template synthesis methods.

Exfoliation Absorption

When a polymer or prepolymer is dissolved in a solvent, the process of exfoliation adsorption, or polymer intercalation from solution, takes place. Before combining the polymer solution with the layered silicate, the silicate is first swelled and dispersed in a solvent. To remove the solvent from between the silicate layers, polymer chains intercalated. When the solvent is removed, the sheets reassemble, trapping the polymer chains, and a multilayered structure is produced. Poly (ethylene oxide), poly (vinyl pyrrolidone), and poly (vinyl pyrrolidone) are all examples of water-soluble polymers that can be utilized to make intercalated nanocomposites (acrylic acid). Due to the enormous amounts of solvents used, melt intercalation is more environmentally friendly than this process. Monomers, such as methyl methacrylate and styrene, are disseminated in water along with an emulsifier and varying quantities of silicates during emulsion polymerization.

In Situ Polymerization

Low-molecular-weight monomer seeps between interlayers during in situ polymerization, causing swelling of the filler in the liquid monomer or monomer solution. Heat, radiation, cationic exchange, or diffusion of initiator are all ways to initiate polymerization. Intercalated or exfoliated nanocomposites are formed

when the monomers polymerize between the interlayers. When compared to melting and exfoliation adsorption approaches, this method provides better exfoliation.

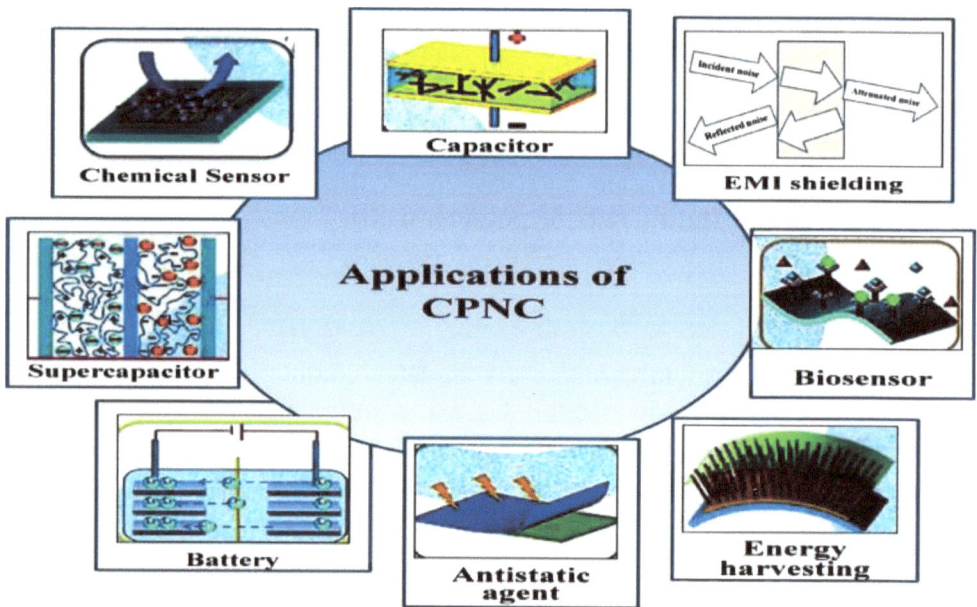

Fig. (3). Novel trends in Conductive Polymer Nanocomposites.

Applications of Polymer Nanocomposites in Electronic Components

Nanotechnology is deeply embedded in the design of advanced devices for electronic and optoelectronic applications. The dimensional scale for electronic devices has now entered the nano range [2, 8]. The utility of polymer-based nanocomposites in these areas is quite diverse, involving many potential applications, and has been proposed for their use in various applications.

1. Chemical sensors, electroluminescent devices, electrocatalysis, batteries, smart windows, and memory devices all rely on these materials.

2. It could also be used to make photovoltaic (PV) cells and photodiodes; supercapacitors; printed conductors; light-emitting diodes; and field effect transistors (FETs).

3. For insulating polymers, the electrical conductivity of carbon nanotubes has also drawn attention. Electromagnetic interference shielding, transparent conductive coatings, electrostatic dissipation of electrostatic charge, and a variety of electrode applications are among the possible uses.

4. Sensors ranging from gas sensors to biosensors to chemical sensors have been developed using conjugated polymers with varied nanoscale filler inclusions. As a part of the manufacturing process, we use a variety of small-scale fillers, such as metal oxide nanowire-coated carbon nanotubes.

5. Grapheme's rise to prominence in electrical applications is reminiscent of the discovery of carbon nanotubes. As the boundaries of Moore's law are reached, grapheme in sheet form may show promise as a replacement for silicon.

6. Using polymer-based solar cells, big flexible panels at a low cost are possible. The sole drawback is that it is significantly less efficient than commercial solar cells. Nanoparticles of Cadmium chalcogenides, such as CdS, have been utilized to create solar cells.

7. A new class of hybrid materials based on polymer semiconductor nanocomposites can be used in a wide range of applications, including optical displays, photovoltaics, and gas sensors, as well as electrical and mechanical devices. These novel trends in Conductive Polymer Nanocomposites are shown in Fig. (**3**).

Advantages of Polymer Nanocomposites

1. As a result of using less high-density material, they are more lightweight than traditional composites [1].
2. When compared to a plain polymer, their barrier characteristics are superior.
3. Superior mechanical properties (modulus and strength).
4. Structural and thermal stability.
5. Promising electrical conductivity.
6. Noise damping and corrosion resistance.

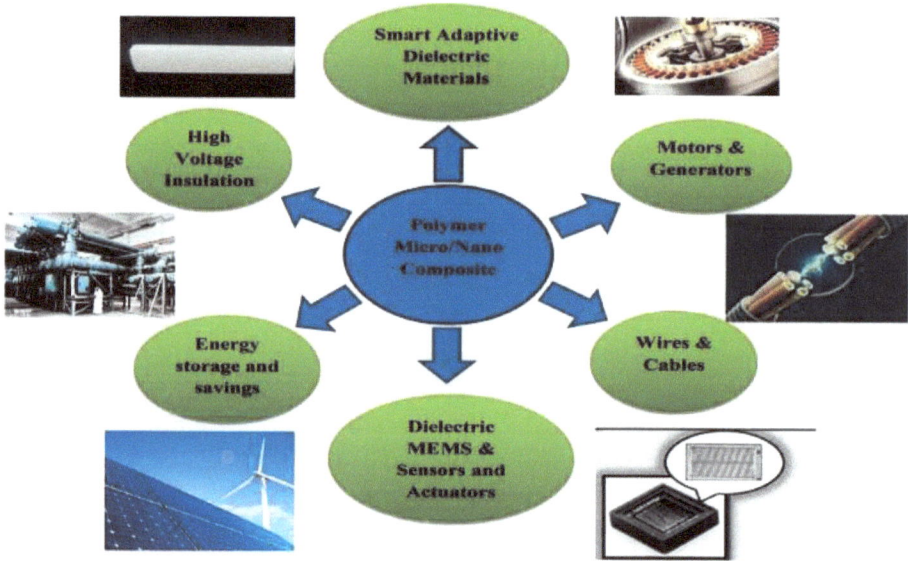

Fig. (4). The next generation of high voltage applications employing polymer-based nanocomposites.

Disadvantages of Polymer Nanocomposites

1. Exfoliation and dispersion of particles must be made simpler due to a lack of knowledge regarding formulation, property, and structural relationships.
2. Cost-effectiveness.
3. Non-uniform distribution.
4. High viscosity.
5. Formation of agglomeration.

Future of Polymer Nanocomposite

Using polymer nanocomposites in packaging will be a game changer for the industry. Companies will use this technology to boost their product's stability and survivability through the supply chain to deliver superior quality to their customers while saving money when production and materials costs are reduced. In the long term, the benefits of nanocomposites exceed the costs and problems. As more research is done into new nanofillers (such as carbon nanotubes), the use of nanocomposite packaging in a wide range of applications is expected to continue to rise [1, 2, 10]. The next generation of high-voltage applications employing polymer-based nanocomposites is shown in Fig. (4).

CONCLUSION

It has just been a decade since smart nanocomposites gained prominence in the field of material research. Even though the development of promising techniques is just a fancy, it has an emerging future in various applications. Nanocomposites made of polymer have been hailed as a hot new research subject in recent years because of their enormous potential. The development and use of nanocomposites will be influenced by our ability to overcome difficulties. Within a decade, it is expected to have a market impact of hundreds of billions of dollars on new materials.

REFERENCES

[1]   S. Thomas, S. Bandyopadhyay, S. Thomas, and R. Stevens, "Polymer Nanocomposites:-preparation, properties and applications", *Gummi Fascrn Kunststoffe,* p. 9, 2006.

[2]   M. Vera, C. Mella, and B.F. Urbano, "Smart polymer nanocomposites: Recent advances and perspectives", *J. Chil. Chem. Soc.,* vol. 65, no. 4, pp. 4973-4981, 2020.

[3]   H. Colquhoun, and B. Klumperman, "Self-healing polymers", *Polym. Chem.,* vol. 4, no. 18, pp. 4832-4833, 2013.

[4]   F. Pilate, A. Toncheva, P. Dubois, and J.M. Raquez, "Shape-memory polymers for multiple applications in the materials world", *Eur. Polym. J.,* vol. 80, pp. 268-294, 2016.

[5]   M. Zare, M.P. Prabhakaran, N. Parvin, and S. Ramakrishna, "Thermally-induced two-way shape memory polymers: Mechanisms, structures, and applications", *Chem. Eng. J.,* vol. 374, pp. 706-720, 2019.

[6]   K.R. Reddy, A. El-Zein, D.W. Airey, F. Alonso-Marroquin, P. Schubel, and A. Manalo, "Self-healing polymers: Synthesis methods and applications", *Nano-Structures & Nano-Objects.,* vol. 23, p. 100500, 2020.

[7]   S.R. Madara, N.S. Raj, and C.P. Selvan, "Review of research and developments in self healing composite materials", In: *IOP Conf. Series: Mat. Sci. Eng.* vol. 346. IOP Publishing., 2018, no. 1, p. 012011.

[8]   M.N. Azlin, R.A. Ilyas, M.Y. Zuhri, S.M. Sapuan, M.M. Harussani, S. Sharma, A.H. Nordin, N.M. Nurazzi, and A.N. Afiqah, "3D printing and shaping polymers, composites, and nanocomposites: A review", *Polymers,* vol. 14, no. 1, p. 180, 2022.

[9]   L. Zhao, B. Jiang, and Y. Huang, "Functionalized graphene-reinforced polysiloxane nanocomposite with improved mechanical performance and efficient healing properties", *J. Appl. Polym. Sci.,* vol. 136, no. 27, p. 47725, 2019.

[10]   S.S. Ray, and M. Okamoto, "Polymer/layered silicate nanocomposites: A review from preparation to processing", *Prog. Polym. Sci.,* vol. 28, no. 11, pp. 1539-1641, 2003.

[11]   C.S. Reddy, R.N. Mahaling, and C.K. Das, "16 Particulate-filled vinyl polymer composites", In: *Handbook of Vinyl Polymers.* vol. 499. , 2009.

[12]   T.P. Sathishkumar, J.A. Naveen, and S. Satheeshkumar, "Hybrid fiber reinforced polymer composites–a review", *J. Reinf. Plast. Compos.,* vol. 33, no. 5, pp. 454-471, 2014.

[13]   B.P. Singh, D. Singh, R.B. Mathur, and T.L. Dhami, "Influence of surface modified MWCNTs on the mechanical, electrical and thermal properties of polyimide nanocomposites", *Nanoscale Res. Lett.,* vol. 3, no. 11, pp. 444-453, 2008.

[14]   B. Kumanek, and D. Janas, "Thermal conductivity of carbon nanotube networks: A review", *J. Mater. Sci.,* vol. 54, no. 10, pp. 7397-7427, 2019.

CHAPTER 20

An Investigation of Constant Amplitude Loaded Fatigue Crack Propagation of Virgin and Pre-Strained Aluminium Alloy

Chandra Kant[1] and **Ghulam Ashraf Harmain**[1,*]

[1] *Department of Mechanical Engineering, National Institute of Technology Srinagar, Jammu, and Kashmir, India*

Abstract: The article examines and explores the impact of pre-strain on resistance to fatigue crack propagation (FCP) *via* analytical models. Most of the materials during service and processing have gone through preexisting strain due to strain-invigorating processes. It is an utmost priority of any low-weight and high strength structural requirement. The numerical study is based on aluminum alloy 7475 with T7375 heat treatment, which is a candidate material for the aerospace industry due to its mechanical properties. In this paper, virgin and pre-strained Aluminum7475, 2.54 & 5% are explored in the time-invariant loading for load or stress ratio (R) of 0, 0.1, and 0.4 (minimum stress/maximum stress) *via* fatigue crack propagation model Paris and Crack annealing model. The model selected for the study is rooted in small-scale yielding theory which is based on linear elastic fracture mechanics (LEFM) without crack closure and accounting crack closure (CL). The emphasis on crack closure behavior before and after the pre-strain of material. Effects of load ratio 0, 0.1, and 0.4 have been studied *via* crack closure models- Elber, Newman, and Virtual crack annealing model. A comparative study of fatigue crack propagation Paris & Crack annealing model forecast has been presented for virgin and strained conditions. The predictions are validated *via* experimental data. Prediction error analysis has been presented in the forecast and actual data.

Keywords: Crack annealing model, Fatigue crack propagation, Pre-strained aluminium.

INTRODUCTION

In the automobile sector, a high strength-to-weight ratio is one of the primary criteria for structural material selection which influence fuel economy. In the case of the aviation industry, the lightweight-to-strength ratio is very crucial to meet

* **Corresponding author Ghulam Ashraf Harmain:** Department of Mechanical Engineering, National Institute of Technology Srinagar, Jammu & Kashmir, India; E-mail: gharmain@nitsri.net

Amar Patnaik, Albano Cavaleiro, Malay Kumar Banerjee, Ernst Kozeschnik & Vikas Kukshal (Eds.)

the floating condition in a less dense medium (air). Structural integrity analysis of such materials considering actual manufacturing process parameter effects and loadings *etc.*, is very much required for reliable design.

Aluminum alloy – 7475 with T-7351 heat treatment is a candidate material for aviation vehicle structures (frames, ribs & spares) as it shows design suitable strength with load and shows aversion to oxidation (corrosion) [1]. Alloy 7474-T7351 is a modified form of Al- alloy 7075, which shows good fracture toughness and corrosion resistance which makes it a candidate material for aviation structures. During service, aviation vehicle continuously encounters cyclic loading, which makes fatigue analysis of such material inexorable.

Plastic deformation accumulated during the manufacturing process of various parts influences the microstructure, mechanical properties, *etc.* accumulated plastic deformation is termed pre-strained material [2, 3]. Accounting for the influence of pre-strain on mechanical properties (tensile strength, compressive strength, fatigue crack propagation resistance, *etc.*) is inevitable for the reliable structural integrity of the structure made of it. Plastic deformation, environmental conditions (temperature, humidity, *etc.*), load ratio (R), microstructure, and grain size influence fatigue crack propagation resistance [4]. Almost 90% of the mechanical structure fails due to fatigue loading; hence analysis of the fatigue crack propagation behaviour of the structure is essential for better reliability of the structural design.

For an accurate assessment of the life of a structure experiencing cyclic loading and rate of fatigue crack growth (FCP) for ductile material, crack closure analysis is inevitable, which has been reported by several researchers [5 - 11]. The crack closure phenomenon is identified and reported in [12 - 14], which incorporates the effect of plasticity, roughness, and oxidation-induced obstructions in crack flank opening and closing. Generally, the effect of plasticity-governed crack closure (PGCC) is influenced by load ratio & environmental conditions (temperature, pressure, humidity, test chamber, and gaseous compositions, *etc.*).

This article presents the influence of pre-strain (0%, 4%, and 5%) on fatigue crack propagation rate *via* an *in-situ* experimental study, and also, the influence of varying stress ratios (0, 0.1, and 0.4) is simulated *via* crack closure models. Crack propagation rate versus stress intensity factor data is modeled *via* linear elastic fracture mechanics based Paris model [15]. The crack annealing model [16], Elber model [12], and Newman model [17] are used to imitate fatigue crack growth rate (FCGR) in the cycle *via* cycle method.

The article is systematized as follows – Section 2 discusses the mechanical properties of Al-7475, the prerequisite process for straining the material, and also discusses the apparatus used briefly. Section 3 gives the details of the experimental procedure, and discusses the process parameters and sample geometry. Next, Section 4 discusses crack propagation models used in this study. Section 5 presents the results. This study is concluded in section 6.

MATERIALS AND METHODOLOGY

For the current study, aluminum alloy 7475-t7375 is used to examine the influence of accumulated plastic strain on FCP behavior in the Paris zone (stable fatigue crack propagation zone). The material has been prominently used in aerospace and automobile structure construction due to its good mechanical properties, such as fracture toughness and corrosion resistance which qualifies the material as an optimal choice for reliable structural design in such industries [18].

Table. **1** gives the Chemical constituents of the material used for this study. Mechanical properties of the material are determined from a tensile test at 27^0C using a 100 KN Tinius-Olsen universal testing machine (UTM) with a 50 mm gauge length extensometer. Table. **2** provides a tensile stress-strain response of the Al-7475 with t-7375 heat treatment. Material has been subjected to pre-strain on UTM, where the strain was measured *via* an axial extensometer.

Table 1. Elements of Al-7475

Alloy wt.%	Si	Fe	Cu	Mn	Mg	Cr	Zn	Ti	Al
7475	0.1%	0.12%	1.4%	0.06%	2.3%	0.22	5.9%	0.06%	Balance

Table 2. Mechanical property of virgin and pre-strained Al-7475-T7531

Material condition	Yield strength (σy) (MPa)	Ultimate tensile strength UT (MPa)
Virgin	396	469.15
2.54% strained	405.18	475.14
5% strained	409.71	476.5

EXPERIMENTAL PROCEDURE

The fatigue crack propagation (FCP) test is conducted on the quarter compact sample in compliance with the standard ASTM-647 [19]. FCP sample, used for this analysis, is prepared *via* electric discharge machining (EDM) from (50*50*6.15) mm sheet.

EDM technique was chosen to minimize the machining-induced residual stress to be able to test the actual behavior of the material. The sample is prepared in compliance with ASTM 647 and ASTM 1820. The quarter compact sample is pre-cracked by fatigue loading on servo hydraulic fatigue testing machine Walter Bai Ag 100 KN under the 50 Hz frequency and 0.1 load ratio. The experimental setup is depicted in Fig. (**1**). During fatigue pre-crack, crack propagation is monitored using digital image correlation.

Fig. (1). Experimental Setup.

The sample is ground and polished using a disk polisher for precise monitoring of fatigue crack growth from the surface. During the experiment, fatigue crack is also monitored using a digital image correlation setup of Imetrum with 2-megapixel camera to ensure cross-validate the crack in a short crack regime.

After pre-cracking, fatigue crack propagation tests are conducted for constant amplitude load increasing stress intensity factor mode as per the ASTM-647. During the fatigue crack propagation test, the growing crack is calculated *via* the elastic unloading compliance method as per ASTM 647 [19].

After completing the fatigue crack propagation test, the sample was torn, and fatigue pre-damage length and damage accumulation in the sample was measured using nine points averaging method as suggested in ASTM E-647 [19]. The crack growth rate is plotted w.r.t. LEFM-based fatigue cracks driving parameter stress intensity factor range (ΔK). Paris model parameter (c, and m) are found from Fatigue crack propagation rate (FCPR) and ΔK plot.

FATIGUE CRACK PROPAGATION MODELS

The present simulation study is based on LEFM-rooted FGC models, which assume the negligible size of the plastic zone during FCP. In this article fatigue, crack propagation rate (FCPR) is simulated using the Paris model [15] and virtual crack annealing model [16]. Paris model is an empirical model for predicting increment in flaw (crack) with crack driving parameter ΔK is shown in Equation 1.

$$\frac{da}{dn} = c * [\Delta K^m] \tag{1}$$

Where da=crack growth increment, dn = change in the number of loading cycles, c, m are empirical relation constants. Value of c and m is found from experimental FCP data.

Crack Closure Phenomena

Crack closure (CC) models take care of the influence of crack tip plastic deformation-induced plastic wake, newly formed fatigue crack surface roughness and oxidation-induced obstructions by introducing effective stress intensity factor (ΔKeff) in place of ΔK, which accounts for the effect of plasticity-induced wake in propagating cracks. The phenomena of crack closure are very complex because of plastic deformation, which shows nonlinear behavior. The inclusion of nonlinearity makes the determination of the effective stress intensity factor tedious [9, 20, 21]. show the crack closure behavior using finite element simulation. An analytical crack closure decomposition model is presented in [22, 23], which decomposes the crack closure effect in constant amplitude loading and interspersed single overload. Crack closure phenomena and various crack closure models are summarized in [11, 24]. For simulating the fatigue crack propagation rate using crack closure models, opening stress (σop) value for each loading cycle is required as shown in Fig. (**2**).

$$da/dn = c\Delta K_{eff}{}^m$$

$$\Delta K_{eff} = K_{max} - K_{op} \tag{2}$$

K_{op} is opening stress intensity factor range

$$K_{op} = \beta \sigma_{op}[\pi a_i] \, , \, K_{max} = \beta \sigma_{max}[\pi a_i]$$

β is a geometric factor, σ_{max} is maximum applied stress, σ_{op} is opening stress.

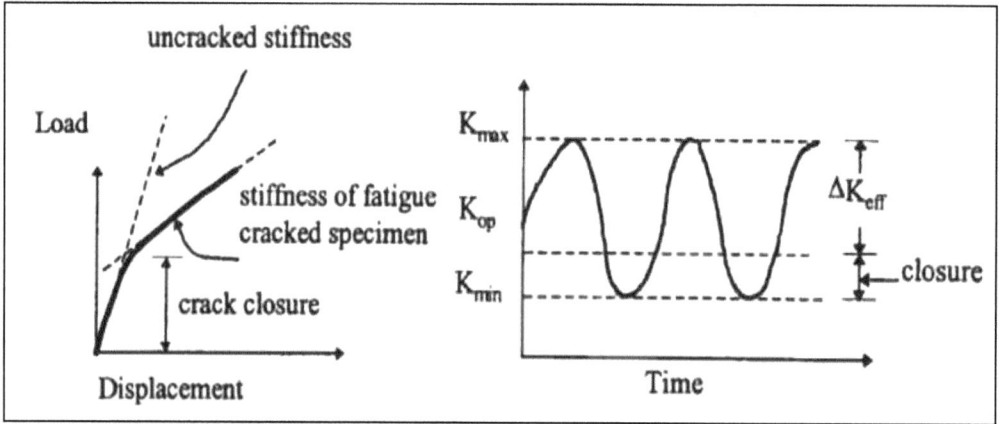

Fig. (2). Crack closure representation.

Primarily ductile material fatigue crack propagation (FCP) is influenced by plasticity-induced crack closure.

Virtual Crack Annealing Model

The virtual crack annealing model (VCA) [16] is a new approach used to calculate crack opening or closure load. The model explains real and virtual crack lengths during loading and unloading, as shown in Fig. (3).

The model seems that during unloading, some part of the crack seems as virgin material, such as no crack as seen in Fig. 3 In the virtual crack reverse plastic zone (drvirtual) is represented *via* Equation 3.

$$d_r virtual = \frac{\pi}{8}\left(\frac{K_{max} - K_{min}}{2\sigma_y}\right)^2 \tag{3}$$

During unloading reversed, plastic zone (dr) if there is no overlapping of crack is determined by Equation 4.

$$d_r = \frac{\pi}{8}\left(\frac{K_{max} - K_{cl}}{2\sigma_y}\right)^2 \tag{4}$$

During loading forward, the plastic zone is given by Equation 5.

$$d = \frac{\pi}{8}\left(\frac{\sigma_{op}-\sigma_{min}}{\sigma_y}\right)^2 (\pi(a-d)) \tag{5}$$

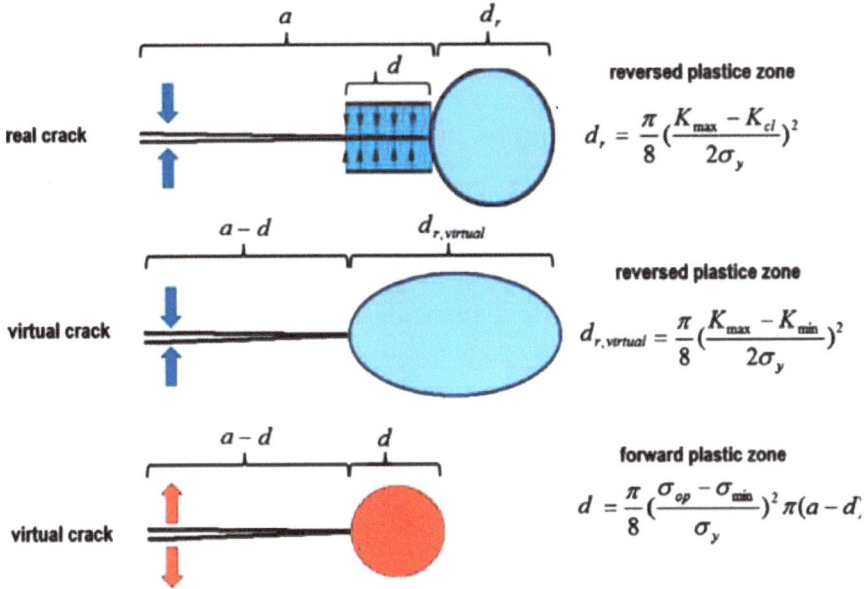

Fig. (3). Pictorial depiction of the plastic zone during loading and unloading.

A pictorial representation of all plastic zones is represented in Fig. (3) Crack closure length d can be determined using Equation 6.

$$d = d_r virtual - d_r \tag{6}$$

Equating equations 5 & 6, and converting the stress intensity into stress terms, we will get a mathematical equation between applied stress and opening stress given by equation 7.

$$\frac{\pi}{8}\left(\frac{\sigma_{max}-\sigma_{min}}{2\sigma_y}\right)^2 (\pi(a-d)) - \frac{\pi}{8}\left(\frac{\sigma_{max}-\sigma_{op}}{2\sigma_y}\right)^2 (\pi(a)) = \frac{\pi}{8}\left(\frac{\sigma_{op}-\sigma_{min}}{\sigma_y}\right)^2 (\pi(a-d)) \tag{7}$$

Most of the metallic material shows cyclic hardening to accommodate the effect of hardening Equation 7 is modified by introducing a hardening factor ç as shown in Equation 8.

$$\frac{\pi}{8}\left(\frac{\sigma_{max}-\sigma_{min}}{2\sigma_y}\right)^2 (\pi(a-d)) - \frac{\pi}{8}\left(\frac{\sigma_{max}-\sigma_{op}}{2\sigma_y}\right)^2 (\pi(a)) = \frac{\pi}{8}\left(\frac{\sigma_{op}-\sigma_{min}}{\sigma_y/\eta}\right)^2 (\pi(a-d)) \tag{8}$$

If the cyclic hardening is insignificant, we take ç=1, in case of cyclic hardening ç>1 & for softening ç<1. If overlaying length d is negligible to the current crack length, we put d = 0, then the simplified equation is

$$(\sigma_{op} - \sigma_{min})(5\sigma_{op} - 2\sigma_{max} - 3\sigma_{min}) = 0 \tag{9}$$

Using Equation 9, crack closure load is determined and used to find an effective stress intensity factor (ΔKeff). For more details on the Virtual crack annealing model derivation given by Equation. 3-9 readers can refer [16].

RESULTS AND DISCUSSION

Propagation of crack in Aluminium-7475 in virgin and pre-plastic deformed condition under time-invariant is evaluated *via* the experimental procedure as ASTM-647 at room temperature 2.54 and 5% pre-strained.

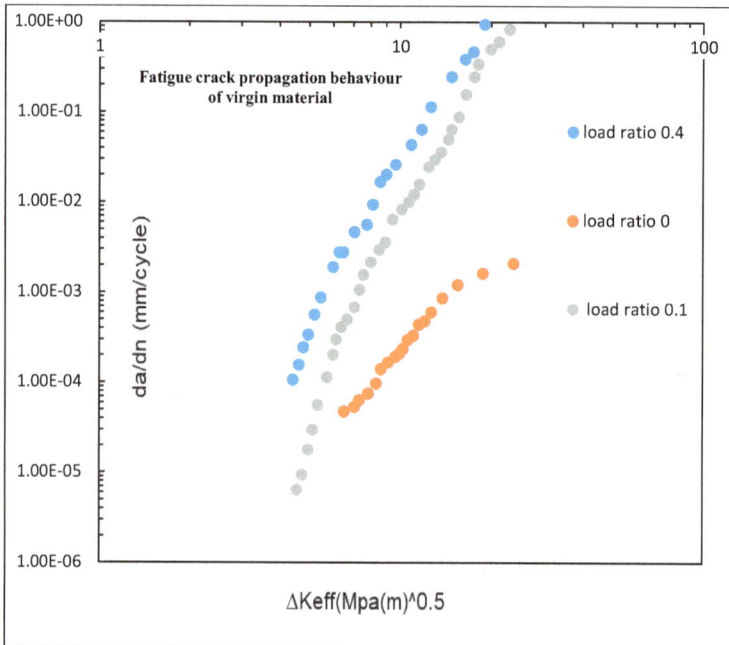

Fig. (4a). Fatigue crack propagation in Virgin and pre-strained condition.

Table 3. Paris model parameter for 5% pre-strained material.

Stress ratio	Paris Model Parameter c	Paris Model Parameter m	R^2
0.1	4×10^{-8}	3.3777	0.9807
0.4	1×10^{-7}	3.3326	0.9874
0.0	3×10^{-5}	1	0.9686

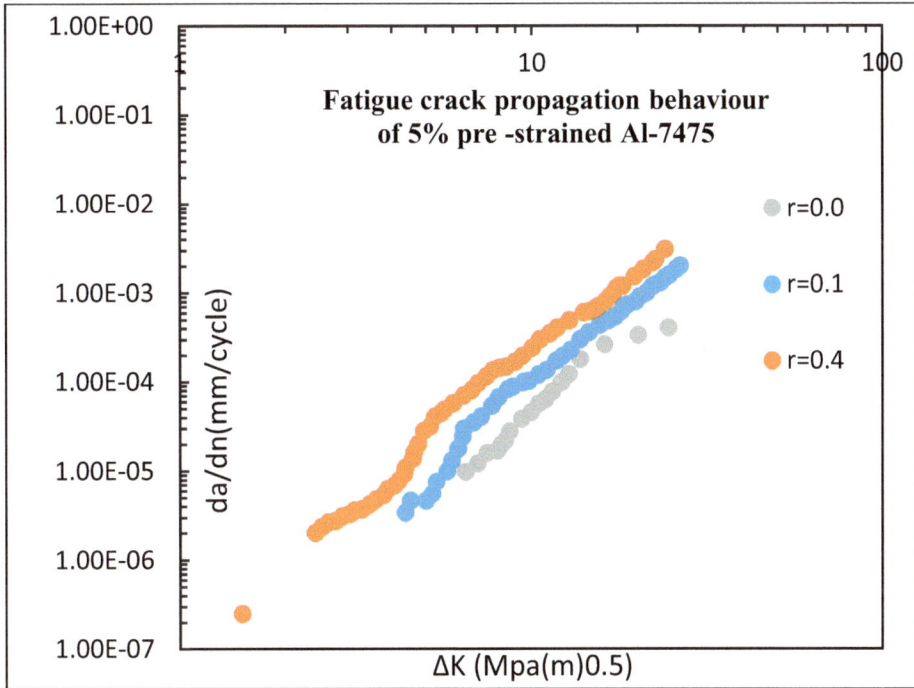

Fig. (4b). Fatigue crack propagation in virgin and pre-strained conditions.

Fig. (4) shows the virgin material trend of growth of fatigued crack and the influence of 5% pre-strain for load ratios 0, 0.1, and 0.4. From Fig. (4a-4b), Fig. 7 and Fig. (9a-9b), it can be observed that FCGR is influenced by load ratio as for each load ratio, a specific trend of the growth rate of a fatigue crack is depicted for both virgin and pre- strained conditions.

Table 4. Paris model parameters for virgin material.

Load Ratio	Paris model parameter, c	Paris model parameter, m	R^2
0.4	4×10^{-8}	5.8663	0.9855

(Table 4) cont.....

Load Ratio	Paris model parameter, c	Paris model parameter, m	R^2
0.0	9×10^{-8}	3.3642	0.9618

Table 5. Paris model parameters for 2.54% pre-strained.

Load Ratio	Paris model parameter, c	Paris model parameter, m	R^2
0.1	1×10^{-8}	3.3777	0.9807
0.4	1×10^{-7}	3.3326	0.9874

Pre-strained material shows increased fatigue crack propagation resistance concerning virgin material. The fatigue crack propagation parameter for both pre-strained and virgin material is presented in Tables **4 & 5** which support the above statement about fatigue crack propagation resistance.

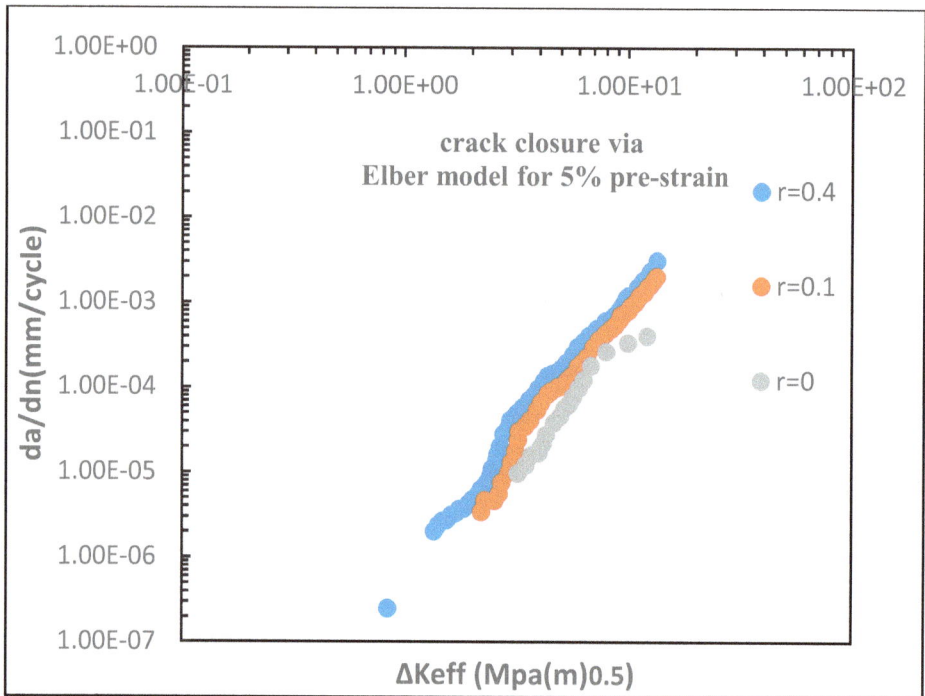

Fig. (5a). Crack closure models simulation.

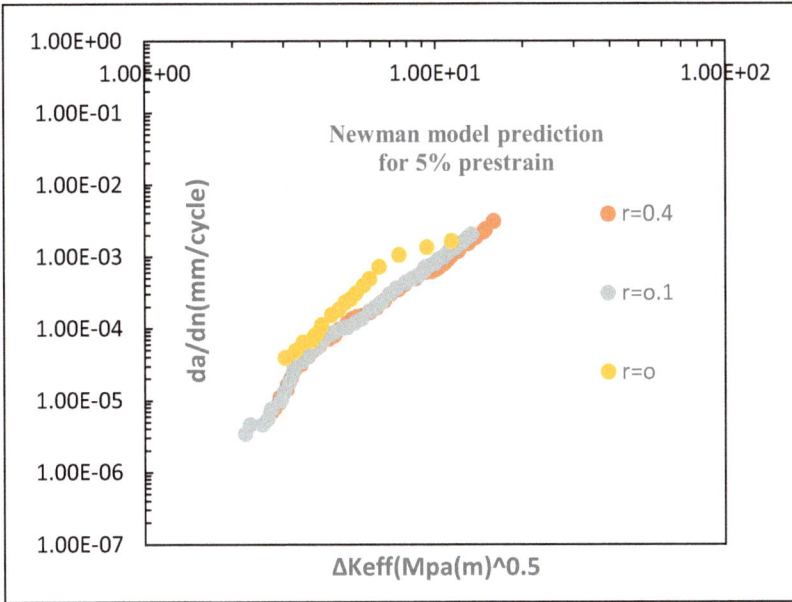

Fig. (5b). Crack closure models simulation.

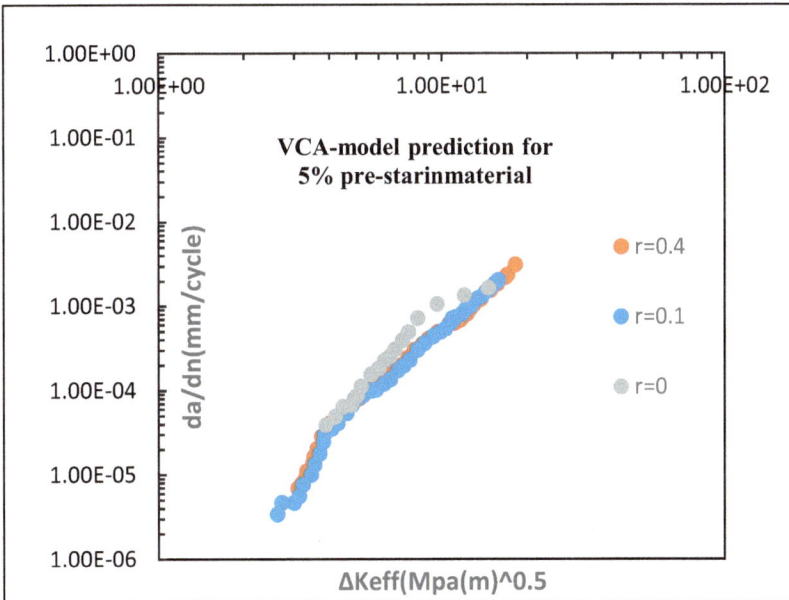

Fig. (6). VCA model prediction.

Load ratio-specific fatigue crack propagation rate is tackled using crack closure phenomena as seen in Fig. (**6**), (Fig. **8**) and (Fig. **9a-9b**) for pre-strain 5% & 2.54%. Various crack closure models are used to simulate the effect of stress ratio as presented in Fig (**5**), (Fig. **6a-6b**), (Fig. **8**), and (Fig. **9a-9b**).

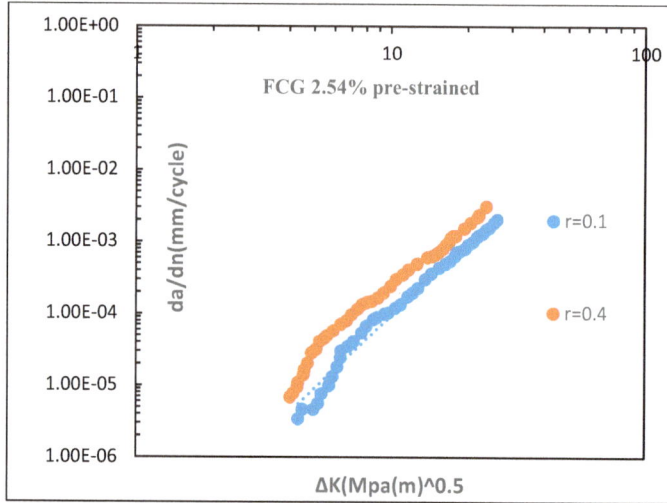

Fig. (7). Fatigue crack propagation of Al-7475.

The used crack closure model shows good results. Fig. (**5a-5b**), Fig. (**6**), Fig. (**8**), and Fig. (**9**) reliability and applicability of studied crack closure models have been verified *via* simulation.

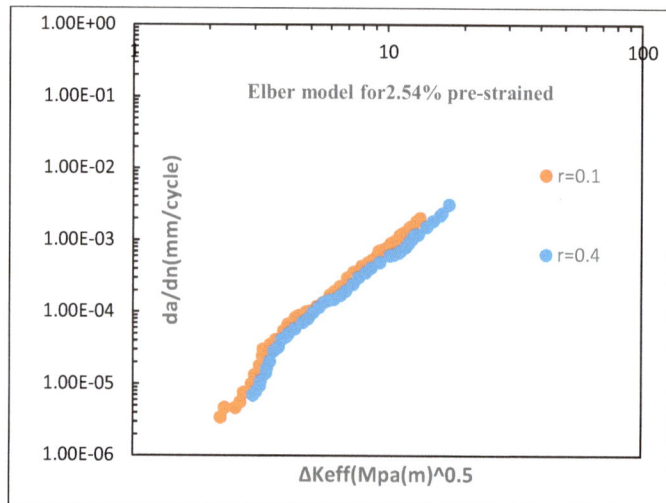

Fig. (8). FCP of *via* Elber CCL model.

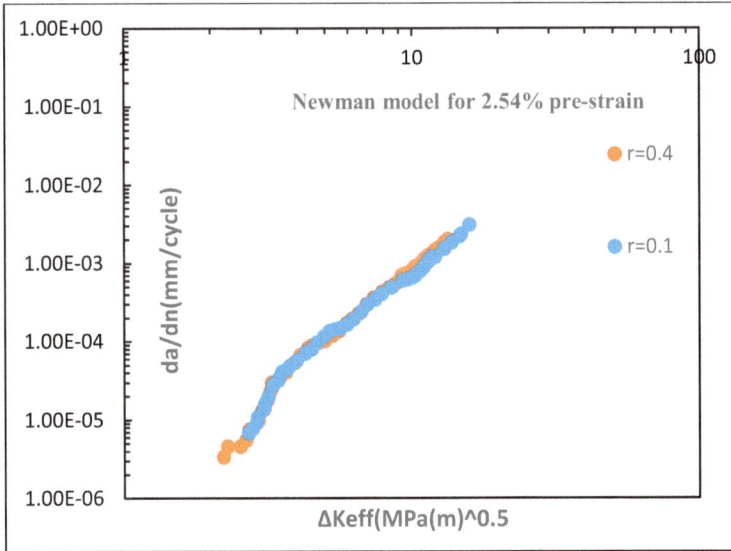

Fig. (9a). Crack closure simulation.

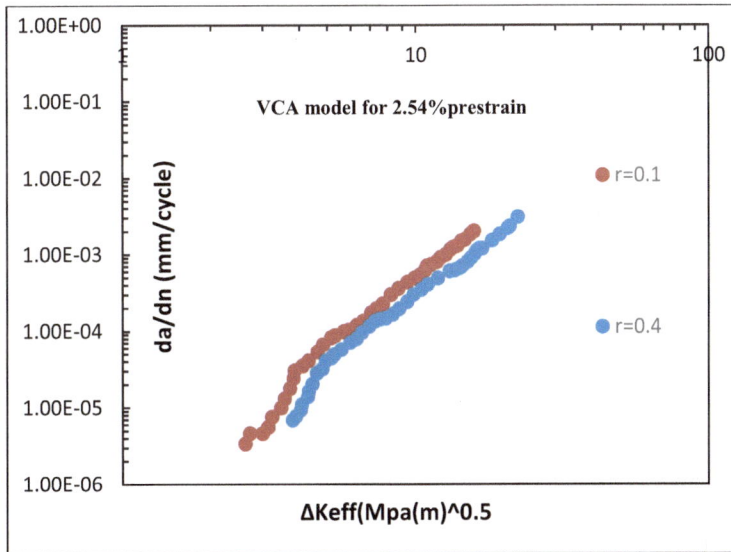

Fig. (9b). Crack closure simulation.

As Elber, Newman, and VCA models prediction is in good agreement for 2.54 & 5% prestrained condition as shown in Fig. (**5a-5b**), Fig. (**6**), Fig. (**8**) and Fig. (**9a-9b**).

Table 6. Crack closure models for 5% pre-strain.

Model	Model Equation	Model parameter, c	Model parameter, m	R^2
Elber	cΔKeffm	6×10^{-7}	3.1775	0.9539
Newman	cΔKeffm	7×10^{-7}	3.1314	0.9383
VCA	cΔKeffm	3×10^{-7}	3.2023	0.9558

The study targets the unification of fatigue crack propagation models under varying load ratios so the designer can use the same fatigue crack propagation model parameters for various load ratios. Elber, Newman, and VCA model show good results for virgin as well as pre-strained Aluminum alloy which is shown in Fig. (**5a-5b**), Fig. (**6**), Fig. (**8**), and Fig. (**9a-9b**). Crack closure model parameters are determined for all three used crack closure models as presented in Tables **6** and **7** for both 2.54% and 5% pre-strained conditions.

Table 7. Crack closure models for 2.54% pre-strain.

Model	Model Equation	Model parameter, c	Model parameter, m	R^2
Elber	cΔKeffm	4×10^{-7}	3.1847	0.9709
Newman	cΔKeffm	5×10^{-7}	3.2324	0.9789
VCA	cΔKeffm	3×10^{-7}	3.1085	0.9535

Paris model parameters are determined and reported in Table **3** for both pre-strained and virgin conditions, which can be used for the life prediction of Al-T7374 for various pre-strained conditions. The study shows that the Paris model is studied for specific sets of conditions for which Paris model parameters are calculated from experimental data. After pre-straining 2.54 and 5% fatigue crack propagation rate decreases by the order of 10-100 for both cases. A decrease in FCPR might be due to pre-tensile strain inducing compressive residual stress in the material.

CONCLUSION

This article analysed and presented the effect of pre-strain on FCP behavior under constant amplitude load (CAL) with varying stress ratios. Fatigue crack propagation rate variation with varying load ratios is studied *via* crack closure models. Consideration of load ratio in fatigue crack propagation models extends the limits of applicability. The experimental and simulation results for the current study reflect the following conclusions:

- Pre-straining 2.54% and 5% increases the residual fatigue life of the structure in order of 10-100, as depicted in Fig. (**4a-4b**) and Fig. (**7**).
- Paris model parameters c and m varies with varying load ratio and under the influence of pre-strain, as shown in Table **3**, **4** and **5**.
- Elber, Newman, and the VCA model are found to be effective in accounting for the load ratio under both virgin and pre-strained conditions.
- The Elber model is observed to be best for the studied parameters.

ACKNOWLEDGEMENTS

The authors acknowledge the Head of Mechanics of material laboratory NIT Srinagar J&K India, for providing the experimental facility for the study.

REFERENCES

[1]　DG Altenpohl, and JG Kaufman, *Aluminum Technology Applications and Env.,* 1998.

[2]　G.H.B. Donato, and F.G. Cavalcante, "Influence of plastic prestrain on the fatigue crack growth resistance (da/dN vs. ΔK) of ASTM A36 structural steel", In: *In ASME International,* 2015.

[3]　J Ferreira, JAFO Correia, G Lesiuk, SB González, MCR Gonzalez, and AMP De Jesus, "Pre-strain effects on mixed-mode fatigue crack propagation behaviour of the P355NL1 pressure vessels steel", *American Society of Mechanical Engineers, Pressure Vessels and Piping Division,* vol. 6, 2018.

[4]　M.N. Babu, and G. Sasikala, "Effect of temperature on the fatigue crack growth behaviour of SS316L(N)", *Int. J. Fatigue,* vol. 140, p. 105815, 2020.

[5]　W. Elber, "Fatigue crack closure under cyclic tension", *Eng. Fract. Mech.,* vol. 2, no. 1, pp. 37-44, 1970.

[6]　J. Newman, A finite-element analysis of fatigue crack closure. in: Mechanics of crack growth. astm international

[7]　M.R. Parry, S. Syngellakis, and I. Sinclair, "Numerical modelling of combined roughness and plasticity induced crack closure effects in fatigue", *Mater. Sci. Eng. A,* vol. 291, no. 1–2, pp. 224-234, 2000.

[8]　K. Solanki, S.R. Daniewicz, and J.C. Newman, "Finite element analysis of plasticity-induced fatigue crack closure: An overview", *Eng. Fract. Mech.,* vol. 71, no. 2, pp. 149-171, 2004.

[9]　N.A. Fleck, "Finite element analysis of plasticity-induced crack closure under plane strain conditions", *Eng. Fract. Mech.,* vol. 25, no. 4, pp. 441-449, 1986.

[10]　R.O. Ritchie, and S. Suresh, "Some considerations on fatigue crack closure at near-threshold stress intensities due to fracture surface morphology", *Metall. trans., a, phys. metall. mater. sci.,* vol. 13, no. 5, pp. 937-940, 1982.

[11]　C. Kant, and G.A. Harmain, A Model Based Study of Fatigue Life Prediction for Multifarious Loadings.*In: Advances in Material Science.* vol. 882. Trans Tech Publications Ltd, 2021, pp. 296-327.

[12]　W. ELBER, "The significance of fatigue crack closure", *ASTM Spec. Tech. Publ.,* pp. 230-242, 1971.

[13]　C. Kant, and G.A. Harmain, Analysis of single overload effect on fatigue crack propagation using modified virtual crack annealing model.*Recent Developments in Mechanics and Design.* Lecture Notes in Mechanical Engineering, Springer Nature, 2022, pp. 1-10.

[14] C. Kant, G.A. Harmain, and C. Kant, "An investigation of fatigue crack closure on 304lss & 7020-t7 aluminium alloy. in: International conference on progressive research in industrial & mechanical engineering (PRIME - 2021)", 2021 pp. 1-10.

[15] P. Paris, and F. Erdogan, "A critical analysis of crack propagation laws", *J Fluids Eng Trans ASME.*, vol. 85, no. 4, pp. 528-533, 1963.

[16] W. Zhang, and Y. Liu, "*In situ* SEM testing for crack closure investigation and virtual crack annealing model development", *Int. J. Fatigue,* vol. 43, pp. 188-196, 2012.

[17] J.C. Newman, "A crack opening stress equation for fatigue crack growth", *Int. J. Fract.,* vol. 24, no. 4, 1984.

[18] N. Ferreira, J.A.M. Ferreira, P.V. Antunes, J.D. Costa, and C. Capela, Fatigue crack propagation in shot peened al 7475-t7351 alloy specimens.*Procedia Engineering.* Elsevier Ltd, 2016, pp. 254-261.

[19] "Amercan society for testing and materials (astm). astm e647: Standard test method for measurement of fatigue crack growth rates. 01 ed", In: *American Society for Testing and Materials* vol. 3. Annual Book of ASTM Standards: West Conshohocken, Pa, USApp. 591-630.

[20] R.C. McClung, and H. Sehitoglu, "On the finite element analysis of fatigue crack closure-2. Numerical results", *Eng. Fract. Mech.,* vol. 33, no. 2, pp. 253-272, 1989.

[21] J. Zapatero, B. Moreno, and A. González-Herrera, "Fatigue crack closure determination by means of finite element analysis", *Eng. Fract. Mech.,* vol. 75, no. 1, pp. 41-57, 2008.

[22] G.A. Harmain, "An investigation on single overload fatigue crack growth retardation, Part-2 (Crack closure decomposition))", *J Metall Mater Sci.,* vol. 47, no. 4, pp. 189-197, 2005.

[23] C. Kant, and G.A. Harmain, "Fatigue life prediction under interspersed overload in constant amplitude loading spectrum via crack closure and plastic zone interaction models - a comparative study", *9th International Conference on Fracture Fatigue and Wear (FFW 2021)* Ghent, Belgium, Springer 2021, pp. 1-10.

[24] J. Newman, *Finite-element analysis of fatigue crack propagation--including the effects of crack closure. - proquest.* Virginia Polytechnic Institute and State University: Blacksburg, VA, 1974.

CHAPTER 21

Principle and Application of Smart Material in the Biosensing Field

Tapan Kumar Patnaik[1,*], Asheem Putel[1], Rakesh Kumar Rout[1] and **Sudhanshu Shekhar Parida[1]**

[1] *Department of Physics, GIET University, Gunupur, Rayagada, Odisha, India*

Abstract: Biosensors are analytical devices that are broadly used for the detection of chemical substances like tissue, organelles, cell receptors, enzymes, antibodies, *etc.* Smart materials respond to the external impulse, and convert the impulse to readable signals. Nowadays, smart materials are used in every requirement of a human being. The various kinds of smart materials are the subject of extensive investigation. This chapter examines the fundamental idea and practical use of smart materials in the biosensing industry.

Keywords: Applications, Biosensors, Smart material.

INTRODUCTION

Smart or Intelligent Materials

Smart materials are a vital research area in the field of material physics. Smart materials exhibit observable changes in response to an external effect or environmental impulse [1]. Temperature, pressure, electric flow, light, mechanical, heat, stress, moisture, electric field, magnetic field, and pH are a few examples of the external impulses or stimuli. Due to the popularity of color-changing shirts and other smart materials, there is an increased use of smart material-based products in everyday life today. Smart materials are used in many fields like aerospace, material, bionics, medical, technologies., civil, engineering and automobiles. A structure that is made from smart material is known as a smart structure.

* **Corresponding author Tapan Kumar Patnaik:** Department of Physics, GIET University, Gunupur, Rayagada, India; E-mail: tapanpatnaik@giet.edu

Smart materials are composite materials, i.e., smart materials are the combination of two or more materials. When compared to standard materials, these materials have various benefits, including reduced weight, immunity to corrosion, and a longer lifespan. Our world's design actively incorporates smart materials. It involves designing things like clothing, buildings, vehicles like cars or planes, bikes, and goods for the house [2].

TYPES OF SMART MATERIALS

There are different types of smart materials. Some of those are as follows:

Biomimetic Materials

Biomimetic materials are capable of sensing their environment, processing the data, and responding instantly. These materials are used in the restoration of natural functions where the original material is not performing well. These materials are used in a very wide range in the biosensing field as it sustains an optimally conducive environment, tissue growth, biomolecular assays, and biotechnology-based manufacturing. Some of these materials are shown in Fig (1).

Fig. (1). Biomimetic Materials.

Smart Gels

Smart gels are the gels that change their structure after getting external stimuli like temperature, pH, light, magnetic field, etc.

Researchers are interested in smart hydrogens, a sort of smart gel, not only for their stunning behavior under external stimuli but also for a broad array of applications, such as industrial or biological, to which they may be used [3]. Drug delivery methods, tissue engineering solutions and injectable biomaterials have all made use of these materials. Some examples of smart gels are shown in Fig 21.2.

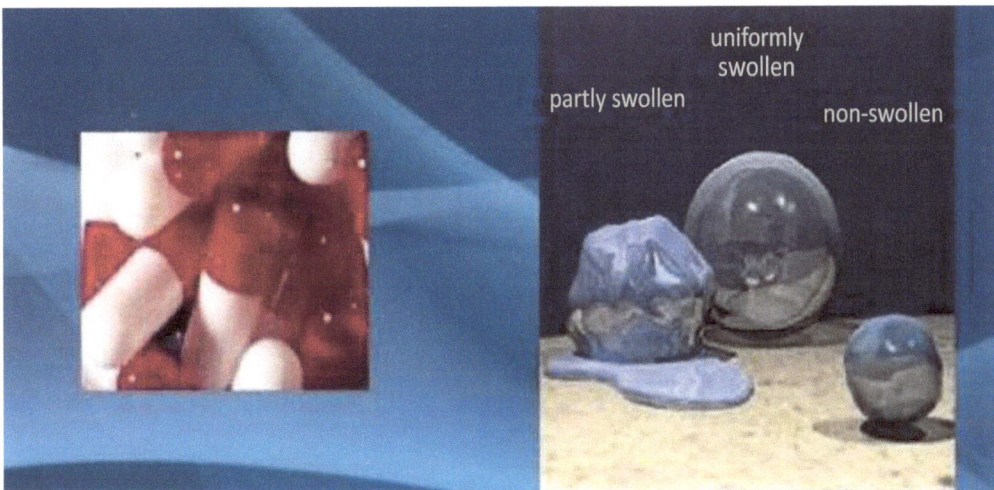

Fig. (2). Smart Gel.

Piezo-electric Materials

Piezoelectric smart materials produce electric voltage when stress is applied and vice versa. Applying the electric voltage throughout this material shows some changes in shape. Some of the naturally piezoelectric occurring materials are berlinite, cane sugar, quartz, Rochelle salt, topaz, and tourmaline [3, 4]. A piezo-electric smart material is shown in Fig. (3).

Piezo-electric materials possess:

• High strain constants, permittivity and coupling constant
• The low mechanical quality factor
• High strain output for large displacement at modest voltage

Fig. (3). Piezoelectric Materials.

These are used for sensing (receivers, knock, acoustic, musical pick-ups, vibration, vortex, and material testing) and also for actuators (valves, positioning, vibrating, AFM) [4]. The recent research on this material is how we can get electricity by walking on the street, dancing on the floor, and charging the laptop by typing, as shown in Fig. (**4**).

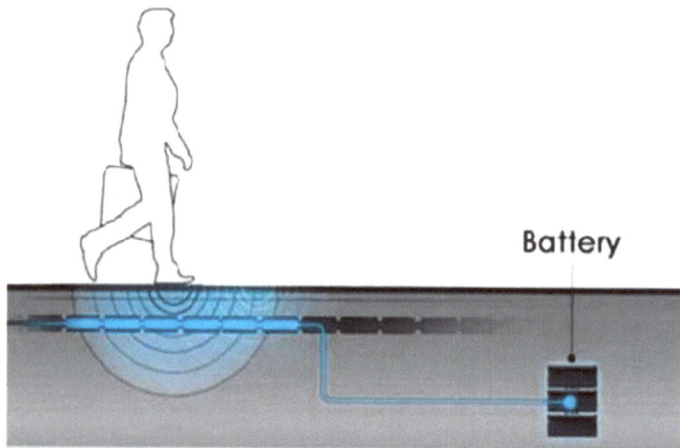

Fig. (4). Electricity generation by walking on the street.

There are some issues with the research work on piezo-electric material as the amount of electricity produced is so small. So, unless a vast installation is set up, it simply would not have the strength to power our latest gadgets.

Electrostrictive Materials

These materials are the same as the piezo-electric material, but the mechanical change is proportional to the square of the electric field. Electrostrictive materials are shown in the dielectric material. This material belongs to the ferroelectric family [5]. These materials are more effective than piezo-electric materials due to some properties. These materials have an electrostrictive constant. An analysis of these materials is shown in Fig. (**5**).

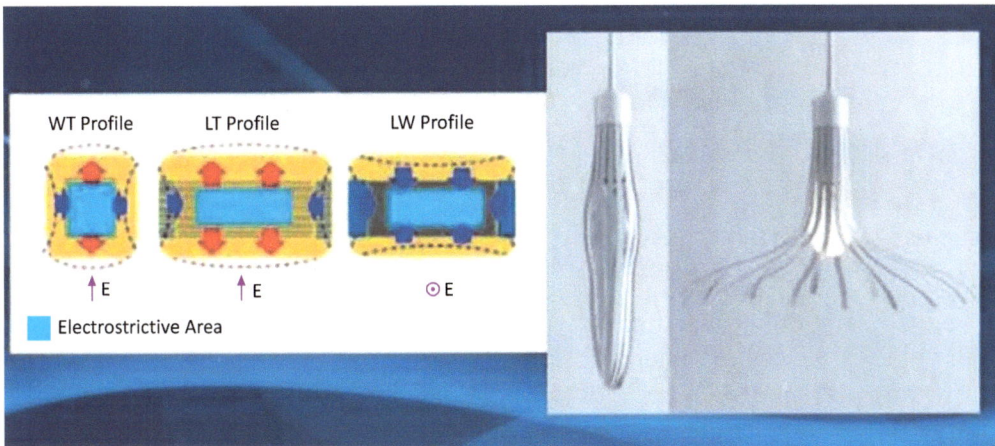

Fig. (5). Electrostrictive Materials.

The most commonly used electrostrictive materials are

• Lead magnesium niobate (PMN)
• Lead magnesium niobate lead titanate
• Lead lanthanum zirconate titanate

These smart materials are used in sonar projectors for submarines and surface vessels. Actuators are used to determine small displacement.

Magnetostrictive Materials

Some ferromagnetic materials change in shape when a magnetic field is applied to them. These materials can transform magnetic energy into kinetic energy. Such substances generate a magnetic field when mechanical force is applied to them. An electric current may be generated in this area, converting mechanical energy into electrical energy [6, 7]. A magnetostrictive material is shown in Fig. (**6**).

Fig. (6). Magnetostrictive materials.

Some uses of these materials are given below

- Used in heavy-duty trucks for active vibrational control system
- For velocity control of pneumatic actuator system
- Low-cost MR sponge damper for washing machines
- Large MR fluid dampers to control wind-induced vibrations in cable-stayed bridges

Rheological Materials

These materials change their states when an electric or magnetic field is applied. This material always changes between a liquid and a solid state. A wide range of industrially significant substances, such as paint and chocolate that have complicated flow properties can benefit from the usage of these materials, such as brakes and shock absorbers for automobile seats [8].

Thermoresponsive Material

Thermoresponsive materials show changes in properties even on a small change in temperature. These are useful in thermostats and parts of automotive and air vehicles. This material can be used in drug delivery systems, tissue engineering, and gene delivery [9]. Thermoresponsive polymers go through a volume phase transition at a certain temperature, which causes the temperature, as shown in Fig. (7). In this procedure, the thermoresponsive polymer is combined with cells at room temperature before being injected into the body. When exposed to a temperature rise (to 37°C) that is higher than the polymer's LCST, the polymer transforms into a physical gel state.

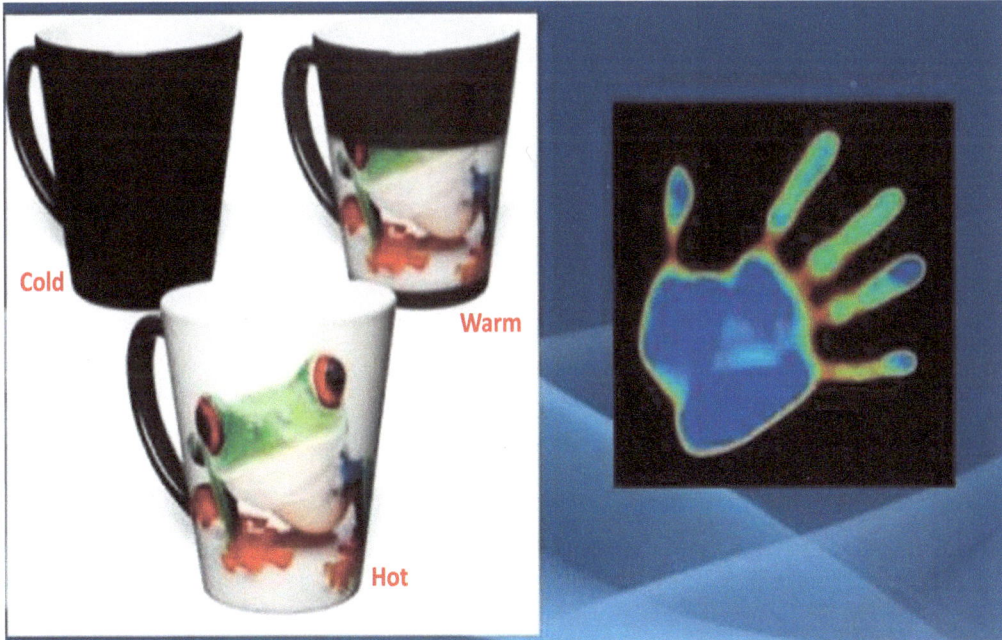

Fig. (7). Thermoresponsive materials.

TYPES OF THERMORESPONSIVE MATERIALS

- TRP, which shows LCST (lower critical solution temperature)
- TRP, which shows UCST (upper critical solution temperature)

Fullerene Materials

Buckyballs and carbon nanotubes are examples of fullerenes, which are any succession of hollow carbon molecules that form a closed cage or cylinder. In terms of tensile strength, the carbon nanotube is one-sixth the weight of steel. Conductors and superconductors are both possible properties of these materials. The C60 molecule is the most frequently synthesized fullerene [6], [8]. It resembles a soccer ball or a truncated icosahedron. Assuming that each vertex is composed of only one Carbon (C), each vertex has two single bonds and one double bond with another Carbon (C), as shown in Fig. (**8**).

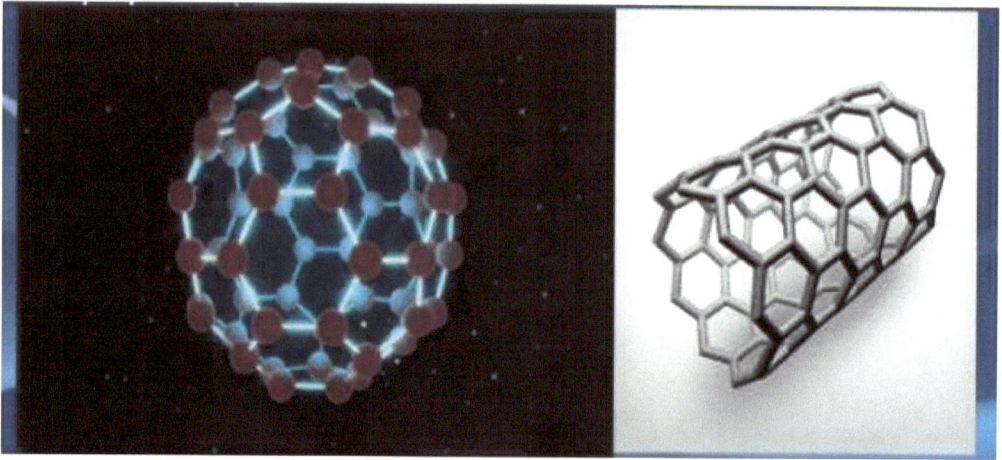

Fig. (8). Fullerene materials.

APPLICATION OF SMART MATERIALS

Smart materials have been applied in various fields like Civil Engineering, Automotive Industry, aircraft, Orthopedic Surgery, Dental Braces, Robotics, Smart Fabrics, Smart glasses, motors and Actuators, Drug Delivery Systems, and Medicines [2, 10]. The smart materials are super elasticity and shape memory, which make them more reliable and applicable

BIOSENSOR

Analytical devices that combine a biological component with a physicochemical detector are known as biosensors. Professor Leland C Clark Jnr's father, Leland C Clark Jnr. Either directly coupled to or incorporated into the transducer, the gadget incorporates a biological sensing element [11]. Biological recognition systems must be linked to a proper transducer to achieve selectivity and specificity. A biological reaction is transformed into an electrical signal by this analytical equipment. Physiological changes or processes can be detected and recorded by this device. The presence and concentration of a certain drug in a test solution may be determined using this technique [12].

Basic Principle of Biosensor

The biosensor's fundamental concept is composed of three parts: - A biological recognition element that is very specific to the biological material produced by analytes, as shown in Fig. (**9**). Second, transducers detect and convert the signal from the receptor molecule in the biological target to an electrical signal as a result of the process taking place [11].

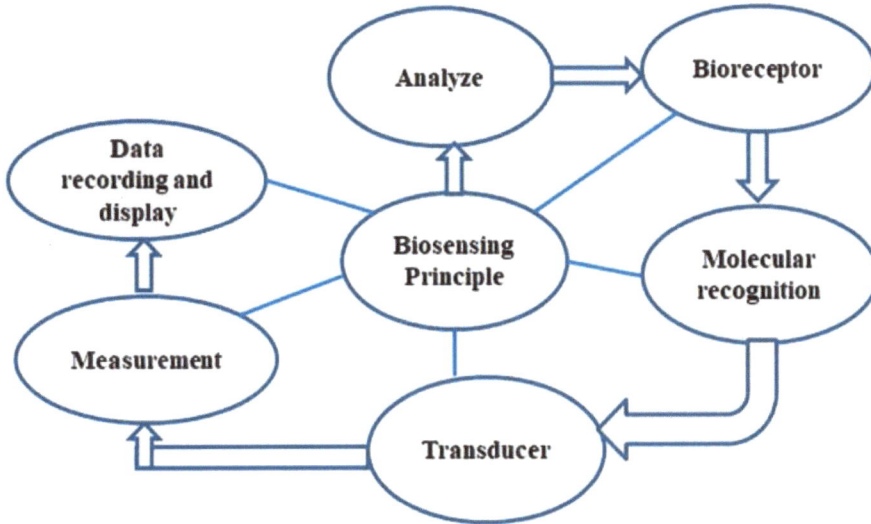

Fig. (9). Principle of biosensor.

Amplification and reading of the detector's output in the form of values for monitoring and regulating the system occur thirdly following signal transduction from biological to electrical signals. The immobilized biological material is brought into direct contact with the transducer. To measure the electrical reaction, the analyte must first bind to a biological substance. The product may be released as heat, gas (oxygen), electrons, or hydrogen ions when the analyte is transformed into another product. An electrical signal is generated by the transducer, which is subsequently amplified and measured [13].

Components of Biosensor

The Biological Element Function is the first component. To identify and interact with a specific chemical, such as a target substance, it has to be able to tell whether or not the test solution contains a certain ingredient. Biosensors are built on the concept of bioelement specificity, or the capacity of a bioelement to interact specifically with a target chemical. The components of the biosensor are shown in (Fig. **10**).

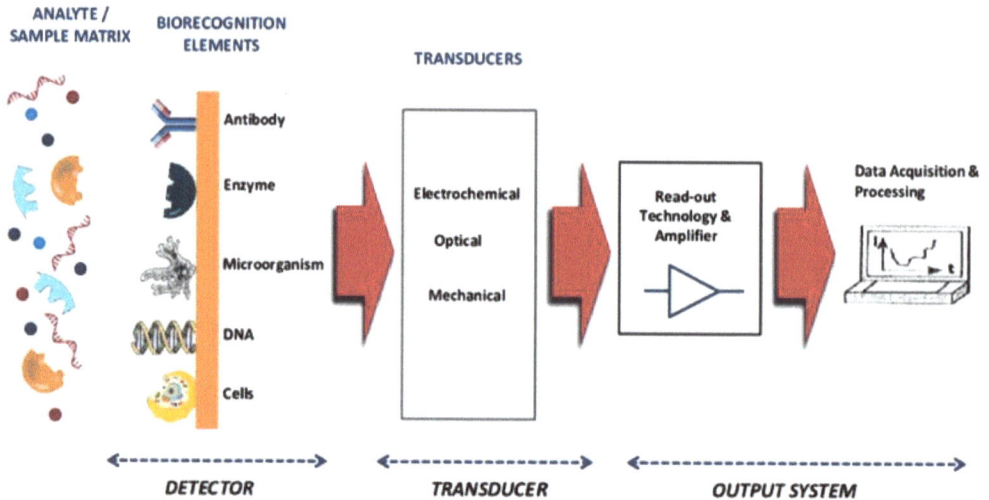

Fig. (10). Component of biosensor [13].

The physiochemical transducer is the second component. As an interface, it measures the physical change that occurs during a response at the bioreceptor and then transforms that energy into an electrical output that can be measured. A detector is the third and last piece of the puzzle. The transducer sends signals to a microprocessor, which amplifies and analyses them. Afterward, the concentration units are transformed to display units and stored on a computer or data storage device

Working of Biosensor

Several different types of biosensors may be used for the quantitative examination of various chemicals. Other factors influencing biosensor performance include the stability of enzymes and the selectivity and sensitivity of biological reactions [10, 13].The working principle of biosensors is shown in (Fig. **11**).

The biological component of the biosensor responds precisely and effectively with the analyte. The surface characteristics of the transducer are altered as a result of this interaction. The optical and electrical characteristics of the transducer surface are also changed. Measurement/conversion of optical/electronic characteristics into an electrical signal is detected.

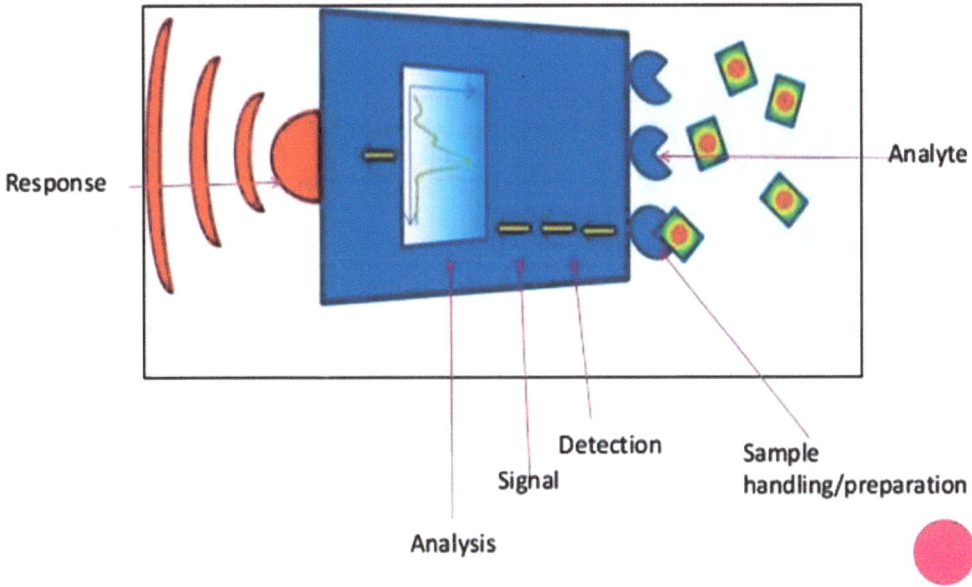

Fig. (11). Working principle of biosensors.

Types of Biosensors

Optical Biosensors

Analyze the change in light absorption when reactants are transformed into products using a colorimetric method. Luminescent or fluorescent processes can be detected using photomultiplier tubes or photodiode devices that use photon output to determine the intensity of light [12].

Calorimetric Biosensors

For exothermic reactions, two thermistors may be used to detect the difference in resistance between reactant and product, and, therefore, to determine whether or not the analyte concentration has been reached [12].

Potentiometric Biosensors

In the case of voltage, ion-selective electrodes, such as pH meters, are used to detect changes in the distribution of charge.

Piezo-electric Biosensors

A piezo-electric device uses gold to detect the precise angle at which electron waves are produced when a substance is subjected to laser light or crystals, such as quartz, that vibrate under the influence of an electric field is exposed to the substance. The frequency shift is related to the amount of material that has been absorbed [10].

Electrochemical Biosensors

The electrical characteristics of a solution can be detected and utilized as a measurement parameter in many chemical processes that make or consume ions or electrons. Amperometric and conductimetric Biosensors are two examples. Biosensors that use potentiometric measurements. The applied current is measured in amperes. When a voltage is supplied between two electrodes, the movement of e- in redox processes may be measured. Ion-selective electrodes, such as pH meters, are used to detect changes in charge distribution in potentiometric voltage measurements. For impedance measurements, conductivity is used.

Amperometric Biosensors

Electric current is the metric that is being measured. Oxidase enzymes manufacture H_2O_2 and use oxygen to produce H_2O. Pt-electrodes can be used to detect the formation of H_2O_2 [14]. The application of biosensors includes food analysis, biomolecular interaction research, drug development, criminal investigation, and medical diagnosis, which are just a few examples of the wide range of disciplines that fall under this umbrella term (both clinical and laboratory use). The applications of amperometric biosensors are shown in (Fig. **12** and **13**).

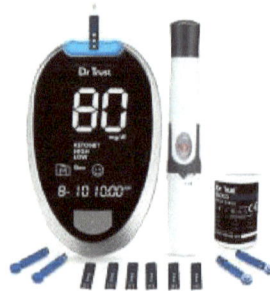

Fig. (12). Glucose monitoring.

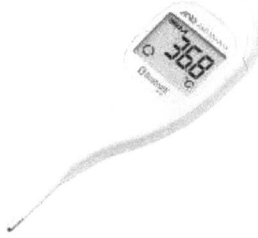

Fig. (13). Digital thermometer.

APPLICATION OF SOME SMART MATERIALS IN BIOSENSING

Some of the smart materials, their responsive principles, and their applications in the field of biosensing are summarized in Table **1** [2, 6].

Table 1. Applications of some smart materials in Biosensing.

Smart materials	Responsive principles	Application
AuNPs	Analytes induced AuNP accumulation and dispersion, resulting in chromogenic alterations in the solution.	Colorimetric testing of glucose, enzyme
Ag+-coated AuNRs	The Ag+/glucose redox process used AuNRs as a substrate.	Naked-eye visualization of glucose
Fe_3O_4 MNPs	Peroxidase substrates were catalyzed by MNPs, which resulted in a noticeable color change on the strip.	Detection of the glycoprotein
QD bioconjugates	Analytes significantly reduced FRET efficiency, which led to the restoration of QD fluorescence in the presence of analytes.	Visual detection of RNA, enzyme
Au/Ag NP pyramids	The reconfiguration or dissociation of pyramids, which modified the CD signals of the pyramids, was made possible by target DNA.	Detection of DNA
GO-decorated PC barcodes	Photoluminescence intensity at the barcode surface is varied for each target miRNA tied to PC barcodes with distinct reflection peaks.	Multiplex miRNA detection
MoS2-integrated PC barcodes	The target miRNA hybridized with the barcode probes and prevented the QDs from coming into contact with the MoS2, allowing for the recovery of the QDs' fluorescence.	Multiplex miRNA detection
Zwitterionic hydrogel	The glucose oxidation was accelerated by GOx and HRP in the hydrogel, resulting in a luminous product.	Optical monitoring of glucose

Smart materials	Responsive principles	Application
pH-responsive hydrogel	A change in hydrogel swelling was caused by a pH shift caused by glucose in the solution.	Visual detection of glucose
PBA-incorporated Hydrogel	The refraction of light through the hydrogel was altered by the reaction between glucose and PBA.	Detection of glucose
Aptamer-based hydrogel	The hydrogel's volume was altered by the protein-aptamer interaction.	Detection of thrombin
Hydrogel integrated with PCs	Hydrogels swelled and contracted in response to analytes, adjusting the encased PCs' diffraction wavelength.	Detection of glucose, DNA, protein, and enzyme
Hydrogel encapsulated with photonic barcodes	Reflection peaks or fluorescence of the PC barcodes were used to identify different types of biomarkers.	Multiplex detection of proteins

CONCLUSION

Today, the most effective and reliable materials are smart materials. These have the potential to impact different fields like science, engineering, medicine, and automotive. The medical field is one of them where smart materials are broadly used by accompanied biosensors. Biosensors have been miniaturized extensively in recent years. Smart materials, due to their properties, have been employed in the construction of biosensors.

REFERENCES

[1] Z. Guo, H. Liu, W. Dai, and Y. Lei, "Responsive principles and applications of smart materials in biosensing", *Smart Materials in Medicine,* vol. 1, pp. 54-65, 2020.
[http://dx.doi.org/10.1016/j.smaim.2020.07.001] [PMID: 33349813]

[2] J.A. Marshall, "Smart materials", *ASEE Annu. Conf. Proc,* vol. vol. 13, 2002pp. 1771-1773
[http://dx.doi.org/10.9790/1684-1305062832]

[3] M.N.O. Sadiku, M. Tembely, and S.M. Musa, "Smart materials: A primer", *Int. J. Adv. Res. Comput. Sci. Softw. Eng.,* vol. 7, no. 3, pp. 43-44, 2017.
[http://dx.doi.org/10.23956/ijarcsse/V7I3/01302]

[4] S.E. Prasad, D.F. Waechter, R.G. Blacow, H.W. King, and Y. Yaman, "Application of piezo-electrics to smart structures", *Eccomas Themat Conf. Smart Struct Mater,* pp. 1-16, 2005. http://www.sen sortech.ca/site/content/2005-01.pdf

[5] S. Sherrit, and B.K. Mukherjee, Electrostrictive materials: characterization and applications for ultrasound
[http://dx.doi.org/10.1117/12.308000]

[6] G. N, and F. M, "Smart materials and structures: State of the art and applications", *Nanotechnology & Applications,* vol. 1, no. 2, pp. 1-5, 2018.
[http://dx.doi.org/10.33425/2639-9466.1015]

[7] S. Kamila, "Introduction, classification and applications of smart materials: An overview", *Am. J. Appl. Sci.,* vol. 10, no. 8, pp. 876-880, 2013.
[http://dx.doi.org/10.3844/ajassp.2013.876.880]

[8]   V. Perumal, and U. Hashim, "Advances in biosensors: Principle, architecture and applications", *J. Appl. Biomed.,* vol. 12, no. 1, pp. 1-15, 2014.
[http://dx.doi.org/10.1016/j.jab.2013.02.001]

[9]   A. Gandhi, A. Paul, S.O. Sen, and K.K. Sen, "Studies on thermoresponsive polymers: Phase behaviour, drug delivery and biomedical applications", *Asian Journal of Pharmaceutical Sciences,* vol. 10, no. 2, pp. 99-107, 2015.
[http://dx.doi.org/10.1016/j.ajps.2014.08.010]

[10]   R. Manikandan, N. Charumathe, and A.F. Begum, "Application of biosensors", *Techniques and Instrumentation in Analytical Chemistry,* vol. 11, no. 100, pp. 291-322, 1992.
[http://dx.doi.org/10.1016/S0167-9244(08)70037-4]

[11]   A. Koyun, E. Ahlatcolu, and Y. Koca, "Biosensors and their principles", *A Roadmap Biomed Eng. Milestones,* 2012.
[http://dx.doi.org/10.5772/48824]

[12]   J. Ali, J. Najeeb, M. Asim Ali, M. Farhan Aslam, and A. Raza, Biosensors: Their fundamentals, "Designs, types and most recent impactful applications: A review", *J. Biosens. Bioelectron.,* vol. 8, no. 1, 2017.
[http://dx.doi.org/10.4172/2155-6210.1000235]

[13]   R. Kazemi-Darsanaki, A. Azizzadeh, M. Nourbakhsh, G. Raeisi, and M. AzizollahiAliabadi, "Biosensors: Functions and applications", *J. Biol. Todays World,* vol. 2, no. 1, pp. 20-23, 2013.
[http://dx.doi.org/10.15412/J.JBTW.01020105]

[14]   I. Journal, A. Issn, B.E.M. Science, and E. Guindy, *Biomedical application of smart materials- An Overview*, 2019.

CHAPTER 22

Experimental Investigation Of Tribological Behavior Of Tin-Based Babbitt And Brass Material

Rohit Kumar Babberwal[1,*] and **Raosaheb Bhausaheb Patil**[1]

[1] *Department of Mechanical Engineering, Army Institute of Technology, Pune-411015, Maharashtra, India*

Abstract: The aim of the experiment is to investigate the tribological behavior of Brass and tin-based babbitt materials. The experiment is conducted on a pin-on-disk wear test machine at room temperature to analyze their effect on tribological behavior. The experiment is conducted at various operating factors like load, sliding time and sliding velocity under dry and lubrication conditions. Application of these materials is mostly found in automobile bearing, precise instrument, railway bearing, aerospace and heavy duty application. After conducting the experiment, it was noticed that tribological behavior slightly changes at elevated temperatures. The use of oil lubricant improves the tribological performance as compared to dry conditions by 18.56 times for tin-based babbitt alloy and by 2.19 times for brass material. Thus the performance of Brass under oil lubrication is superior to tin-based Babbitt due to its hardness. Under dry conditions, the wear rate of brass material is approximately four times that of wear in tin-based Babbitt; thus, the service life of tin-based Babbitt is longer than brass material under dry conditions.

Keywords: Brass, Temperature, Tin based babbitt, Tribology, Wear.

INTRODUCTION

Nowadays, various materials with different compositions are used in industry, such as steel, Aluminum, Copper, Nickel, Titanium, Bronze, Nickel, etc., to reduce wear and friction and improve the performance or service life of the component. But these materials and their combination have wear and friction problems, and their available tribological data is also limited, so it is difficult to predict the actual tribological behavior of these materials. The Selection of the correct material is of great concern in the design of components [1]. In this experi-

---

* **Corresponding author Rohit Kumar Babberwal:** Department of Mechanical Engineering, Army Institute of Technology, Pune-411015, Maharashtra, India; Email: rkbabberwal01@gmail.com

Amar Patnaik, Albano Cavaleiro, Malay Kumar Banerjee, Ernst Kozeschnik & Vikas Kukshal (Eds.)

mental investigation, tin-based Babbitt and Brass are selected to analyse their tribological performance. These materials show good compatibility with other materials such as steel; also, they have a good anti-frictional properties and the ability to embed

foreign particles. These materials are mostly used in various industrial applications, such as bearing, gear, cam, and follower. Earlier studies have shown that wear and friction depend on various factors such as normal load, sliding speed, surface condition, specimen geometry, system rigidity, sliding time, type of material in contact, etc. Out of these factors, sliding speed and normal load are the main factors that play a significant role in the change in tribological performance. So, this experimental investigation focuses on examining the tribological performance of babbitt alloy and brass material with different working parameters under dry and oil lubricating (SAE 50) conditions. This research paper aims to investigate tribological behavior of tin-based Babbitt and brass material under various operating conditions.

Dongya *et al.* [1] studied the tribological performance of Babbitt (ZSnSb11Cu6) alloy with PU polymer coating with dry sliding and oil lubricating conditions. It is observed from the result that Babbitt with polyurethane coating shows better tribological performance than bare Babbitt. The coefficient of friction of bare Babbitt is more than Babbitt with PU coating. Goudarzi *et al.* [2] study about tribological properties of white metal (Sn-Sb-Cu) as a journal bearing. It also investigates the heating effect and cooling rate on white metal. The result shows that under heavy load application, the tribological performance of WM5 is better than WM2 because of alloying materials. It also noticed that under the same working condition, the amount of wear in WM2 is more compared to WM5. Feyzullahoglu *et al.* [3] investigated the tribological behavior of Brass (CW 619), SAE (7% Sb), and Sn-Sb-Cu (20% Sb) for heavy duties application. The experiment was conducted using Tecquipment HFN type 5 journal-bearing equipment. It notices that increase in tin content in WM5 increase results in an increase in hardness. Chowdhury *et al.* [4] investigated the effect of normal load, sliding velocity, and relative humidity on wear and friction of Brass (disc) sliding against mild steel and stainless steel (pin). It has been observed from the experiment that the wear rate of SS 202, SS314 and MS is affected by sliding velocity and perpendicular load. Also, study the wear properties of different grade steel [5, 6] and other material pairs [7].

MATERIALS AND METHODOLOGY

The pin specimen used in the experiment is made up of tin-based Babbitt and brass material, whereas the disc is made up of E8 steel material. Pin specimen is cylindrical in shape with a flat surface. It is manufactured by extraction process followed by machining on a lathe machine to get the desired dimension of 6mm diameter and 28mm height. Chemical analyses using a spectrometer show that the composition of tin-based Babbitt is 81.74% tin, 12.13% antimony, 5.05% copper, and 0.22% lead, whereas brass chemical composition is 55.77% copper, 41% zinc, 2.42% lead, and 0.33% tin. The experimental setup with pin specimen is shown in Figs. (**1-2**) [8].

Fig. (1). Tin based babbitt pin.

Fig. (2). Brass pin.

Experiment

The experiment was performed on Pin on disc (POD) machine according to the G99 ASTM test standard to find wear and friction properties of babbitt alloy and brass material. The experiment was conducted, by considering working parameters like normal load (10N, 20N, and 30N), and sliding velocity (200, 300, and 400rpm) into the account. The experiment was carried out for 20 minutes with both dry and oil lubricating conditions. Test Pin specimen weight is measured before and after the wear, an experiment is to find out the wear loss with the help of an electronic microbalance weight machine [9, 10]. Experimental results are shown in Table **1**.

Table 1. Experimental result for tin-based Babbitt and brass materials under dry and lubricating conditions.

Experiment number	Normal Load (N)	Sliding speed (rpm)	Dry condition				Libration condition			
			Tin based Babbitt		Brass		Tin based Babbitt		Brass	
			Wear (micron)	COF	Wear (micron)	COF	Wear (micron)	COF	Wear (micron)	COF
1	10	200	34	0.33	43	0.166	34	0.023	1.8	0.02
2	10	300	37.21	0.21	8.44	0.063	-3	0.005	-5	0.003
3	10	400	30	0.207	98.53	0.068	-4.75	0.029	19	0.055
4	20	200	57	0.028	376.56	0.158	98	0.003	19	0.002
5	20	300	47.4	0.27	67	0.192	-0.75	0.027	20	0.042
6	20	400	238	0.48	643	0.226	-.05	0.014	11.69	0.048
7	30	200	86.6	0.251	465.45	0.155	-11.85	0.019	0.66	0.042
8	30	300	215	0.452	895	0.228	46.59	0.012	20.47	0.044
9	30	400	149	0.345	1119.2	0.163	6	0.007	51.78	0.006

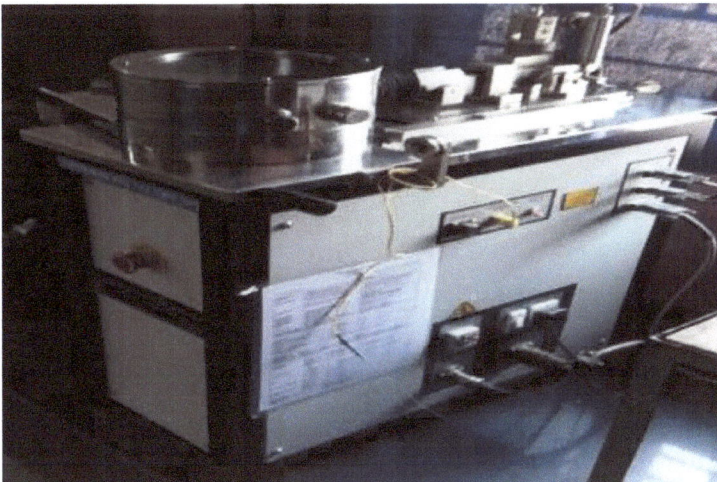

Fig. (3). Pin-On-Disc Apparatus.

RESULTS AND DISCUSSION

Wear Analysis

Figs. (**4** and **5**) show the relation of wear with sliding duration at different sliding speeds and normal load with a dry condition for babbitt alloy and brass material.

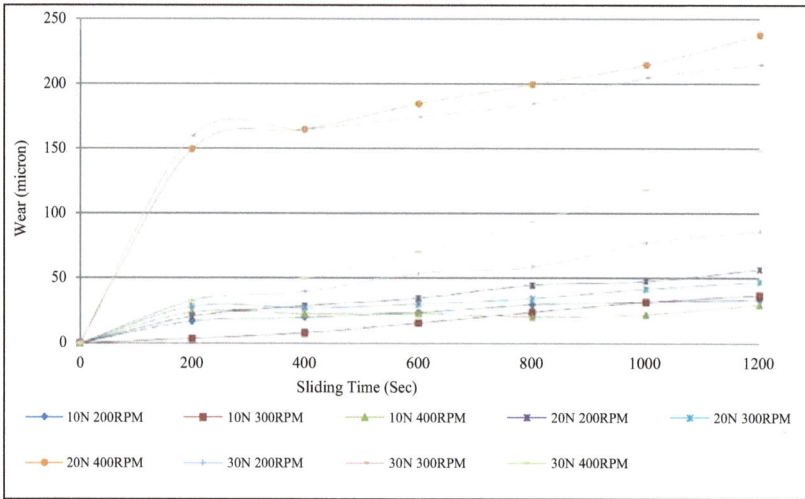

Fig. (4). Wear variations with sliding time for Tin based Babbitt under dry condition.

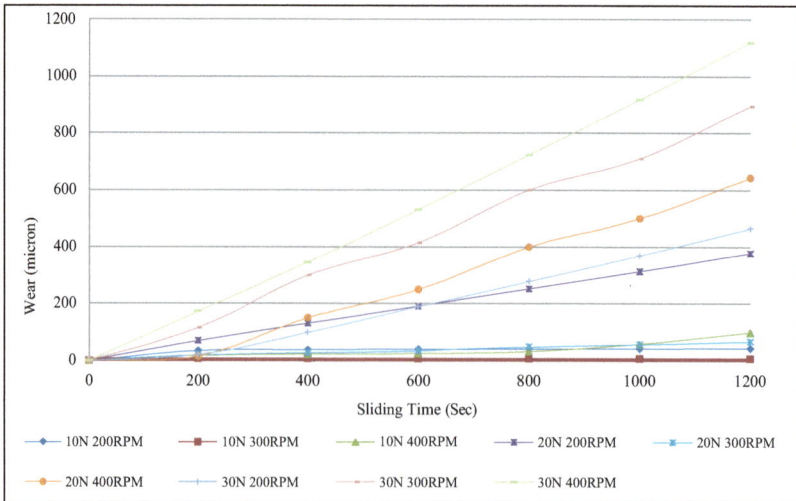

Fig. (5). Wear variations with sliding time for Brass under dry condition.

It has been noticed from the graph that wear depends on the sliding velocity and perpendicular load for both materials. Under dry conditions, tin-based Babbitt shows a sharp increase in wear initially up to 200sec, then wear increases gradually with sliding time. But in the case of brass material, wear increases at a constant rate. In dry conditions, Tin-based Babbitt shows maximum wear loss at 20N and 400rpm, while in the case of brass material, wear loss is maximum at 30N and 400rpm. The minimum wear loss for tin-based Babbitt is at 10N, 400rpm, and for brass material, it is at 10N, 300rpm. It is noticed from the curve that for both materials, wear increases as the normal load increases. While maintaining constant sliding speed under dry conditions. As sliding distance (sliding time) increases, there is a decrease in wear rate. Thus, it has been concluded that the wear rate changes with sliding distance, and does not remain constant throughout for Brass and tin-based Babbitt. While performing the experiment, it was observed that wear occurs through two processes- adhesive and abrasive wear. At low sliding speed and normal load, abrasive wear occurs. But at high sliding speed and normal load, both adhesive and abrasive wear occurs.

Similarly, for oil lubricating conditions, wear characteristics depends upon normal load and sliding speed. Figs. (**6** and **7**) show the wear variation with the sliding duration at different sliding speeds and normal loads with oil lubricating conditions. This graph shows that the wear raises linearly up to 200 sec, after which it remains almost stable for the rest of the experiment. Initially, because of the emergence of lubricating film between the pin and disc surface, the graph shows some negative value for wear. The result shows that the tribological performance of babbitt alloy and brass material changed according to sliding operating conditions. As sliding distance (sliding time) increases, there is a decrease in wear rate. Thus, the wear rate changes with sliding distance, not remaining constant throughout for Brass and tin-based Babbitt.

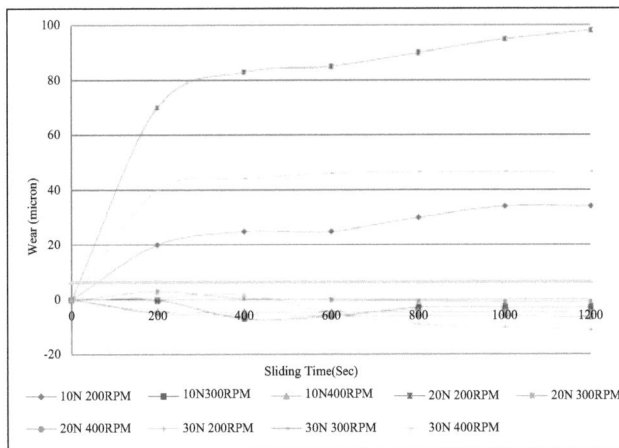

Fig. (6). Wear variations with sliding time for Tin-based Babbitt under oil lubrication conditions.

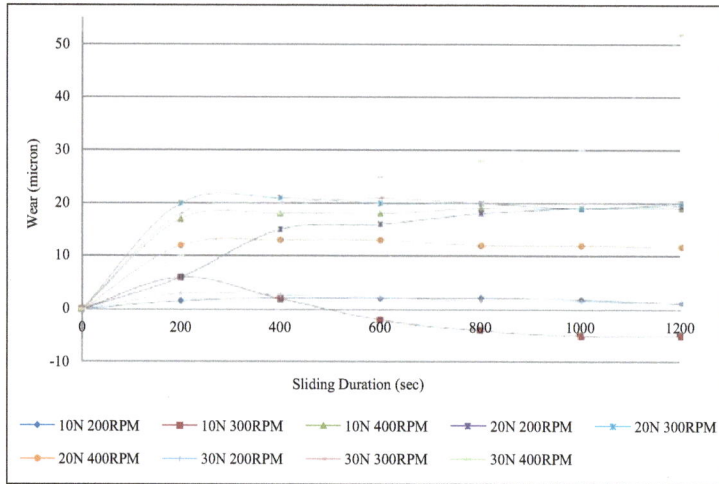

Fig. (7). Wear variations with sliding time for Brass under oil lubrication condition.

Under the dry condition (Fig. **8**), wear loss in Brass is more than in tin-based Babbitt material, whereas in oil lubricating condition (Fig. **9**), wear loss is more in tin-based Babbitt. The amount of wear in brass material is approximately higher than in babbitt material with dry condition. Thus, the service life of Babbitt is longer than brass material under dry conditions.

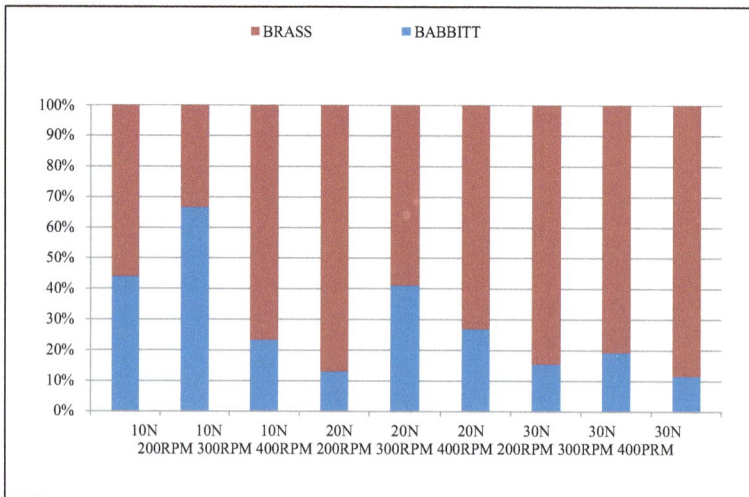

Fig. (8). Comparison of wear behavior of Tin-based Babbitt and Brass material under dry conditions.

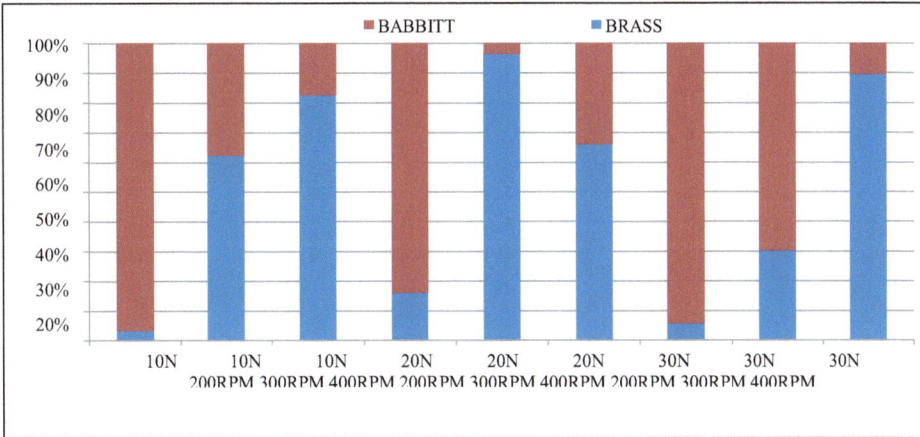

Fig. (9). Comparison of wear behavior of Tin-based Babbitt and Brass material under oil lubrication conditions.

Friction Coefficient Analysis

Figs. (**10** and **11**) show the friction coefficient variation with rubbing duration for tin-based Babbitt and brass material under dry conditions at different sliding speeds and normal load conditions.

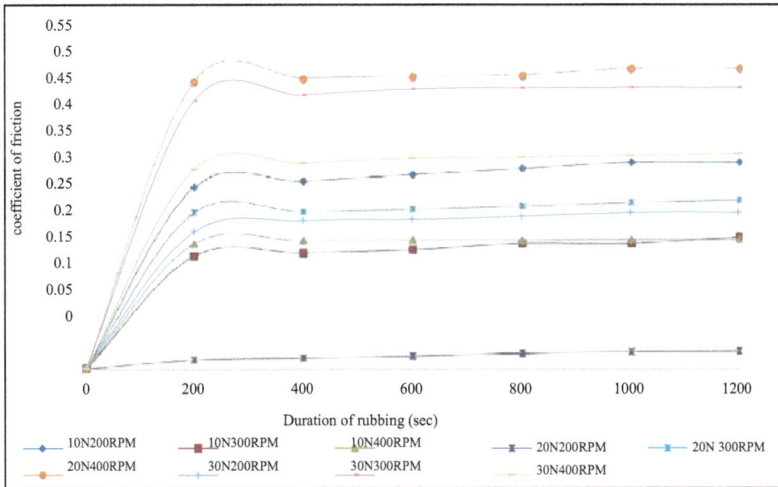

Fig. (10). Relationship between coefficient of friction and duration of rubbing for Tin based babbitt alloy under dry condition.

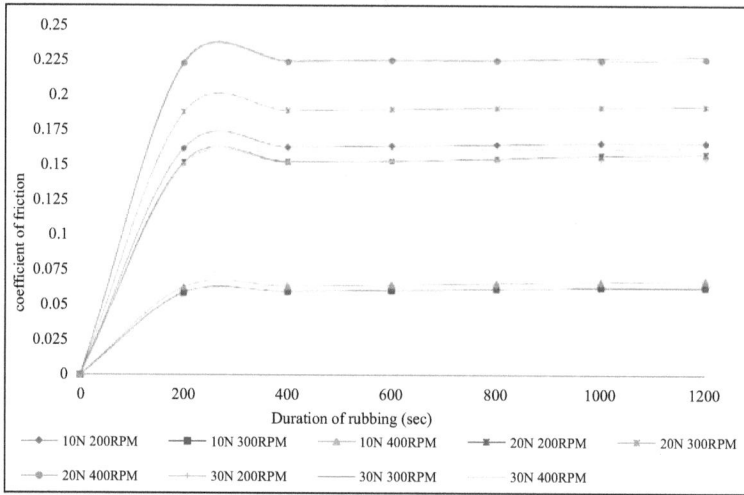

Fig. (11). Relationship between the coefficient of friction and duration of rubbing for Brass alloy under dry conditions.

It is observed from the graph that tin-based Babbitt at different sliding speeds and normal load takes 200sec to reach a steady state. The tin-based Babbitt friction coefficient (cof) varies between 0.028 to 0.48, and the brass material friction coefficient (cof) varies between 0.063 to 0.228 for dry conditions. As observed from (Figs. **14** and **15**) under lubrication conditions, the friction coefficient varies between 0.003 to 0.029 for tin-based Babbitt, while for brass material friction, the coefficient varies between 0.002 to 0.055. For both materials, the friction coefficient increases with the increase in sliding time under different working conditions. Initially, the friction coefficient is higher due to the higher frictional force between the pin and disc surface. With the help of frictional force, the friction coefficient is calculated for the test specimen during a wear test experiment. In oil lubricating conditions, an initial value of the friction coefficient is low due to the emergence of the fluid layer. For tin-based Babbitt friction coefficient value is higher at a normal load of 10N and sliding speed 400rpm. While brass value is also higher at 10N and 400 rpm. As observed from the graph, the initial value of the friction coefficient increased sharply, and then it became almost constant for the rest of the experiment for dry conditions, and a similar trend was seen for the lubrication condition also.

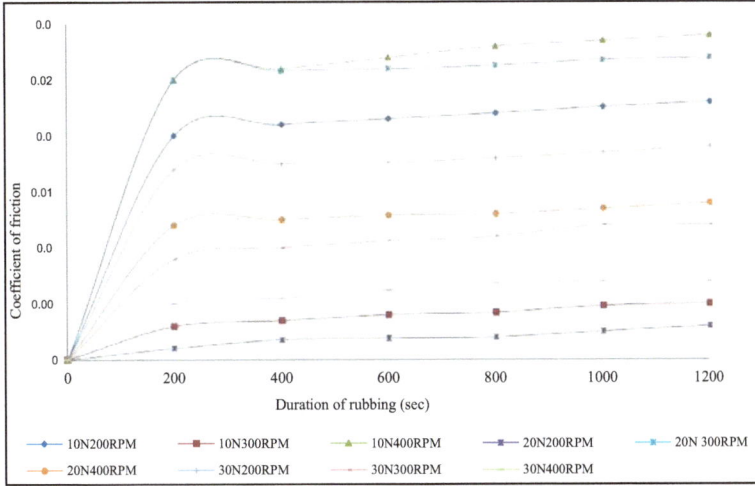

Fig. (12). Relationship between the coefficient of friction and duration of rubbing for Tin based Babbitt alloy under lubrication condition.

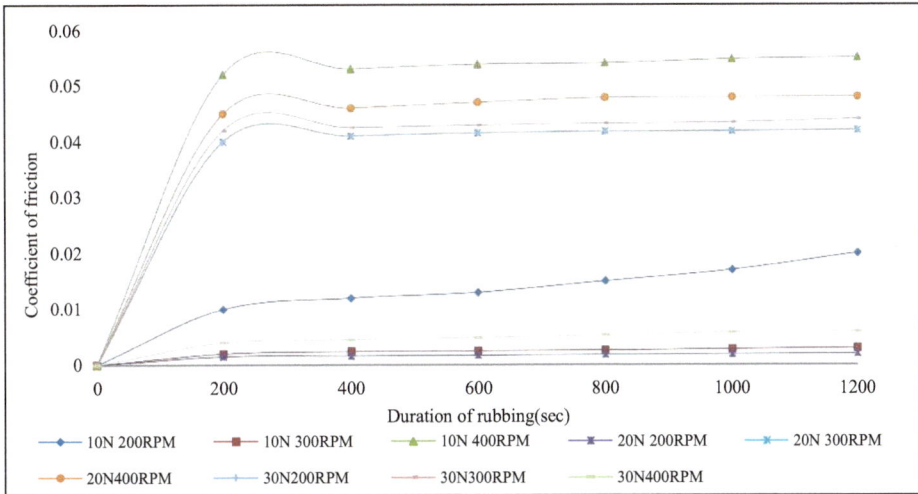

Fig. (13). Relationship between the coefficient of friction and duration of rubbing for Brass alloy under lubrication condition.

From the result obtained through the experiment, it was observed that the friction coefficient has a higher value for tin-based Babbitt and a lower value for brass material under dry conditions (Fig. **14**). While in oil lubrication condition, brass material has a higher value for friction coefficient as compared to tin-based Babbitt, as shown in Fig. (**15**).

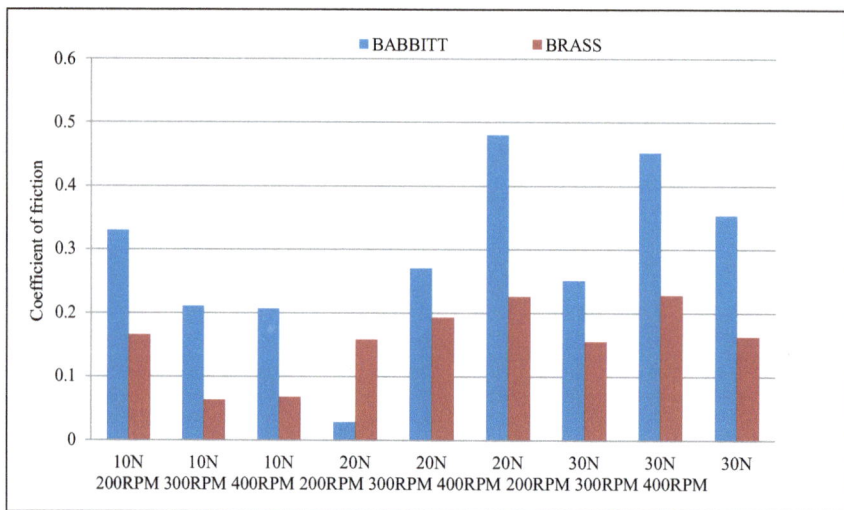

Fig. (14). Comparison of friction coefficient of Tin-based Babbitt and Brass material under dry condition.

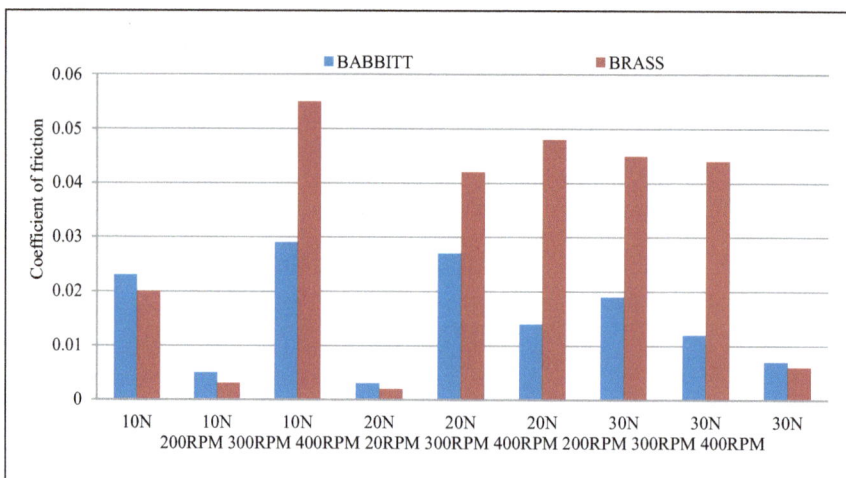

Fig. (15). Comparison of friction coefficient of Tin-based Babbitt and Brass material under oil lubrication condition.

CONCLUSION

In this experimental work, the tribological behavior of two different materials sliding against E8 steel under dry and lubrication (SAE 50) has been studied. The use of oil lubricant (SAE50) improves tribological performance relative to dry conditions. For tin-based Babbitt, performance improves by 18.56 times. At the same time, brass material performance improved by 2.19 times. At higher load (30N) and higher sliding speed(300rpm)-the friction coefficient (cof) of tin-based

Babbitt is 14.28% times higher compared to the friction coefficient (cof) of brass material under oil librating (SAE50) conditions, while under dry condition, coefficient of friction is 52.75% higher in tin-based Babbitt compared to Brass. Hence, if Babbitt is replaced by brass material, then power lost in a system can be reduced, and it helps in improving the overall performance of the system. While performing the experiment, if the speed has been kept constant on a wear test machine and the load is increased in increments from 10N, 20N to 30N, then it is observed that the wear loss increases under dry conditions for both babbitt alloy and brass materials. Under similar experimental operating conditions, the wear loss in Babbitt is more than in brass material; therefore expected service life of Brass is longer than tin-based Babbitt.

REFERENCES

[1]   A. Zeren, E. Feyzullahoglu, and M. Zeren, "Mater Des, A study on tribological behavior of tin based bearing material in dry sliding", *Materials & Design,* vol. 28, pp. 318-323, 2007.

[2]   D. Zhang, K.L. John, D. Guangneng, H. Zhang, and I. Tribol, "Tribological properties of tin based babbitt bearing alloys with Polyurethane coating under dry and starved lubrication condition", *Tribology International,* vol. 90, pp. 22-31, 2015.

[3]   M. Goudarzi, S. Jenabali Jahromi, and A. Nazaraboland, "Mater des, investigation of characteristics of tin based white metal as a bearing material", *Materials & Design,* vol. 30, pp. 2283-2288, 2009.

[4]   E. Feyzullahoglu, A. Zeren, and M. Zeren, "Tribological behavior of tin based materials and Brass in oil lubricated conditions", *Materials & Design,* vol. 29, pp. 714-720, 2008.

[5]   M.A. Chowdhury, and D.M. Nuruzzaman, "Experimental investigation on friction and wear properties of different steel materials", *Tribol Ind,* vol. 35, pp. 42-50, 2013.

[6]   A. Chowdhury, M. nuuruzzaman, and A. Hannan, "Effect of sliding velocity and relative humidity on friction coefficient of brass sliding against different steel counter faces", *IJERA,* vol. 2, pp. 1425-1431, 2012.

[7]   M.A. Chowdhury, D.M. Nuruzzaman, A.H. Mia, and M.L. Rahaman, "Friction Coefficient of Different Material Pairs Under Different Normal Loads and Sliding Velocities", *Tribol Ind.,* vol. 34, pp. 24-32, 2015.

[8]   R.K. Babberwal, and P.B. Patil, "Tribological behavior analysis of tin based babbitt alloys and brass materials", *IOSR JEN ,* vol. 1, pp. 28-33, 2019.

[9]   K. Deore, L. Aage, and M. Hajara, "Testing of Wear rate, Frictional force and Coefficient of friction computation using pin on disk", *IJETT,* vol. 20, pp. 256-259, 2015.

[10]   H. Wu, Q. Bi, and S. Zhu, "Friction and wear properties of babbitt alloy 16-16-2 under sea water", *Tribol Int,* vol. 44, pp. 1161-1167, 2011.

SUBJECT INDEX

A

Acid 147, 151
　polylactic 151
　reinforced-polylactic 147
　stearic 147
Air pressure 9, 188
Aluminium 16, 20, 169
　matrix composites (AMCs) 16, 20
　metal matrix composites 169
ANOVA analysis 119
Applications 2, 141, 144, 145, 147, 150, 159,
　225, 248
　automobile 144, 145, 150
　automotive 141, 147, 159, 225
　electrical 248
　electrode 248
　stress 2
Approaches, exfoliation adsorption 247
Artificial neural network (ANN) 115
Automobiles 22, 58, 114, 128, 144, 145, 155,
　157, 170, 184, 268, 283
　fuel-efficient 58
Automotive vehicle 155

B

Ball milling method 134
Binder jetting 28, 33, 34, 38
　process 28, 38
　system 33
　technique 38
　technology 34
Biomolecular assays 269
Biosensors 248, 268, 275, 276, 277, 278, 279,
　281
　amperometric 279
　conductimetric 279
BLA reinforcements 171, 177, 178, 179, 180,
　181
Box-behnken design (BBD) 114, 118
Brass material friction 291

coefficient 291

C

Calorimetric biosensors 278
Cancer 128, 135
　intestinal 135
　therapy 128
Carbon 17, 229
　nanoparticles 17
　nanotube density 229
Carbon nanotubes 242, 248
　nanowire-coated 248
　single-wall 242
Ceramic(s) 38, 187, 198, 199
　moulds 187
　non-ferroelectric oxide 199
　powder 38, 198, 199
Ceramic magnets 82, 84
　ring-shaped Fe-based 82
Charge 248, 278
　electrostatic 248
Chip deformation coefficient 102, 104, 105,
　106, 107, 108, 109, 110, 111, 112
Composite(s) 3, 14, 15, 19, 142, 145, 149,
　151, 154, 157, 170, 248
　aluminium-based 15, 19
　biopolymer 151
　ceramic 170
　design technique 3
　energy absorber 157
　fiber-hybrid 145
　fiber-reinforced 154
　reinforced thermoplastic 149
　straw-reinforced polymer 149
　synthetic fiber-reinforced polymer 142
　traditional 248Composite materials
　fiber-reinforced 14
　hybrid 145
Computation technique 63
Conductive polymer nanocomposites 247, 248
Control, emission 127, 128, 134

www.ingramcontent.com/pod-product-compliance
Lightning Source LLC
Chambersburg PA
CBHW050810220326
41598CB00006B/167